Microfluidic Biosensors

Microfluidic Biosensors

Edited by

Wing Cheung Mak

Department of Biomedical Engineering, The Chinese University of Hong Kong, Shatin, Hong Kong; Division of Sensor and Actuator Systems, Department of Physics, Chemistry and Biology (IFM), Biosensors and Bioelectronics Centre, Linköping University, Linköping, Sweden

Aaron Ho Pui Ho

Department of Biomedical Engineering, The Chinese University of Hong Kong, Shatin, Hong Kong

ELSEVIER

ACADEMIC PRESS

An imprint of Elsevier

Academic Press is an imprint of Elsevier
125 London Wall, London EC2Y 5AS, United Kingdom
525 B Street, Suite 1650, San Diego, CA 92101, United States
50 Hampshire Street, 5th Floor, Cambridge, MA 02139, United States
The Boulevard, Langford Lane, Kidlington, Oxford OX5 1GB, United Kingdom

ISBN: 978-0-12-823846-2

For Information on all Academic Press publications
visit our website at https://www.elsevier.com/books-and-journals

Publisher: Mara Conner
Acquisitions Editor: Carrie L. Bolger
Editorial Project Manager: Joshua Mearns
Production Project Manager: Kamesh R.
Cover Designer: Vicky Pearson Esser

Typeset by MPS Limited, Chennai, India

Contents

List of contributors

Asrulnizam Abd Manaf
Collaborative Microelectronic Design Excellence Center, Universiti Sains Malaysia, Sains@USM, Pulau Pinang, Malaysia

Katrina Campbell
Institute for Global Food Security, School of Biological Sciences, Queen's University, Belfast, United Kingdom

Sandro Carrara
Integrated Circuits Laboratory, School of Engineering, Institute of Microengineering, École Polytechnique Fédérale de Lausanne (EPFL), Neuchâtel, Switzerland

Stefano Cinti
Department of Pharmacy, University of Naples "Federico II", Naples, Italy; BAT Center—Interuniversity Center for Studies on Bioinspired Agro-Environmental Technology, University of Napoli "Federico II", Naples, Italy

Christopher T. Elliott
Institute for Global Food Security, School of Biological Sciences, Queen's University, Belfast, United Kingdom

Aroosha Faheem
State Key Laboratory of Agricultural Microbiology, College of Life Sciences, Huazhong Agricultural University, Wuhan, P.R. China

Sara S. Ghoreishizadeh
Department of Electrical and Electronic Engineering, University College London, London, United Kingdom

Tuan Guo
Institute of Photonics Technology, Jinan University, Guangzhou, P.R. China

Jana Hajslova
Department of Food Analysis and Nutrition, Faculty of Food and Biochemical Technology, University of Chemistry and Technology Prague, Prague, Czech Republic

Aaron Ho Pui Ho
Department of Biomedical Engineering, The Chinese University of Hong Kong, Shatin, Hong Kong

Georges Humbert
CNRS, XLIM Research Institute, UMR 7252, University of Limoges, Limoges, France

Itthipon Jeerapan
Center of Excellence for Trace Analysis and Biosensor, Prince of Songkla University, Hat Yai, Songkhla, Thailand; Division of Physical Science, Faculty of Science, Prince of Songkla University, Hat Yai, Songkhla, Thailand; Center of Excellence for Innovation in Chemistry, Faculty of Science, Prince of Songkla University, Hat Yai, Songkhla, Thailand

Tugba Kilic
Center for Systems Biology, Massachusetts General Hospital, Harvard Medical School, Boston, MA, United States

Supatinee Kongkaew
Center of Excellence for Trace Analysis and Biosensor, Prince of Songkla University, Hat Yai, Songkhla, Thailand; Center of Excellence for Innovation in Chemistry, Faculty of Science, Prince of Songkla University, HatYai, Songkhla, Thailand; Forensic Science Innovation and Service Center, Prince of Songkla University, Hat Yai, Songkhla, Thailand; Division of Health and Applied Sciences, Faculty of Science, Prince of Songkla University, Hat Yai, Songkhla, Thailand

Kaiwei Li
Key Laboratory of Bionic Engineering of Ministry of Education, Jilin University, Changchun, P.R. China

Warakorn Limbut
Center of Excellence for Trace Analysis and Biosensor, Prince of Songkla University, Hat Yai, Songkhla, Thailand; Center of Excellence for Innovation in Chemistry, Faculty of Science, Prince of Songkla University, HatYai, Songkhla, Thailand; Forensic Science Innovation and Service Center, Prince of Songkla University, Hat Yai, Songkhla, Thailand; Division of Health and Applied Sciences, Faculty of Science, Prince of Songkla University, Hat Yai, Songkhla, Thailand

Yanting Liu
Department of Applied Physics, The Hong Kong Polytechnic University, Hong Kong, P.R. China

Jacky Fong Chuen Loo
Department of Biomedical Engineering, The Chinese University of Hong Kong, Shatin, Hong Kong; Department of Neuroscience and Biomedical Engineering, School of Science, Aalto University, Aalto, Espoo, Finland

Wing Cheung Mak
Department of Biomedical Engineering, The Chinese University of Hong Kong, Shatin, Hong Kong; Division of Sensor and Actuator Systems, Department of Physics, Chemistry and Biology (IFM), Biosensors and Bioelectronics Centre, Linköping University, Linköping, Sweden

Lingyin Meng
Biosensors and Bioelectronics Centre, Division of Sensor and Actuator Systems, Department of Physics, Chemistry and Biology (IFM), Linköping University, Linköping, Sweden

Joost L.D. Nelis
Institute for Global Food Security, School of Biological Sciences, Queen's University, Belfast, United Kingdom

Anis Nurashikin Nordin
Department of Electrical and Computer Engineering, Faculty of Engineering, International Islamic University Malaysia, Kuala Lumpur, Malaysia

Kiattisak Promsuwan
Center of Excellence for Trace Analysis and Biosensor, Prince of Songkla University, Hat Yai, Songkhla, Thailand; Center of Excellence for Innovation in Chemistry, Faculty of Science, Prince of Songkla University, HatYai, Songkhla, Thailand; Forensic Science Innovation and Service Center, Prince of Songkla University, Hat Yai, Songkhla, Thailand; Division of Health and Applied Sciences, Faculty of Science, Prince of Songkla University, Hat Yai, Songkhla, Thailand

Jana Pulkrabova
Department of Food Analysis and Nutrition, Faculty of Food and Biochemical Technology, University of Chemistry and Technology Prague, Prague, Czech Republic

Panote Thavarungkul
Center of Excellence for Trace Analysis and Biosensor, Prince of Songkla University, Hat Yai, Songkhla, Thailand; Center of Excellence for Innovation in Chemistry, Faculty of Science, Prince of Songkla University, HatYai, Songkhla, Thailand; Division of Physical Science, Faculty of Science, Prince of Songkla University, Hat Yai, Songkhla, Thailand

Aristeidis S. Tsagkaris
Department of Food Analysis and Nutrition, Faculty of Food and Biochemical Technology, University of Chemistry and Technology Prague, Prague, Czech Republic

Jiajia Wang
College of Agricultural Equipment Engineering, Henan University of Science and Technology, Luoyang, P.R. China

Yuye Wang
Department of Biomedical Engineering, The Chinese University of Hong Kong, Shatin, Hong Kong; CNRS, XLIM Research Institute, UMR 7252, University of Limoges, Limoges, France; Laboratory of Biomedical Microsystems and Nano Devices, Bionic Sensing and Intelligence Center, Institute of Biomedical and Health Engineering, Shenzhen Institute of Advanced Technology, Chinese Academy of Sciences, Shenzhen, P.R. China

Lei Wei
School of Electrical and Electronic Engineering, Nanyang Technological University, Singapore, Singapore

Wu Yuan
Department of Biomedical Engineering, The Chinese University of Hong Kong, Shatin, Hong Kong

Shuwen Zeng
CNRS, XLIM Research Institute, UMR 7252, University of Limoges, Limoges, France; Light, Nanomaterials and Nanotechnologies (L2n), CNRS-ERL 7004, Université de Technologie de Troyes, Troyes, France; Department of Applied Physics and Applied Mathematics, Columbia University, New York, NY, United States

Xuming Zhang
Department of Applied Physics, The Hong Kong Polytechnic University, Hong Kong, P.R. China

Wenchao Zhou
State Key Laboratory of Applied Optics, Changchun Institute of Optics, Fine Mechanics and Physics, Chinese Academy of Sciences, Changchun, P.R. China

Shaodi Zhu
Department of Biomedical Engineering, The Chinese University of Hong Kong, Shatin, Hong Kong; CNRS, XLIM Research Institute, UMR 7252, University of Limoges, Limoges, France; Light, Nanomaterials and Nanotechnologies (L2n), CNRS-ERL 7004, Université de Technologie de Troyes, Troyes, France

Foreword

According to Scopus, over a hundred thousand papers mentioning biosensor(s) in their title, keywords, or abstracts have been published to date. Most of these focus on the mechanism of detection and the effectiveness of devices in meeting basic analytical objectives such as sensitivity and specificity. A body of work is also dedicated to evaluating the performance of instruments in meeting real-world analytical challenges. However, the academic literature pays relatively little attention to designing and manufacturing practical devices as reliable means of measurement in the hands of the everyday user. Despite this lack of research, companies have successfully commercialized biosensors, ranging from the hugely successful glucose sensors produced for home use by people with diabetes to their effective use in automated analytical machinery based on surface plasmon resonance (SPR) in the pharmaceutical industry. The recent pandemic has brought technologies such as PCR and lateral-flow devices to limelight, but the question is: Why have such approaches as those aforementioned been so successful? This book focuses on the key underpinning enabler, microfluidics.

The mediated electrochemical sensors that catapulted home blood glucose biosensors to fame in the 1980s were novel in several respects, but it was innovations in microfluidics that enabled ground-breaking features, such as self-filling, fill detection, and hematocrit correction, making the products so enduringly successful. Also, in the same decade, SPR was shown to be an exciting new way to monitor biological affinity reactions such as antibody–antigen and cell-receptor interactions, but it was the commercial exposure of the supporting microprocessor-controlled microfluidics that turned this scientific discovery into practical devices with such huge commercial impact in the pharmaceutical industry. Lateral-flow devices, initially commercialized for pregnancy testing, paved the way for simple immunoassays whereby fluidic separations could be achieved and visualized on an extremely robust and easy-to-use platform, which was later developed for a range of analyses, including, most notably, the detection of the coronavirus, for use by general public. The recent pandemic highlighted the value of high-throughput microfluidic PCR machinery in facilitating tracking and tracing of the virus and also monitoring the emergence of new strains. All these examples share a common feature in that the information about the bulk of the key design features can be found in the patent literature or in the hands of commercial engineers rather than in the academic literature. This book helps to address this paucity of information by reviewing the role of microfluidics in modern biosensor design.

Professor Mak and his coworker Professor Ho have called together an impressive array of colleagues to work with them to write this book. This talent pool consisted of some of the most exciting new and emerging teams in the field of biosensors from across the globe, including nations spearheading biosensor innovation. We see scientists from 3 continents and 13 countries working together in various combinations targeting clinical chemistry, food analysis, and environmental monitoring. Classical technologies such as fiber optics and analytical electrochemistry are blended with new ways to design fluidics for improved performance, and interfaces with smartphones to improve the processing of results. Smartphone technology intuitively offers considerable promise in terms of improved quantification, on-line result processing, and data communication. This exciting approach is evaluated in terms of both opportunities and issues that are hindering progress. Nanoliter drop-based analyzers are described, whereby discrete nanoliter droplets deliver a "digital assay," while the confinement of the assay within a small volume accumulates signal to deliver improved sensitivity. Throughout the book, a range of electronic components to probe, detect, and process signals derived from the electrochemical and optical properties of biomolecules are explored for incorporation into truly integrated analytical systems.

The idea of this volume is to explore the integration of microfluidics to create more practical and inexpensive biosensors that use less reagents, require smaller samples, and are simpler to use, while still delivering reliable results. The two most widespread types of biosensor transducer, electrochemical and optical, are covered in detail, and a variety of other formats are explored. Considerable attention is paid to appropriate microfabrication techniques such as the use of advanced polymers and polymer composites, including those containing graphene and 2-D inorganic and metal composite materials, to support a wide range of printing and other manufacturing technologies. Professor Sandro Carrara and his colleagues cover the integration of electronics with analytical systems, that is, CMOS-enabled biosensors. Other methods described cater to the development of wearable and flexible microfluidics and sensors, along with the burgeoning interest in paper-based devices. The various contributions show us how to immobilize biomolecules along with appropriate reagents and then integrate these with transducers such as electrodes and optical fibers. The chapter on fiber-optic biosensors, in particular, focuses on bringing solutions, sample treatment, and assay components that are typically handled in large laboratory settings into miniaturized 1-D devices that integrate multiple functionalities.

Very often in this field, we see detailed descriptions of the analytical performance of a technique in the lab, but with scant detail about how relevant samples are delivered from the point of interest to the analytical instrument. We often see impressive lab-based limits of detection quoted, but they bear little relevance to the clinical objective or environmental need when applied in a real-life situation. Biosensors are nothing if they cannot deliver relevant analytical information, and appropriate means to acquire, deliver, and process samples are fundamental to the construction of realistic tools. The main editor of this volume, Professor Mak, leads with the essential premise that "there always needs to be fluidics in front of the biosensor."

I trust that you will enjoy this refreshing perspective on the field microfluidic biosensors and that this is just the first of several volumes to emerge from Professor Mak, whom I have had the great pleasure to work with for a number of years when we were in Sweden together.

Anthony P.F. Turner
Department of Biotechnology, SATM, Cranfield University, Bedfordshire, United Kingdom

CHAPTER

1 Printed microfluidic biosensors and their biomedical applications

Jacky Fong Chuen Loo[1,2], Aaron Ho Pui Ho[1] and Wing Cheung Mak[1,3]

[1]*Department of Biomedical Engineering, The Chinese University of Hong Kong, Shatin, Hong Kong*

[2]*Department of Neuroscience and Biomedical Engineering, School of Science, Aalto University, Aalto, Espoo, Finland*

[3]*Division of Sensor and Actuator Systems, Department of Physics, Chemistry and Biology (IFM), Biosensors and Bioelectronics Centre, Linköping University, Linköping, Sweden*

1.1 Introduction

1.1.1 The emerging need of microfluidic biosensors

Global health issues remain major challenges across the world, and rapid disease diagnostics is one of the keys to disease control and surveillance. Take the 2014–2016 Western African Ebola virus epidemic as an example. It is essential to differentiate the Ebola virus from other viruses which share similar symptoms and exclude false detection of co-circulating diseases, such as malaria, to curb the spread of this highly contagious virus. This called for an effective on-site diagnostic for multiplexed detection of the Ebola virus and other viruses. Since medical centers, resources, and transportation are scarce in West Africa, sending samples to centralized laboratories for tests is not always feasible. This hampered large-scale diagnosis while addressing the epidemic. Even though conventional rapid test kits were available, e.g., ELISA kit, handling of biological fluid samples, and processing of the reaction posed a risk to healthcare workers administering the assay or monitoring the readout. Commercial rapid tests with paper-based lateral flow assays on other diseases, such as HIV, tuberculosis, influenza, and *Escherichia coli (E. coli)*O157:H7 infection, are designed to be simple test strips. Still, they only provide a yes/no answer and test for a single disease [1]. Therefore, a more reliable diagnostic method for multiple diseases is desired [2].

Biosensor is an integrated self-contained receptor–transducer system, which is capable of providing quantitative or semiquantitative analytical information. The main components include (1) a recognition element or bioreceptor, which is responsible for the specific targeting of the target analyte to support a high selectivity and specificity; (2) a transducer, which converts the interaction between the

Microfluidic Biosensors. DOI: https://doi.org/10.1016/B978-0-12-823846-2.00001-8

bioreceptor and the target into a signal, such as optical and electrochemical signals; and (3) electronics and display, which convert the signal obtained from the transducer into a processed, organized format to be read by human. Biosensors are often used in self-diagnosis and monitoring of disease progression. The earliest example of point-of-care testing (POCT) biosensors is the glucose biosensor that monitors the blood glucose level of diabetic patients. It is portable and generates results rapidly. To make the POC diagnostics with biosensors practical for use in the developing world, the World Health Organization issued criteria under the acronym ASSURED, i.e., Affordable, Sensitive, Specific, User-friendly, Rapid and robust, Equipment-free, Deliverable to end users [3]. In order to achieve ASSURED, biosensors are now designed to be inexpensive, portable, and perform as a standalone device. The desired way of performing the POC diagnostics is sample-to-answer biosensing. The user injects the sample into the inlet. The testing procedures, including reactions and signal acquisition, are automated, and the result readout show on the display without the need for further judgment or post-analysis. In order to perform automation inside a biosensor of limited space and weight, microfluidics handle fluid actuation and hence its reaction is the best option to be incorporated into the biosensor.

1.1.2 Printed microfluidics biosensors

Microfluidics is the process of fluid in micron scale, in the range of pico- to micro-liter. This process includes:

1. actuation of fluid of a particular volume to a specific location;
2. separation of the components in the fluid based on different properties, e.g., viscosity, density, etc.;
3. mixing of several kinds of fluid in a controlled temporal and spatial manner, causing bioreactions in a controlled manner.

Compared to conventional laboratory methods, it reduces the need for large volume of the sample, and labor-intensive works to process samples and reagents. Microfluidics is often applied in basic life science and applied science research [4]. In basic life science research, biomolecular and cell analysis, especially single-cell analysis, makes assay systems based on mammalian cells for developing drugs as individualized medicine [5–8]. In high-throughput screening in medical diagnosis requiring multiplex reactions, as microfluidic chambers are equivalent to multiple test tubes, microfluidics guides the sample through microchannels to simultaneously control fluidic flow to the sequential reaction sites and eventually to the sensing chambers [9].

Commercial printing technologies have been used for decades, with flexographic printing on a plastic sheet to fabricate printed sensing electrode or noncontact liquid line printing on paper to fabricate a lateral flow paper strip. Other printed components, such as printed circuit boards (PCBs) and 3D-printed machine exterior, have been used to construct biosensors [10]. It is, therefore, reasonable to extend the concept and advantages of printing into microfluidics fabrication and integration.

The COVID-19 pandemic showed us the importance of local production. Since transportation is limited or hauled due to restrictions, the import of medical supplies is insufficient to meet the skyrocketing rise in demand. The lack of medical supplies further impaired the already strained healthcare system. In this context, 3D printing of medical supplies, such as masks, face shields, and ventilators, was approved in many medical centers and became standard practice during emergency periods. Besides, visits to medical centers by patients with chronic illnesses and noncommunicable diseases were also disrupted due to risks of infection with the SARS-CoV-2 and limited medical resources. Similar to 3D printing of medical supplies, demand of POC diagnostics for the COVID-19 and other tests during the pandemic can be tackled by printing technology [11].

Printed microfluidics can be fabricated in 2D to 4D scale, depending on the complexity of the biosensing reaction. Table 1.1 summarizes some of the examples of sensing targets and the choices of microfluidic dimension. For example, immunoassay requires simple interaction between target molecules and recognition elements for visual readout and can therefore be performed in 2D microfluidics, such as lateral flow paper strip. For reactions requiring multiple procedures, e.g., target extraction and purification, washing, reagent mixing, before detecting the target, such as *E. coli* detection from food, 3D microfluidics are commonly employed. 4D or dynamic microfluidics possess reversible properties and respond to external stimuli, or responsive signals, such as light and heat, although there is a lack of example in biosensing target currently. More importantly, by leveraging the integration with sensor components, which can also be printed, the printed microfluidics can be translated into a printed biosensor to be practically used without compromising the advantages of how printing technology tackle the emergency [12].

Fig. 1.1 illustrates how the printed microfluidic biosensors tackle the current and upcoming emergency needs around the globe. Printed microfluidics have exhibited attractive unique advantages, including fast turnaround time of production, flexibility in customization, and sustainability in fabrication, to fit this need. In conventional production methods (routes colored in red), the design of microfluidic biosensors is sent to the factory for mass production, stored in a warehouse, shipped to the destination, and transported to the target site. This kind of industrial production is less flexible in customizing the product design and the production schedule and is confined to factories due to various requirements of equipment and technicians. On the contrary, printed microfluidics can be printed anywhere, after obtaining the design in a digital format via the internet or cloud storage, anytime by anyone, as the printer is usually automated and printing can be easily controlled via software with the imported pre-optimized setting. They are also highly flexible in customization, and the design can be tailored or modified according to conditions and requirements. Printing microfluidic biosensors

Table 1.1 Examples of the sensing target by the use of different printed microfluidic approaches.

Sensing target	Dimension	Transducer	Printing materials	Printing methods	Integration (recognition element)	Integration (signal acquisition)	Reference(s)
HIV (HIV-1 p24 antigen)	2D	Optical (absorptive measurement)	Wax on NC membrane	Wax pencil	Anti-HIV-1 p24 antibody	UV–VIS spectrometer	[13]
Bisphenol A	2D	Electrochemical (voltammetric measurement)	Wax on cellulose fiber paper	Wax solid-ink printing	Carbon black	Screen-printed paper electrode	[14]
Glucose	Pseudo-3D	Optical (colorimetric observation)	Wax on chromatography paper	Wax solid-ink printing and stacking	Glucose oxidase	Photo scanner	[15]
Four tumor markers (AFP, CA153, CA199, and CEA)[a]	Pseudo-3D	Optical (chemiluminescence)	Wax on chromatography paper	Wax solid-ink printing and folding	AFP, CA153, CA199, and CEA-targeting antibodies	Digital camera	[16]
Malaria DNA marker	Pseudo-3D	Optical (fluorescence observation)	Wax on glass-fiber paper	Wax solid-ink printing and folding	LAMP primers	Handheld UV lamp (365 nm) and naked eye	[17]
Lead ion (Pb^{2+}) and mercury ion (Hg^{2+})	Pseudo-3D	Electrochemical (electro-chemiluminescence)	Wax on chromatography paper	Wax solid-ink printing and folding	Pb^{2+} and Hg^{2+}-binding DNA	Paper electrodes	[18]
Escherichia coli	3D	Optical (absorptive measurement)	ABS plastic polymer	FDM	*E. coli*-targeting antibody	Light absorption spectrometer	[19]
E. coli crooks strain	3D	Electrochemical (impedimetric measurement)	Polyacrylate resin	MultiJet printing	*E. coli* outer membrane proteins-targeting aptamer	Screen-printed gold electrodes	[20]
E. coli O157:H7	3D	Optical (luminescence)	Photocurable resin (acrylate-based photopolymer)	DLP	*E. coli* O157:H7-targeting antibody	Portable luminometer	[21]
Light wavelength	4D	Optical	Shape-memory polymers	MultiJet printing	/	/	[22]

ABS, Acrylonitrile butadiene styrene; DLP, digital light processing; FDM, fused deposition modeling.
[a]Key: AFP: fetoprotein, CA153: carcinoma antigen 153, CA199: carcinoma antigen 199, CEA: carcinoembryonic antigen.

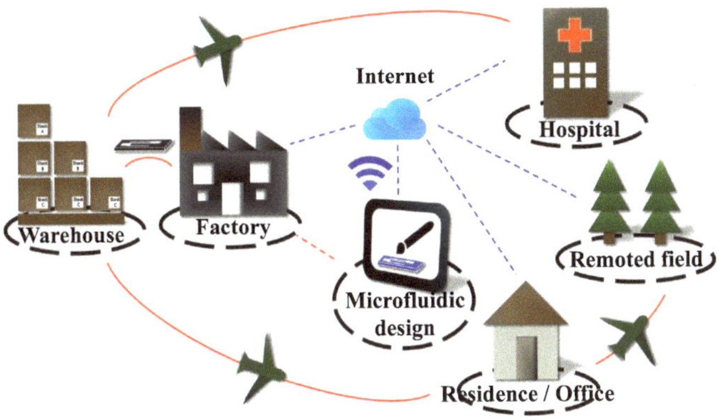

FIGURE 1.1

The illustration shows the logistic of fabricating printed microfluidic biosensors (*blue route*) and their advantages over conventional biosensing systems (*red route*).

also promotes sustainability by reducing unnecessary waste from excessive production, extra storage, and power to maintain warehouses.

1.2 Technologies in fabrication and assembly of printed microfluidics

This section describes printing technologies, especially advanced methods investigated and demonstrated in the past few years, for microfluidic fabrication and assembly. It also discusses the design, materials, and methods of printing and possible steps for post-processing of 2D, 3D, and 4D-printed microfluidics.

1.2.1 2D microfluidics

Cellulose-based paper has long been one of the most popular printing substrates. The advantages of paper are that it comes with a wide range of thickness and weight for commercial and industrial use at a low cost and that its ease of functionalization supports modifying the paper surface properties, e.g., hydrophilicity, permeability, piezoelectricity, and reactivity. Its high biocompatibility allows the printing of recognition elements on paper. Paper is also an environment friendly material which degrades quickly without polluting the environment and does not create toxins upon incineration. Paper-based microfluidics usually is more straightforward and more affordable. They rarely require an external power source, since fluid movement through the device can easily be facilitated by capillary forces. Cellulose fibers, chromatography paper, e.g., Whatman grade 1 chromatography paper, are commonly employed for printing, thanks to the detailed specification of well-defined properties in terms of pore size, water absorption property, hydrophobicity, etc. [23,24].

Nitrocellulose (NC) membrane is another popular printing substrate in 2D microfluidics. NC and polyvinylidene fluoride (PVDF) membranes have been extensively used in immunoblot for protein and nucleic acid analysis since the 1970s, thanks to their high protein- and nucleic acid-binding affinity as well as the availability of different pore sizes for capturing particular molecular size protein [25]. Although PVDF membrane is more popular given its better retention and less brittle nature than NC membrane, it is less desirable in POCT as it has to be pre-wetted with bipolar solvents, such as methanol or ethanol, prior to interaction with biomolecules in biosensing. Since NC membrane has better intrinsic protein and nucleic acid-binding properties than chromatography paper [13,26], it has been popularly used in conventional NC paper-based strip assay where antibodies can be immobilized on the NC surface to form a "test line" and "control line" to enable the end users to visualize the results more easily in comparison to color-change based assay. Similar to the chromatography paper, the well-defined properties of NC can define the flow rate in a 2D microfluidic, and geometric flow control, in terms of width, length, input angle, and output angle of the microchannel, in an NC membrane provides a new dimension of additional dynamic flow rate control of fluidic actuation [27,28].

To design 2D microfluidics, office software as simple as Microsoft (Ms) Paint, as long as they can generate the print layout to the printer, can be theoretically used for the drawing, although the accurate control of scaling is compromised. Therefore, professional computer-aided design (CAD) software are generally considered better at drawing the fine details of the micrometer resolution. Open-source software, e.g., Sketchup, Blender, and commercial software, such as AutoCAD or Solidwork, are commonly used for 2D-printed microfluidics design. 2D microfluidics contains microchannels of various lengths and widths and microchambers of various sizes and shapes, surrounded by the hydrophobic barrier such as wax. Therefore, after the dimensions and positions of the channels and chambers are determined in the drawing, the drawing is converted into the "inverted" image, i.e., the regions of channels and chambers are left blank while the surroundings are shown in "black" color, so that the printer head will move to the "black" region to print the wax materials according to the image.

Wax is the most common printing material. Printing wax channels on paper or NC membrane and then subjecting the paper to heat treatment causes wax penetration to form the wax barrier, of which the resolution and the channels can be as low as approximately 500 μm [24,29,30]. The percentage of wax in the printing material determines the hydrophobicity and fluidic properties, e.g., flow rate of the fluid. Commercial 2D inkjet printers commonly used in households can print 2D microfluidics with other materials instead of toner and ink. Fig. 1.2A demonstrates the use of a commercial printer (Xerox ColorQube color printers) for routinely printing paper-based microfluidics of different wax percentages with the addition of distinguishable color dyes, i.e., bright cyan (C), magenta (M), yellow (Y), black (K) that

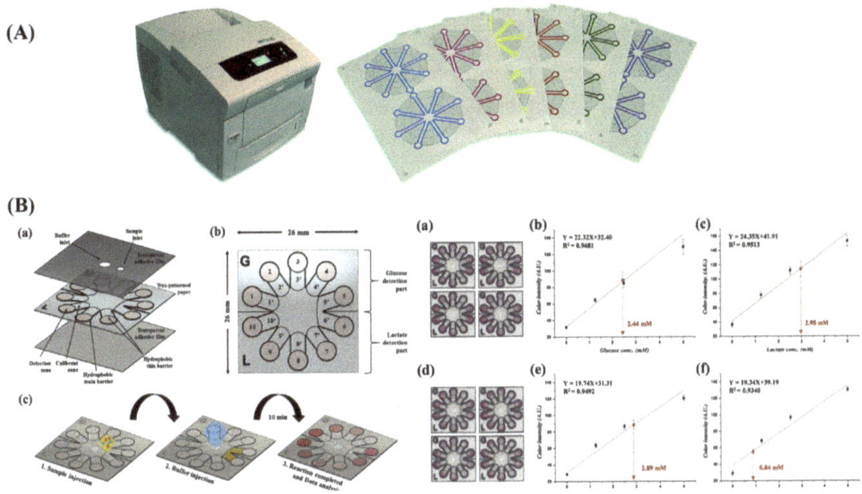

FIGURE 1.2

2D-printed microfluidics. (A) The Xerox ColorQube 8570 model wax-based color printers for creating paper microfluidics of different color combinations of the wax ink, i.e., cyan, magenta, yellow, magenta + yellow, cyan + yellow, and cyan + magenta. (B) Fabrication and preparation of the self-calibrating microfluidic paper-based analytic device (left): its assembly, schematic diagram and flow of procedures, and (right) the image of the microfluidics for color intensity determination and plot fitting for detection in standard concentration solution (5 mM glucose and 5 mM lactate) and plasma samples [31,32].

Reprinted with permission from J. Potter, P. Brisk, W.H. Grover, Using printer ink color to control the behavior of paper microfluidics, Lab Chip 19 (2019) 2000–2008. https://doi.org/10.1039/C9LC00083F; S. Kim, D. Kim, S. Kim, Simultaneous quantification of multiple biomarkers on a self-calibrating microfluidic paper-based analytic device, Anal. Chim. Acta 1097 (2020) 120–126. https://doi.org/10.1016/j.aca.2019.10.068.

can produce a full range of colors, to various percentages of paraffin wax and resin ranging 50%–60% and 10%–20%, respectively. The color based on the combination of CMYK colors on the paper microfluidics are shown with the defined hydrophobicity, so that the behavior of fluid can be controlled conveniently [31].

However, wax printing requires additional post-processing steps for melting wax into the paper to make the final product, e.g., heating at 120°C for 90 seconds and then cooling at 25°C. The heating and cooling need to be controlled accurately to guarantee the consistency of the end product. Prolonged heating might cause the melted wax to diffuse into the surrounding area. Due to its instability at high temperatures, storage and operation of the printed microfluidics should be kept at a controlled temperature. Also, printing quality depends on the smoothness of the paper or membrane surface. Inconsistent deposition of wax affects its hydrophobicity and hence its functionality.

2D-printed microfluidics can be inserted into a detector, such as potentiometric, fluorimetric, or colorimetric sensors, for visualization or signal acquisition [33]. Integration of 2D-printed microfluidics is relatively easy thanks to its properties such as thin layering, flexibility, and compatibility with daily-use adhesive materials. The microfluidics can be joined to any contact area of a detector at any orientation and geometry with simple adhesion such as adhesive tape and glue.

Another practical use of printed microfluidics in real-life applications is quantification accuracy. Calibration with standard concentrations is often used in the laboratory when determining the concentration to minimize the effect of environmental conditions (i.e., ambient light intensity, temperature, humidity, and pressure). However, on-site testing with calibration of printed biosensors can be challenging. Some 2D microfluidics introduces self-calibration and reserve areas for pre-loading and actuating calibrants. Fig. 1.2B shows an example of the microfluidic that supports automatic partition of the injected sample into different zones, pre-loaded with varying concentrations of the sensing elements such as glucose and lactate. The microfluidic is then imaged to determine the color intensity of each detection zone. Post-processing of the data by fitting the data plot with linear interpolation is used to simultaneously determine the glucose and lactate concentration [32].

1.2.2 Pseudo-3D microfluidics

Pseudo-3D microfluidics is defined as the 3D microfluidics in which fabrication of microfluidics is in a 2D manner, with extra procedures transforming multiple 2D microfluidics into 3D microfluidics. Stacking and folding are the two main methods to produce pseudo-3D microfluidics. Since most of the 2D microfluidics are printed on paper or membrane, its flexible feature enables folding without breaking the microfluidic bases.

Stacking multiple 2D microfluidics is popular because it extends the dimensional scale of microfluidics. It also reduces the size of the microfluidics while retaining a similar fluidic path distance. The difference in the diffusing speed of various molecules across layers of 2D microfluidic chambers, mostly with filter paper of defined pore size as a barrier, generates functions like molecule separation, and filtration, which is not conveniently done in conventional 2D microfluidics. Fig. 1.3A shows the schematics of pseudo-3D-printed well-based microfluidics based on stacking of multiple printed layers with laminate support and components to hold the assembled chip [15]. In detail, five layers of wax-printed chromatography paper were stacked together. Hydrophobic parafilm layers with hole-punch sites, creating a sample loading site, were then placed on both the top and the bottom of the stack layer. After proper alignment, all these were assembled manually with polyvinyl chloride enclosure and clips.

On the other hand, folding and sliding action can join printed compartments of channels and chambers to build functional microfluidics. This simplifies the fabrication because printing is completed on a single surface instead of printing

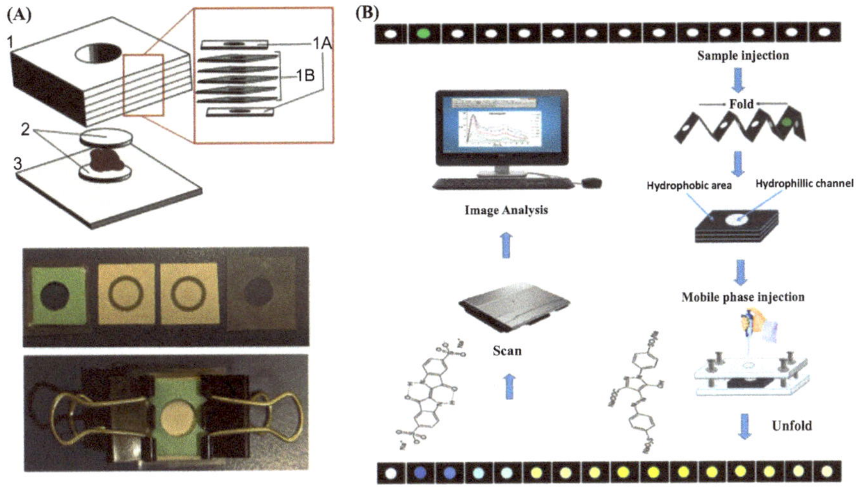

FIGURE 1.3

Pseudo-3D-printed microfluidics. (A) Stacking of multiple layers comprises five laminate sheet layers of wax-printed chromatography paper to form a paper wall. (B) Schematic of folding the 2D-printed microfluidic to generate a hydrophilic channel and function as separation of food colorants in the mixture [15,34].

Reprinted with permission from G.C. Ilacas, A. Basa, K.J. Nelms, J.D. Sosa, Y. Liu, F.A. Gomez, Paper-based microfluidic devices for glucose assays employing a metal-organic framework (MOF), Anal. Chim. Acta 1055 (2019) 74–80. https://doi.org/10.1016/j.aca.2019.01.009; F.M. Gharaghani, M. Akhond, B. Hemmateenejad, A three-dimensional origami microfluidic device for paper chromatography: application to quantification of Tartrazine and Indigo carmine in food samples, J. Chromatogr. A 1621 (2020) 461049. https://doi.org/10.1016/j.chroma.2020.461049.

on multiple surfaces and then assembling them. The alignment by folding is more feasible than by stacking since the alignment can be well defined in the microfluidic design, while stacking requires external position markers, such as labels at particular corners, and positioning tools to facilitate the stacking process [16–18,34,35]. Fig. 1.3B shows origami-based three-dimensional paper chromatography (3D-PC) as a micro-chromatographic microfluidic platform. In this microfluidic, repeated mountain and valley folds are created during the folding and microchambers are formed. Chromatographic separation is possible only when the folding is completed to form a fluidic path across the printed layers. After the separation, the paper microfluidic is unfolded to reveal the color of all regions, and it is then followed by a colorimetric detection by determining the colors and their positions to identifying the target molecules [34].

However, the disadvantages of stacking and folding include the post-processing of stacking multiple printed materials to make it 3D, creating the folding lines that make the crease pattern to facilitate accurate folding upon usage, and the tedious process of folding during the assay. Their lack of robustness and automation hinders mass production and extensive use in large-scale screening tests. Therefore, researchers started moving to 3D-printed microfluidics in the last decade when the technology became widely available.

1.2.3 3D printed microfluidics

3D printing has recently become a popular area in the additive manufacturing process, as its technology has matured over decades and turned into commercial products widely available in the last decade. Its popularity can be observed from the availability of the 3D printing design and viewing software, education materials, and access to the equipment. In the 3D design, Microsoft changed its default drawing software from MS Paint to Paint 3D in the latest Windows operating system (Windows 10) on personal computers. It accelerates the education of basic 3D drawing. Open-source designs for 3D printing are free on online platforms under the "Creative Commons – Attribution" license, such as Thingiverse, Sketchfab, GrabCAD with digital design for physical objects and 3D printing services on demand, encouraging people to participate in the design and exploitation of 3D printing. Apart from the widely available commercial 3D printers that can produce high resolution, more and more open-source 3D printers are available for less than 1000 USD while restraining its printing resolution below submillimeter scale. Free online tutorials illustrate the procedure of building a 3D printer from scratch using another 3D printer, i.e., the replication of a 3D printer in-house, and it makes possible urgent repairing of a printer by replacing the damaged parts with 3D-printed parts. On the other hand, high-resolution 3D printing services on demand are now available at an affordable price. This section focuses on illustrating the practical procedure from design to production of 3D-printed microfluidics, particularly designing with CAD software to production and post-processing with various types of 3D printing.

Similar to 2D microfluidics, the design heavily relies on CAD software to produce an accurate 3D geometry with high resolution in the micrometer scale. However, the dimensions are not handled in Paint 3D, and more professional software are needed for accurate microfluidic drawing. Since different file extensions can be exported from various software, the drawing drafts need to be converted into a generally readable format for printing. In 3D printing, STL file, i.e., "Standard Triangle Language" or "Standard Tessellation Language," is the most popular file extension to transform the drawing into a 3D illustration, readable on other computers without 3D drawing software. It can also be uploaded to the online platforms mentioned. Up till early 2021, more than 50 3D microfluidic models are available for download. Detailed model information and print settings are also attached to the models to help the user modify and print the 3D microfluidics. The STL file is then converted into a g-code file according to the type of printer. These g-codes are later on used to guide the printer head position and its printing sequence. According to the G-code command, the microfluidic designs are formed by moving the extrusion print head in the XY plane.

Fused deposition modeling (FDM), stereolithography (SLA), and digital light processing (DLP) are the most common 3D printing approaches. FDM, based on the layer-by-layer disposition of printing materials, is the most widely employed 3D printing method currently because of the simplicity of operating the 3D printer and large varieties of choice of the plastics to fabricate microfluidics of different properties with minimal effort of adjusting the machine or optimizing the printing parameter. At the same time, SLA and DLP require the use of specific printing materials, which are light-responsive and able to be solidified under light illumination, most often UV light, as well as the machine is more sophisticated, where mirrors, lens, and UV lights have to be implemented and aligned well, compared to FDM one which usually requires a printing head that disposes the printing materials at a particular location guided by several actuation motors. Despite research and development of printing materials, choices of light-responsive resin materials are still limited, and hence the few tunable properties. In contrast, many thermoplastic types with melting point up to 280°C, such as acrylonitrile butadiene styrene (ABS), PLA, can be used in FDM printing.

3D FDM printing with wax is the simplest way to turn 2D wax-printed microfluidics into 3D without changing the fluidic properties and sample compatibility, unlike using a new printing material [24]. Fig. 1.4A shows the extension of wax printing from 2D to 3D for patterning hydrophobic wax barrier on the filter paper with the commercial Prusa i3 desktop printer with a modified extruder. The patterns of the solidified wax printed on filter paper matched the fluidic patterns designed by SolidWorks CAD software, and the loading of methylene blue dye solution clearly labeled the fluidic channels [24]. Apart from wax, microfluidics made of elastic plastic, the rigidity of which will affect the fluid actuation performance inside the fluidic channel, with a tunable hole as a tunable flow valve, was printed and coupled with a resistive pulse sensor (RPS) as a printed microfluidic RPS sensor [36].

Other strategies to increase the resolution of 3D microfluidics include stereolithographic 3D, which relies on optic and polymerization kinetics instead of melted plastic flow dynamics, which relies on a piezo printhead to further enhance the resolution in FDM printing and enable the printed microfluidic channels and chambers of a high resolution as low as 40 μm in the horizontal plane of the 3D print [38,39]. Combination of the mechanism of FDM and SLA to leverage their advantages of the versatile printing head and high resolution has turned into the inkjet printing method, which is also called MultiJet or PolyJet printing [40]. Jetting layers of photocurable plastic resin, such as UV curable liquid photopolymer, are printed with FDM process onto the surface, and instantly cured voxel by voxel to form a microfluidic building structure with resolution as low as 50 μm in the lateral direction [40].

However, 3D-printed microfluidic chips are often subjected to post-processing to preserve fine details of microchannels and microchambers and enhance their

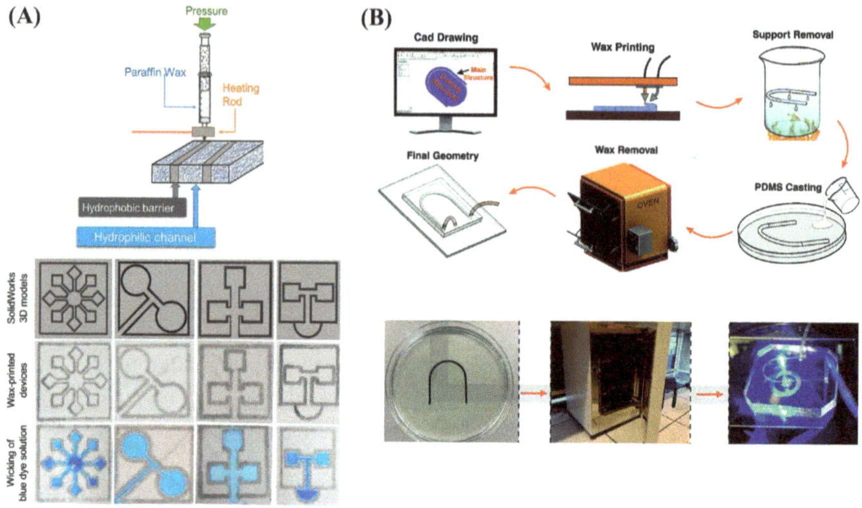

FIGURE 1.4

3D-printed microfluidics. (A) (Top) 3D printing process with wax for patterning a hydrophobic barrier on the filter paper with a commercially available 3D printer; (bottom) drawing and images of the wax-printed microfluidics with arbitrary designs. (B) (Top) Schematic illustration of the workflow for the fabrication of microchannels from drawing, 3D printing with wax complex 3D geometry, removal of the supporting structure, curing of PDMS from liquid to solid on the fabricated wax channel, removal of wax; (bottom) the photograph of pouring PDMS for curing, melting of a wax channel in an oven, and connection of injection tubing as the inlet and outlet [24,37]. *PDMS*, Polydimethylsiloxane.

Reprinted with permission from C.-K. Chiang, A. Kurniawan, C.-Y. Kao, M.-J. Wang, Single step and mask-free 3D wax printing of microfluidic paper-based analytical devices for glucose and nitrite assays, Talanta 194 (2019) 837–845. https://doi.org/10.1016/j.talanta.2018.10.104; M.A. Raoufi, S.R. Bazaz, H. O. Niazmand, Rouhi, M. Asadnia, A. Razmjou, et al. Fabrication of unconventional inertial microfluidic channels using wax 3D printing, Soft Matter 16 (2020) 2448–2459. https://doi.org/10.1039/C9SM02067E.

cleanness and functionality. There are physical and chemical methods for post-processing. For physical methods, SLA- and DLP-printed microfluidic plastic are treated with UV irradiation as a 30-s post-curing stage on each side to make the harder chips and more stable 3D-printed microfluidic device [41]. For the wax-printed 3D microfluidics, complex 3D structures can be fabricated with dummy structured printing. The schematic procedures and images of the key steps are shown in Fig. 1.4B. After 3D wax printing and curing with polydimethylsiloxane (PDMS) silicone elastomer, the microfluidic chip is immersed into a dissolving solution and heated in the oven at 115°C to melt and displace the wax from the PDMS structure [37]. This turns the disadvantages of wax-printed 2D microfluidics mentioned previously into an advantage. The high-temperature heating causes the microchannels to appear gradually. For chemical methods, although acetone has been used for smoothing in 3D-printed materials made with ABS, it is less often used for 3D-printed microfluidics as it will dissolve or deform parts of the microfluidic structure. Instead, rinsing with isopropanol and water are commonly used to remove the water-soluble polymer template, plastic residues, and impurities, and increase hydrophilicity prior to biomedical use [40,42].

For mass production, the printing of individual microfluidics can be substituted by soft lithography replica methods. The hard mold of a microfluidic chip is first printed with FDM and PDMS silicone elastomer solution, e.g., Dow Corning Sylgard 184, and a chemical curing agent is injected into the mold and solidified with thermotreatment. Compared to the conventional injection molding used in the industry, this method creates microfluidic replica with superior resolution ranging from millimeter to micrometer. Nevertheless, micro-molding technology has improved the molding-based mass manufacture of microfluidics with an accuracy of microchambers and microchannels in micron size [43]. Although the metallic molds are currently made by computer numerical control (CNC) milling, they can potentially be produced by 3D printing in the future.

1.2.4 4D microfluidics

4D microfluidics are also called dynamic or stimuli-responsive microfluidics because the fourth dimension, i.e., the temporal dimension, is usually governed by external factors such as simulants. Having the temporal dimension equips microfluidics with new functionalities, such as responsivity to external physical or chemical stimuli, and dynamically tuning the dimension of microchannels to create new connections or block existing connections and microchambers to vary the fluid volume accumulation. Despite scaling issues, 3D-printed microfluidics can also be turned into 4D simply by using printing materials that bring the response in the temporal dimension. Generally speaking, the shape or 3D geometry needs to change upon external environment or stimuli, such as light and temperature, so responsive polymer or hydrogels are often chosen as the raw materials for fabrication. Shape memory materials, such as shape memory polymers (SMPs) or swellable hydrogels, can be another option as these provide the reversible feature of the microfluidic to extend the ability of shape responsivity from one-off to multiple times [22,44].

Temperature responsivity is another common property in microfluidics, primarily used as temperature-controlled valves to control the opening/closing of certain microchannels that control the fluidic actuation. The responsive action can be reversible thanks to the temperature sensitivity in HBC-MA, which influences the mechanical strength of the 4D structure [45–47]. Fig. 1.5 shows the temperature response in the hydrogel of the microfluidics. In the responsivity test, the printed 4D structure changes the swelling ratios with temperature. Tiny pores are formed by HBC-MA between the pore-free walls made of photopolymerized poly (ethylene glycol) (PEG)-MA, and the pore structure changes after bending or

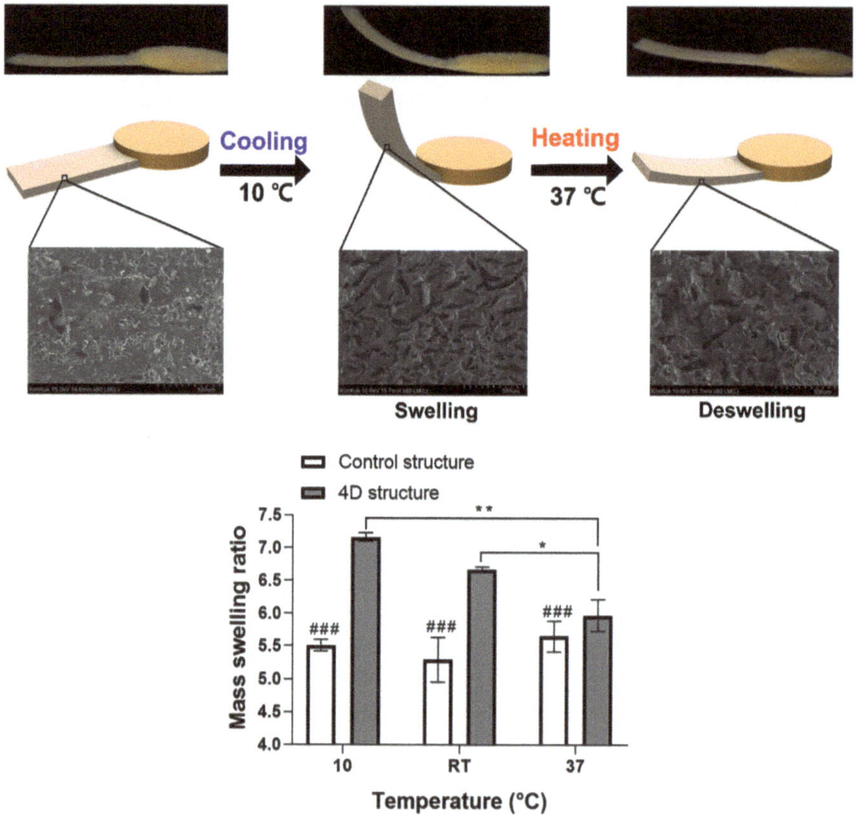

FIGURE 1.5

4D-printed microfluidics. (Top) Schematic shows the entire process of the reversible dynamic movement of 4D printing, showing bending movements upon cooling and unbending movements upon heating in the printed 4D structure. The zoom-in from the schematic shows the microscopic images of the pore structures. (Bottom) Analysis of the equilibrium swelling characteristics of control regarding different swelling ratios with temperature in the 4D printing structure [47].

Reprinted with permission from J.W. Seo, S.R. Shin, Y.J. Park, H. Bae, Hydrogel production platform with dynamic movement using photo-crosslinkable/temperature reversible chitosan polymer and stereolithography 4D printing technology, Tissue Eng. Regen. Med. 17 (2020) 423–431. https://doi.org/10.1007/s13770-020-00264-6.

cracks are formed from bending during swelling. Also, thermal cross-linked pores with deswelling were observed at a temperature higher than room temperature. Although they are not extensively used in microfluidics currently, microfluidic regulation and performance are expected to further improve by introducing these materials to printed microfluidics.

1.3 Integration of printed microfluidics into functional biosensors

The integration of other components, i.e., recognition elements, signal acquisition, processing and display units, to be an operational biosensor without the need for additional equipment or laboratory could turn the printed microfluidics into functional POC biosensor for biomedical use. This chapter covers the integration with biological sensing elements, i.e., molecular recognition elements, actuators for fluidic and temperature regulation, which is critical to control the bioreaction, and the transducers, including components for optical and electrochemical detection.

1.3.1 Conjugation of molecular recognition elements

In a molecular recognition system, the biological recognition element, or the bioreceptor, captures the analyte of interest, i.e., target analyte. These can be small molecules such as ions, drugs, toxins, nucleic acids, proteins, or even cells. For example, in the biosensing of viral infection, the presence of the viral RNA and surface proteins, such as spike protein, indicates the presence of the virus in a sample. Serum biomarkers, such as microRNAs that regulate gene expression post-transcriptionally, cytokines and T cells that participate in the immunity, will serve as the fingerprint of the diseases and their infectious status, e.g., what is the endocrine signaling and immune response of an individual to the viral infection [48,49].

Biological recognition elements that can specifically bind to the target analyte can be categorized into nucleic acid-based and protein-based elements. Nucleic acid- and protein-based recognition elements targeting nucleic acids and proteins or other small molecules are commonly employed in microfluidic biosensor. There are other types of recognition elements, such as metal ions, that target metal ion-binding analytes, but they are seldom used for biomedical applications.

For nucleic acid-based recognition elements, the specificity relies on hybridization between complementary bases, i.e., A to T or U, G to C. The nucleic acid probe of a specific sequence will hybridize to its reverse complementary strand DNA or RNA. Thanks to the convenience of designing a probe sequence for specific targeting, it has been used for decades in most nucleic acid-based biosensing assays, which can be classified as nucleic acid amplification tests (NAAT) and microarray gene chips.

Polymerase chain reaction (PCR) is one of the NAAT relying on thermocycling reaction to amplify primer-bound target DNA sequence. With the help of DNA-binding dyes or sequence-specific DNA fluorescence probes, such as molecular beacons, the presence of a biomarker can be detected and quantified as they emit light at a particular light excitation wavelength. PCR can also be applicable for RNA detection, such as viral identification based on viral RNA sensing, where reverse transcription (RT) with enzyme reverse transcriptase will be needed to convert the RNA into cDNA before PCR. The advantages of RT-PCR include high sensitivity that can target only a few copies of DNA and RNA and high specificity to be multiplexable for many targets. Therefore, it has become the most common NAAT in the past few decades. There are also many reagent cocktails, pre-mixed with all RT and PCR ingredients, available to be incorporated into microfluidics [50,51]. However, the need for accurate thermocycling and high energy consumption discourage its translational use in POC biosensors.

Various isothermal amplification methods, such as rolling circle amplification (RCA), recombination polymerase amplification (RPA), which can occur as low as at room temperature, has drawn additional attention since they can be performed without thermocycler [52]. They have therefore established an instrument-free nucleic acid test in the last decade [53,54]. Although isothermal amplification also yields high sensitivity, its specificity is often compromised by self-priming or mispriming since annealing of primers to partially complementary sequences other than targets happens at a lower temperature. In this context, the coupling with CRISPR reaction, which relies on enzyme Cas13 that cleaves reporter probes when bound on the target sequence, has a single-nucleotide specificity that enables it to detect single-nucleotide polymorphisms [55,56]. Loop-mediated isothermal amplification (LAMP) is the solution to the complicated heating and cooling procedures as the reaction can be completed at a constant temperature, ranging from 50°C to 70°C, which is higher than other isothermal amplification methods but ensures a higher specificity. Although isothermal amplification methods usually involve more reagents and enzymes than that of PCR, the reagent cocktails have been commercially available in both dry lyophilized pellet format and solution to be pre-inserted or injected into the microfluidics [57,58]. For example, a reagent cocktail containing enzymes for CRISPR has been integrated into a lateral flow read-out format and tested its possibility for field-deployable assay [59]. Printing-based immobilization of these to the printed microfluidics will not be affected by the variation of the NAATs as long as the solutions or lyophilized powder are made available to fit the printer.

For protein-based recognition elements, antibody, protein receptor, and enzyme are the three most common elements for integration. The antibody has been employed in immunoassay since the 1950s for biosensing assay in the clinical laboratory, and tags, including radioactive, fluorescent, and enzymatic tags, have been added as the signal reporter. However, the stability of antibodies can be challenging: heat, pH values, and ionic strength induce denaturation of antibodies and hamper the functionality of antibodies. To prevent this denaturation and subsequent dysfunctionality, stabilizing agents, such as sucrose and trehalose, can be used when printing on a surface. For example, a pullulan-trehalose sugar mixture contains 8% pullulan-based ink, a reversibly gelling polysaccharide, and 0.25 M trehalose in a HEPES buffer of 5 μL that can be spotted at 100 Hz rate with 1.2-ms drop deposition time on top of an NC membrane paper and dried prior to use [60]. Besides stabilizing agents, antibody fragments, e.g., scFv antibody, Fab antibody, may replace animal-raised antibodies. They are the minimization of a functional antibody with similar affinity but higher stability and can be produced in genetically engineered bacteria at a lower cost [61,62]. Another solution is using an engineered binding protein with resistance to thermal denaturation, i.e., hyperthermophilic property [63]. This property also brings the benefit of storing ingredients on site before printing.

Instead of the attempt to increase the stability, there are researches in using DNA that can substitute protein-based recognition elements to the target protein. Aptamer, a single-strand nucleic acid that can fold itself into a unique 3D configuration to act similarly to antibody, has been introduced [64]. The advantages of nucleic acid-based protein-based recognition elements, such as low cost, low log-to-log variation, and on-demand production, also apply to the aptamer, in addition to its higher thermostability and chemical stability in comparison to antibody. Apart from its functional feature in molecule recognition, the aptamer can also act as a signal reporter as a nucleic acid sequence that can be bound with a primer of complementary sequence in NAATs [65,66]. It can also be linked to an aptamer that converts 3,3′,5,5′-tetramethylbenzidine (TMB) into color products, such as the enzyme in ELISA bioassay as the visual signal output [67]. For example, a biotechnology company "Aptagen" has intensively tailor-made linked target-binding aptamers and TMB-converting aptamers into apta-beacons. The TMB-converting aptamer will only be folded into an active form for TMB color conversion upon the presence of the target. As a result, the limited stability of TMB-converting enzyme peroxidase, e.g., horseradish peroxidase, will no longer be an obstacle to implementing the current TMB-based assay into a reliable POCT.

Nanomaterials have been popular since the 1990s for enhancing biosensing signals upon recognition element and analyte interaction. Recent developments have greatly improved the sensitivity of optical sensors based on metal nanoparticle arrays and single nanoparticles [68,69]. Metallic nanomaterials, e.g., gold nanoparticles, silver nanoparticles, iron oxide nanoparticles, possess excellent absorptive and scattering optical properties, a high surface area to volume ratio for recognition element attachment. The visual color is tunable by simply adjusting the size and shape of the nanoparticles, and these products are easily accessible in the market nowadays. Moreover, there are well-established protocols for surface functionalization, such as click chemistry, biotin–avidin interaction, and thiol ssDNA, to attach different types of recognition elements. Fluorescence detection is also possible by the use of fluorescent nanoparticles, e.g., fluorescent-doped polymeric nanosphere, quantum dots (QDs), and upconverting and down-converting nanoparticles (UCNPs and DCNPs) [70,71]. However, because of the biological and environmental toxicity of QDs, they are less sustainable choices for printed microfluidic biosensors, which will be disposed of outside the laboratory after use [72]. Nevertheless, these nanomaterials are commercially available and direct inkjet printing with the inks containing nanomaterials can be used to print them onto microfluidics [73,74].

There are a lot of printing-based immobilization methods for microfluidics nowadays. The most accessible printing principle is the method for making lateral flow assay, although the printing is only 1D, i.e., line printing [75]. Recognition elements can be printed on site with the inkjet dispersion technique, usually used for paper test strips. In this process, the solution containing the recognition element and other reagents, such as stabilizers and viscosity regulators, is sprayed onto the paper test line and control line. Additional movement in dimensions will support printing in a 2D or 3D manner. Commercial products are also able to print recognition elements on various types of surfaces such as plastic and paper. Dispensing equipment, e.g., BioDot printers, HP D300e Digital Dispenser, has recently demonstrated its precision in dispensing solutions from pico- to microliter volumes of spots, as a 3D bioprinting. Meanwhile, the possibility of omitting the step of immobilization has been explored by incorporating the immobilized materials into the printing materials, e.g., the filaments used in 3D printing. For example, a filament made of polyvinyl alcohol (PVA), mixed with *o*-phenylenediamine (OPD), a chromogenic substrate that converts into color products upon interaction with H_2O_2, and glucose oxidase used in blood glucose test, and another filament made of ABS resin, incorporating peroxidase—mimicking iron oxide (Fe_3O_4) nanoparticles and forming a Fe_3O_4-filament, have been used for one-step fabrication of micro-wells [76]. Since OPD color conversion is common in conventional bioassays, such as ELISA, this printing approach can be utilized for other bioassays with the same principle, i.e., converting OPD into color products as signal visualization, by shifting the recognition element from, in this case, enzyme glucose oxidase to other products.

1.3.2 Integration of actuators

The actuator is particularly important in printed microfluidic biosensors than in other biosensors because the fluid flow in micron-scale or nanoscale can be challenging. In the conventional printed lateral flow assay, only simple passive-based diffusion is needed. However, for lab-on-a-chip, i.e., conducting all the laboratory works inside one chip, and chip-in-a-lab, the microfluidic chip must be connected to sophisticated equipment to control its fluidic actuation properly in order to perform the desired biosensing functions in the correct sequential order.

Efforts have been made to miniaturize this sophisticated equipment into a portable format to integrate with microfluidics as a biosensor. In the lateral flow assay, the fluidic flow is unidirectional, one-dimensional from the sample chamber, or reservoir, to the reaction lines and toward the end of the strip, jointed with an absorbent pad. Since the lateral flow assay was designed as a simple diagnostic device for general end users with no technical experiences, the control of fluidic speed is not critical in the assay and there is a better tolerance to volume variation in the specimen. For example, in urine pregnancy tests, no exact volume of urine is needed, as long as it is sufficient to reach the reaction lines. Similarly, in the 2D paper- and membrane-based microfluidics, the fluidic driven force is based on passive capillary action, and therefore properties of different base materials, i.e., different types and features of paper and membrane in the 2D microfluidics, should be well-characterized to understand how fluids of different viscosities and compositions actuate inside the materials in order to control fluidic actuation properly.

To guide the fluid to move from one direction to another, materials of different water retention capacities can be used. For example, NC has a better water retention capacity than glass fiber. The overlap between NC and glass fiber will cause the solution content in the latter to drain completely. To dynamically adjust the fluid flow speed, the end terminal of the microfluidic can be connected to absorbent pads of larger water retention capacity with different surface areas exposed to the outer environment. The evaporation force provides an additional fluidic driving force for active actuation. On the other hand, the flow speeds of microfluidics made by the same base material can be made tunable by adjusting the channel's length and width, and the capillary force will change accordingly. Start- and stop-flow of the fluid can be regulated by the separation of fluidic channels into various partitions, i.e., discrete elements, where they can be reconnected with origami folding in a microfluidic as described in pseudo-3D microfluidics. However, dynamic actuation of fluidic direction and speed cannot be easily controlled, as a pump cannot be easily integrated in 2D-printed microfluidic. Paper or the membrane itself absorbs the liquid so that large-volume processing will be challenging and the predicted model may not fit all the different types of 2D microfluidics [77]. Another way of controlling the flow rate is centrifugation. It is an indirect method that applies a centrifugal force to the fluid. By integrating the paper strip in a microfluidic disk and subjecting it to spinning action, the fluidic speed can be regulated, either acceleration or deceleration, in addition to the capillary action of the lateral flow assay [78–80].

Dynamic actuation of fluidic direction and speed can be continuously done in 3D microfluidics when each channel is connected to a commercialized pump with high precision. However, these pumps are not readily employed for in-field use because of their high cost and bulky size. Besides redesigning these microfluidics to reduce the complexity in fluidic actuation and increasing the tolerance of variation in controlling fluidic volume, 3D-printed pumps, such as peristaltic pumps, have offered a low-cost solution of on-site rapid fabrication and integration into the microfluidic channels [81,82]. In contrast to the convention pump, centrifugal microfluidic is an approach relying on spinning action in a microfluidic for active fluidic actuation. The microfluidic modification is minimal, as long as the direction of movement is not altered. Centrifugal microfluidics, including the microfluidic disk and centrifugal platform, i.e., the holder that allocates the microfluidics, embedded with PCB for additional regulation, such as heating, can be easily constructed by 3D printing technology [83–85].

Besides, finger-powered fluidic actuation was introduced for elastic microfluidic materials. In this actuation, positive and negative pressures are created by the pressing action and are applied to the microfluidic chamber. The channels connected to this chamber respond to the pressure change by pushing the solution out of or pulling them into the chamber. By using flexible materials in 3D printers, finger-powered fluidic actuation and mixing units were produced by 3D printing [86].

Although finger-powered actuation is practical in a low-resource setting, it is not suitable for fine control of the fluidic actuation. One solution is to redesign the microfluidic platform to imitate the shape of a pipette tip and its adapters, which can fit commercial pipettes. When this microfluidic with a pipette tip connects to a pipette, it can attain high accuracy in its two-directional fluidic actuation, i.e., suction or release. The actuation force is represented by the pipette volume unit and can be adjusted conveniently.

Antibodies coated on the inner walls of the 3D-printed tip enable the visualization of signal output as well as charge-coupled device (CCD) camera quantification [87]. The fine control of fluid actuation, i.e., start- and stop-flow, can be further enhanced with printed valves, e.g., cyclo-olefin copolymer (COC) and polyethylene terephthalate (PET) films. The valves, responsive to the optical energy, can be installed in the microfluidics and controlled by the laser beam signal, which turns on the valve and guides the fluid actuation from stop flow to continuous flowing [88].

1.3.3 Integration with optical and electrochemical signal acquisition units

Optical and electrochemical methods have been most frequently used in POC rapid biosensing tests because the integration of optical and electrochemical sensors is easy and their stability for prolonged use has been well evaluated. The stability, accuracy, and sensitivity of optical and electrochemical sensors in detecting optical and electrochemical signals have improved over time [89]. Usually built for laboratory use, conventional sensors are bulky and sensitive to collision, thus hindering their use as portable signal acquisition units. To overcome this barrier, printing technology has simplified their construction and miniaturized them for POC use, and in turn facilitated the fabrication of sensors, e.g., optical lens and electrodes, and their integration with microfluidics for optical and electrochemical sensing.

Optical measurement based on naked-eye colormetric detection has been popular for POC use, since minimal equipment is needed to record the signal. Among the approaches to obtaining the optical signal, the use of smartphone cameras for smartphone-based imaging has been increasingly important [90]. This can be seen as the miniaturization of a laboratory imaging device or even microscope when coupled to an objective lens, compatible and aligned with the smartphone camera. In addition to imaging, thanks to the built-in multifunctionality, e.g., data processing with own CPU power and communication, such as cloud computing, post-processing of the captured images and videos, such as storage, analysis, and distributing, can be carried out on the same phone, unlike other imaging devices that may require accessories. Because of the prevalence of smartphones, mobile phone repair stores are numerous and therefore hardware maintenance is easily accessible.

3D printing has been used to fabricate a smartphone platform to fit into the detection site of a microfluidic chip for on-site monitoring [91]. 3D-printed stage has been demonstrated for positioning the phone and the microfluidic holder and fitting different microfluidics into the image area steadily. High resolution of the smartphone camera makes possible microscopic imaging with a submicron scale resolution when it is coupled with a microscopic objective lens. To reduce the cost, various methods were developed to replace the expensive microscopic objective lens. For example, embedding a glass bead, acting as a ball lens, in a 3D-printed holder connected to a smartphone has been made with detailed tutorials and commercialized.

Furthermore, printing optical lenses can reduce the barrier of obtaining commercial lens in remote regions. Printed products should have excellent optical properties, e.g., transparency and accuracy in light-guiding property, and their fabrication should be robust. Glass has also been 3D-printed in high-temperature settings, similar to general glass production. However, in conventional 3D FDM printers, the highest temperature is usually below 300°C, making it challenging to print glass. Therefore, other polymers have been examined for potentially 3D-printing lenses in the last decade. Among them, an inverse droplet of resin PDMS liquid on a glass slide has been characterized with the microscopic imaging property with high resolution. PDMS is produced with 3D printing technology and the product is optically transparent and has high strength comparable to glass. This enhances its extensive use in optical biosensing, so that the signal visualization part of microfluidics can be further minimized [92]. Also, IP-S photopolymer, a photoresistant that can create smooth surfaces within a short period of light illumination, has been used to fabricate a high NA compact catadioptric lens that is even suitable as a microscopic lens [93].

Apart from printing high-quality optical lens, it is essential to make the printed microfluidics highly transparent for optical detection. Instead of simply using optically transparent materials, e.g., silica glass and SU-8, commonly exploited in photolithography for fabricating traditional microfluidics, plastic is another choice of material. But the trouble in printing plastic with high transparency lies not only in the materials, such as PDMS, poly(methyl methacrylate) (PMMA), being used, but also in plastic heating profile and actuation of the printed head during the FDM process [94]. For example, PMMA is fused with ABS by the solvent bonding method, creating a transparent upper layer for 3D microfluidics without re-characterizing the fluidic actuation properties of a microfluidic with new printing material [95]. It is noteworthy that PMMA and PDMS have different light-guiding properties than glass, so the optical setting should be realigned when the microfluidic is modified to be 3D printed. Also, VeroClear, a novel transparent 3D printing material with a refractive index similar to silica glass (1.52 at 589 nm), can be an alternative to PMMA and PDMS [96]. Fig. 1.6B shows a highly transparent 3D-printed microfluidic using printing methodology to improve

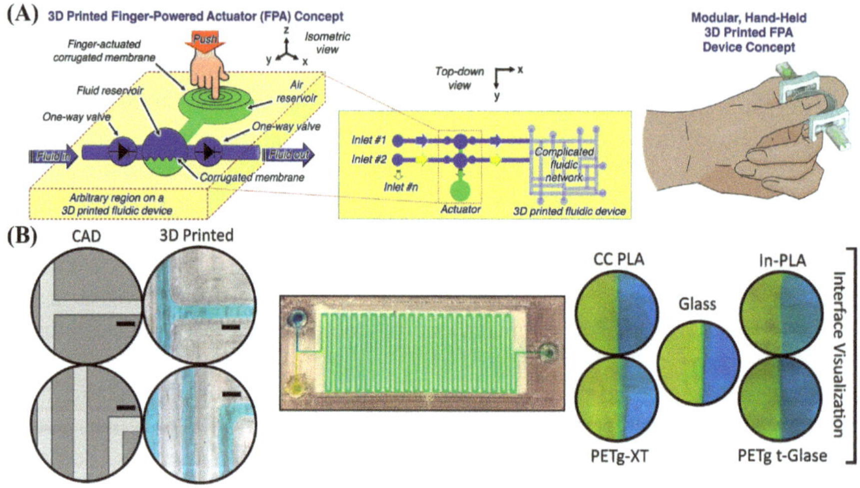

FIGURE 1.6

Integration of the printed microfluidic biosensor. (A) With a 3D-printed fluidic actuator. Illustrations of the 3D-printed FPA design, which can be incorporated into (left) the interfacing between a source fluid reservoir and a discrete device at any arbitrary location where fluid actuation is desired; (middle) an arbitrarily complicated 3D printable microchannel network for the simultaneous actuation of multiple fluid channels; (right) a modular format for human-powered actuation; (B) optical imaging with a highly transparent microfluidic channel. (Middle) A 3D-printed transparent microfluidic using printing methodology for improving the clarity of the microfluidic device. Transparency characterization on (left) the tracking intensity changes; both the CAD design and images of 3D-printed channels filled with a solution of blue dye are shown; (right) visualization of two touching laminar streams (*yellow and blue dye*) interface inside the printed microfluidic [86,97]. *CAD*, Computer-aided design; *FPA*, finger-powered fluidic actuator.

Reprinted with permission from E. Sweet, R. Mehta, Y. Xu, R. Jew, R. Lin, L. Lin, Finger-powered fluidic actuation and mixing via MultiJet 3D printing, Lab Chip 20 (2020) 3375–3385. https://doi.org/10.1039/D0LC00488J; V. Romanov, R. Samuel, M. Chaharlang, A.R. Jafek, A. Frost, B.K. Gale, FDM 3D printing of high-pressure, heat-resistant, transparent microfluidic devices, Anal. Chem. 90 (2018) 10450–10456. https://doi.org/10.1021/acs.analchem.8b02356.

the microfluidic device's clarity. To characterize transparency of clarity in visualizing the microfluidic channel, the color intensity of a blue dye solution filled in the channels was tracked and two touching laminar streams of yellow and blue dyes inside the printed microfluidic were distinguished [97]. The apparent color yellow and blue dye interface supports the transparent 3D-printed microfluidic for optical-based biosensing measurement.

Post-processing the 3D-printed transparent plastic can further improve their optical smoothness, i.e., surface smoothness, which is critical when evaluating the percentage of light dispersion to yield a low error in reflection, refraction, and scattering during sensing. Although acetone vapor is often used in polishing and spraying of polymer coatings in commercial 3D-printed objects, it may not achieve high optical smoothness in the case of plastic. Wet-sanding and polishing are used to reduce the overall surface roughness in nanoscale and the polished glass behaves similarly as conventional lens made from BK7 or SF2 glass, which can support high clarity demanded by optical sensing methods, such as surface plasmon resonance (SPR) sensing [98]. Coating the printed optics with a UV curable polymer mixture consisting of methacrylates, acrylates, and urethane-based polymers could also increase its surface smoothness to be reflective mirrors for the downstream procedure for metallization [99,100].

Overall, improvement in transparency of printed microfluidics enables visual observation and detection. To obtain a more accurate reading in quantification, the choice of printing materials should be taken into consideration, so as to match absorption and transmission properties with the requirements of biosensing. For spectrum analysis of the visible region, transparent plastic is acceptable, as long as its intrinsic absorptive spectra do not entirely block the light of a particular wavelength or affect the spectra measurement in the samples inside the microfluidics. For fluorescence detection, the plastic has a different optical properties than that of the glass materials. For example, the plastic has a lower transmission in the UV wavelength region, may affect the biosensing assays which need UV excitation light.

Electrochemical sensing is another major stream of signal acquisition, as the signal obtained, e.g., from square wave voltammetry measurements, is an intrinsic digitalized value. Many portable electrochemical sensing instruments are well connected with mobile devices, such as laptops or smartphones [101,102]. In electrochemical sensing, the sensing electrode is one of the critical elements that affect sensing sensitivity and reliability. Conventionally, the sensing electrodes are prepared in laboratories or factories before sealing, storage, and shipping. In industrial-scale manufacture, a screen printer with designated mesh polyester screens is used to deposit a carbon layer onto flexible films, e.g., PET substrates, to produce the electrodes and electrical contact. Since these electrodes are often embedded as a sensing electrode cell, on top of which test samples will be loaded directly, they are considered nonreusable consumables. Therefore, on-site printing of electrodes with a rapid prototyping printing method can tackle centralized production and delivery issues described previously.

Thanks to the improvement in the resolution of printer head-actuation-based printing and conducting inks made by conducting polymers, such as polypyrrole, poly(3,4-ethylenedioxythiophene) (PEDOT), inkjet-based printing has become common for printing electrodes. The office laser printer and inexpensive home craft printer have long been used to fabricate the electrodes on transparent films purchased from local stationery stores. They are then stacked with double-sided adhesive polystyrene cards for constructing a disposable microfluidic device [103,104]. The default carbon printing cartridge is also flexible for further modification to print other materials, such as silver [105]. Nanoparticle imprinting on the electrode is known to enhance the electrochemical signal. Traditionally, the nanoparticles were synthesized in a solution format and then disposed of on the electrode by dipping the electrode into the solution of nanoparticles in a controlled coating environment, with respect to temperature and time. Conductive graphite-based ink containing graphene oxide has been a popular choice to coat an additional layer as a graphene oxide-modified electrode [74,101,102]. Electrode modified with photoactive nanomaterials, such as CdS nanoparticles and $ZnIn_2S_4$ (ZIS) nanostructure, could enable the signal acquisition based on photocurrent, instead of conventional electrochemical oxidative/reductive signal, and the latter nanostructure further triggers the amplification of photocurrent for a better signal output [106,107].

1.4 Biomedical applications with printed microfluidic biosensors

Disease and food safety are two main biomedical issues that are inseparable in their relationship with human health. Accurate, timely biosensing and proper real-time monitoring hold the key to realizing a proactive action for interventive and preventive care. This section covers how printed microfluidic biosensors target biomarkers of cancers for disease screening, foodborne pathogens, additives, and contaminants for food safety.

1.4.1 Disease screening

Cancer is one of the leading causes of death worldwide. Early diagnosis and adequate treatment are known to reduce mortality. Biomarkers useful in screening of cancer include (1) oncogene or mutation of a gene, (2) overexpressed protein, and (3) circulating tumor cells (CTC). For example, epidermal growth factor receptor (EGFR), a transmembrane protein that which acts as a receptor to bind with EGF family involving signaling pathways that promote cell proliferation and protein synthesis, is regarded as an oncogene when overexpressed or mutated [108].

1 The detection of the relative expression level of EGFR mRNA and its mutation, which can be point mutation or deletion, is often used in diagnosis for adenocarcinomas and glioblastomas [109]. NAAT is the standard approach to the detection of EGFR mRNA and mutation for rapid cancer diagnosis. As described previously, isothermal nucleic acid amplification is a better choice of NAAT. The development of a rapid test with it has been demonstrated to determine EGFR mutation from extracted genomic DNA samples under 10 minutes in a test tube-based format [110], and it is feasible to be incorporated into printed microfluidics for complete sample-to-answer POCT.

2 EGFR protein biomarker is more often reported in proteomic analysis than POCT since the detection of mRNA and mutation have already provided adequate diagnostic indicators for further therapy. Therefore, there is lack of development of microfluidic-based POCT to target EGFR protein. On the other hand, carcinoembryonic antigen (CEA) is a glycoprotein involved in cell adhesion, which is normally present in the bloodstream. Its increased level has been related to various cancers, e.g., colon cancer. Therefore, it is useful as a tumor biomarker used in POCT in the rapid screening of cancer. High detection sensitivity and specificity are more often stressed for detecting protein cancer markers when compared to NAAT as NAAT has shown an intrinsic high sensitivity and specificity by amplifying the target gene region with sequence-specific primer. In the case of detecting CEA protein, an electrochemical aptasensor and optical sensing with a smartphone were reported, with the detection limit in the picomolar range. It also demonstrated a high sensitivity of more than 95% and specificity of 95% in the detection of human samples for evaluation of the clinical performance [111,112]. Apart from diagnosis for the discovery of the disease, other biomarkers, such as lactate dehydrogenase (LDH) related to prognosis, were commonly detected together with other biomarkers, such as uric acid and urea, to reveal the efficiency of anticancer drug treatment [113,114], and the practical implementation in microfluidics are relatively easy as multiplex assays from a sample can be designed with sample partition from a single microchannel into different zones during sample actuation.

3 Circulating tumor cells (CTC) are the cancer cells circulating in the bloodstream, and their appearance are usually correlated to cancer metastasis, where the cancer cells, in the form of either single cells or clusters, are released from a solid tumor site and migrated through the bloodstream [115]. The relationship between CTC and cancer was discovered more than a decade ago, and it was used as a biomarker in liquid biopsy, e.g., blood, to detect cancer [116–118]. Since their concentration is low in circulating blood, enrichment or isolation prior to sensing is often required to minimize false-positive signals or inhibit the sensing effect due to other blood cells [119,120]. Prior to biosensing, the enrichment of tumor cells is performed by injecting a whole blood sample, which contains blood cells and other residues, into the 3D-printed microfluidic device. In detail, a membrane filter is used to retain nucleated cells while eliminating anucleated blood cells. The white blood cells are captured in the multilayered immunocapture channels, which consist of 200 μm-diameter microposts within the microfluidic layers, and arrayed micropillars that maximize cell–micropost interactions within each layer [121,122]. Common 3D printing materials, such as ABS, may hinder prolonged cell survival, including cell seeding and culture inside the microfluidics. The versatility of printing materials should be taken into consideration in facilitating cell culture in microfluidics [123,124]. In addition, image-based CTC analysis was proposed. With a high-resolution smartphone

camera of a small dimension, counting cells on paper was also made possible and showed a limit of detection (LOD) of one cancer cell per μL. This type of portable, smartphone-based fluorescence imaging may be helpful in future POC CTC detection [125].

1.4.2 Food safety

Foodborne diseases impede socioeconomic development by straining healthcare systems and harming national economies, tourism, and trade [126]. Food poisoning caused by bacterial contamination in raw and processed food is a common cause of fatality. For instance, the toxins released from some bacteria, such as *E. coli* O157:H7 and *Listeria monocytogenes*, do not degrade and are not detoxified upon washing and heating in standard food preparation procedure. Thus, identification of bacteria and toxins before distribution and consumption of food material is a public health priority.

Both 2D- and 3D-printed microfluidics have been used to detect *E. coli* O157:H7, and the anti-*E. coli* O157:H7 antibody or aptamer are recognition elements for detection via impedimetric sensing or paper-based ELISA colorimetric assay. In this case, the microfluidics do not need to do much sample processing, such as bacterial lysis, since the recognition elements are used to target the surface antigen of the bacteria. The LOD could reach 10^4 CFU/mL in standard samples and 1 CFU per 25 g beef sample [20,127]. Without further sample treatment, it is already more sensitive than the conventional ELISA biochemical assay used in the centralized laboratories. The 3D-printed device is capable of conducting more complicated procedures, such as size-based separation in a helical channel with a trapezoid cross section, which cannot be conducted in 2D microfluidics [19]. In addition to bacteria, toxins produced by toxin-producing algae are also common contaminants. Algae and their toxins in aquatic ecosystems can detrimentally affect the water quality. 3D-printed smartphone platform integrated with a light-driven microfluidic chip operated by optoelectrowetting (OEW) provided on-site monitoring of algae in water [91]. In contrast to relying on surface antigens to detect bacteria, algae are large enough to be visible under the microscope after dye staining. Surprisingly, a smartphone camera with a digital zoom of 45 × is enough to count the algae stained in fluorescence dye for further quantitative analysis. Although this biosensing principle is not highly specific to the algae species due to the nonselective feature of the dyes used, other existing solutions, such as antibody–fluorescent dye complex, help the specific labeling on the algae.

The emergence of multidrug-resistant (MDR) bacteria, which are often fatal due to the lack of effective drugs, is linked to misuse or abuse of antibiotics. There is an increasing demand for detecting antibiotic residues and an antibiotic susceptibility test of bacteria in samples. The antibiotics found as residues in pork or the bloodstream could be determined with the 2D-printed paper-based microfluidic biosensor [128,129]. However, the biosensing principle relies on the chemical reaction between antibiotics and the reactive reagent to lead to the metal ion–antibiotic complex precipitate formation. Cross-reactions may take place for other antibiotics of the same family that share the same functional groups and may result in false-positive if the sample contains other biological compounds that also react with the reactive reagent.

Food additives, such as dyes and nitrites, have long been used as preservatives and enhancers to appeal to customers. The intake of commonly used food dyes raises concerns about their toxicity. The colorimetric sensing method, which requires minimal sample preparation and equipment, is the conventional detection and quantification method of food additives. Although it might not be accurate enough to identify the exact chemical formula of the food additives compared to more supplicated laboratory equipment, such as HPLC and mass spectrometry, the colorimetric sensing method can be easily incorporated into microfluidics for a quick assessment. Pseudo-3D-printed micro-chromatographic paper origami microfluidics demonstrated the detection of two azo food dyes, tartrazine (E102) and indigo carmine (E132). Based on the result in the chromatographic separation of the food colorant mixture, it achieved a LOD of 0.620 and 0.060 g/L, respectively, which is sufficient for evaluating whether the food dyes exceed the threshold for safe consumption [34]. Besides, nitrite is used as a strong oxidant to produce the bright red color in meat and imitate fresh meat, and as an inorganic fertilizer in farms from where it may leak into drinking water source. It can form carcinogen metabolites inside the human body [130]. The coffee-ring effect, the result of varying evaporation rates of analytes moving across a surface, was introduced to enhance visual detection, thus improving the detection limit [131]. As shown in Fig. 1.7A, the 2D-printed microfluidic biosensor was placed in an oven at 45°C, the temperature below the melting point of the printed wax channels, to produce a distinct coffee-ring effect to enhance visual detection. The signal was then recorded with a mobile phone camera for image-based analysis. The LOD of nitrite in pork was determined to be 19.2 mg/kg by analyzing the inner chamber reaction, while it could be as low as 1.1 mg/kg if the coffee-ring regions were analyzed [24,132]. With a more complex design with the PCB-based palm-size digital microfluidics (DMF), the LOD can be reduced to 5 μg/L, which is 60-fold more sensitive than the lowest acceptable limit of 300 μg/L nitrites in drinking water [130,133].

Hormones and heavy metals are common food contaminants. Bisphenol A (BPA), commonly found in drinking water, especially inside plastic bottles, is considered the priority hormone to tackle. Both 2D- and 3D-printed microfluidics have been utilized in the electrochemical sensing of BPA with an LOD of 30 nM. The former microfluidics (Fig. 1.7B), conveniently used for on-site sensing, employs the entire electrochemical cell printed on filter paper for the reagent-free reaction, while the latter takes part in more downstream analyses, such as coupling with and laser desorption ionization mass spectrometric detection (LDI-MS) that improves the precision of BPA detection [14,134]. The disadvantage of the latter is its portability, since the LDI-MS detector is yet to be made into a small-

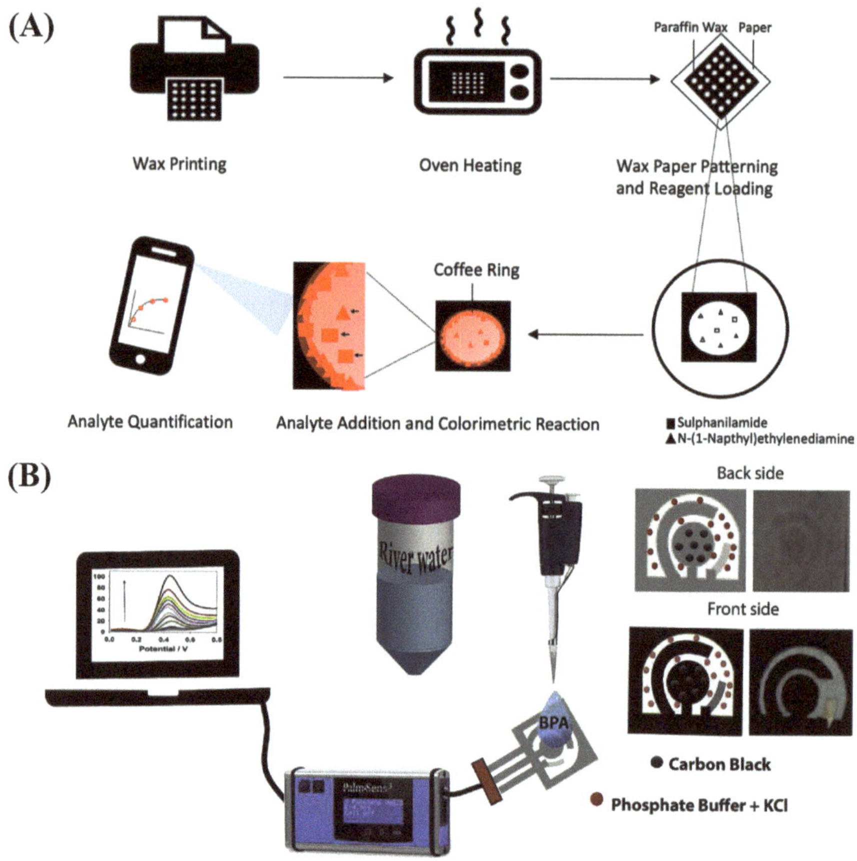

FIGURE 1.7

Biosensing with the printed microfluidic biosensor. (A) Schematic diagram of the paper-based microfluidic biosensor fabricated through the wax printing technique and integrated with reagents to quantify nitrite concentration in pork. Camera-based signal acquisition with a smartphone was used to record the color and measure the color intensity of the reaction chamber. (B) Schematic diagram of the configuration of the paper-based microfluidic biosensor and procedures for electrochemical measurement [14,132].

Reprinted with permission from D. Jemmeli, E. Marcoccio, D. Moscone, C. Dridi, F. Arduini, Highly sensitive paper-based electrochemical sensor for reagent free detection of bisphenol A, Talanta 216 (2020) 120924. https://doi.org/10.1016/j.talanta.2020.120924; E. Trofimchuk, Y. Hu, A. Nilghaz, M.Z. Hua, S. Sun, X. Lu, Development of paper-based microfluidic device for the determination of nitrite in meat, Food Chem. 316 (2020) 126396. https://doi.org/10.1016/j.foodchem.2020.126396.

sized, lightweight format. Heavy metal contamination, particularly toxic heavy metals such as arsenic (As), cadmium (Cd), chromium (Cr), lead (Pb), and mercury (Hg), in food and drinking water is a serious health threat because these are nonbiodegradable, able to accumulate in soft tissues, and are classified as human carcinogens [135]. The epitaxial property of graphene exhibits the preferential ability to bind with Pb than Cd ions, so it was integrated into a 3D-printed microfluidic and it showed high selectivity in the sensing of specific heavy metals [136]. However, its selectivity with other nontoxic metal ions has not been evaluated with this graphene recognition element. Therefore, it is questioned how specific is the capturing ability of graphene and its signal output upon specific capture.

1.5 Conclusions and outlooks

The readily available printing technology supports the mass production of printed microfluidic biosensors to be used at the POC. The strategic use of printed microfluidic biosensors could make possible large-scale testing in future disease outbreaks. In spite of the contribution of printed microfluidic biosensors to translational use, there are still doubts over the development and application of printed microfluidic biosensors, including their feasibility in mass production, their long-term sensing ability, biocompatibility for human use, and their reliable utilization in practical use in epidemic control.

The previous sections discuss that the advancement in printing technology and more printing material choices facilitate printing microfluidics from raw ingredients. The forthcoming focus will be the reliability of the integration process of printing microfluidics and other printed components to produce consistent and repeatable results, similar to the mass-produced sensors from the factory. In fact, the current integration by manual procedures or automated steps can be further simplified with one-step multicomponent printing to cut down the downstream assembly of different printed materials; although there is an uncertainty about how these innovative methods can be effectively included in the printed transducers during the printing of the microfluidics [137]. Although there are researches on this one-step multi-material printing, more effort is still needed to materialize it into industrial practice.

Despite the latest trend in printing on a flexible surface and functioning as a wearable sensor for daily healthcare monitoring, the new printing materials' biocompatibility in prolonged contact with the human body is still questionable [138,139]. Another related issue is the toxicity of these materials if used in tailored drug screening, in which living cells are loaded into the printed microfluidic biosensors. Noncytotoxic resin, e.g., biocompatible poly(ethylene glycol) diacrylate (PEGDA) resin, has been well evaluated for cytotoxicity and a suitable material in 3D printing [140], so its incorporation into printed microfluidics will be expected.

This pandemic sparks interests in the practical use of advanced biosensing technologies in disease control and surveillance. The continuous biosensor development will address their robustness to fulfill the POC standards, i.e., increasing the capacity and flexibility and aiming for rapid sample-to-answer sensing. Real-time map-based epidemic control can benefit from the proper integration with computing networks, such as big data analysis and artificial intelligence (AI). The latter suggestion has, in fact, already showcased in AI-based image processing and analysis to identify some diseases such as tumors from histological images, which outruns the accuracy of judgment by medical experts as a potential tool in medical diagnostics [141]. Therefore, it reduces laborious signal recording and analysis from the result generated and improves the sensitivity and accuracy of biosensors. Overall, as their development grows, printed microfluidic biosensors will play a more prominent role in biomedical sensing and POC health monitoring.

References

[1] W.C. Mak, V. Beni, A.P.F. Turner, Lateral-flow technology: from visual to instrumental, TrAC Trends in Analytical Chemistry 79 (2016) 297–305. Available from: https://doi.org/10.1016/j.trac.2015.10.017.

[2] A.K. Yetisen, M.S. Akram, C.R. Lowe, Paper-based microfluidic point-of-care diagnostic devices, Lab Chip 13 (2013) 2210. Available from: https://doi.org/10.1039/c3lc50169h.

[3] C.S. Kosack, A.-L. Page, P.R. Klatser, A guide to aid the selection of diagnostic tests, Bulletin of the World Health Organization 95 (2017) 639–645. Available from: https://doi.org/10.2471/BLT.16.187468.

[4] G.M. Whitesides, The origins and the future of microfluidics, Nature 442 (2006) 368–373. Available from: https://doi.org/10.1038/nature05058.

[5] J. Sibbitts, K.A. Sellens, S. Jia, S.A. Klasner, C.T. Culbertson, Cellular analysis using microfluidics, Anal. Chem. 90 (2018) 65–85. Available from: https://doi.org/10.1021/acs.analchem.7b04519.

[6] D. Gao, F. Jin, M. Zhou, Y. Jiang, Recent advances in single cell manipulation and biochemical analysis on microfluidics, Analyst (Lond.) 144 (2019) 766–781. Available from: https://doi.org/10.1039/C8AN01186A.

[7] Q. Huang, S. Mao, M. Khan, J.-M. Lin, Single-cell assay on microfluidic devices, Analyst (Lond.) 144 (2019) 808–823. Available from: https://doi.org/10.1039/C8AN01079J.

[8] J.F. Loo, H.P. Ho, S.K. Kong, T. Wang, Y. Ho, Technological advances in multiscale analysis of single cells in biomedicine, Adv. Biosyst. 3 (2019) 1900138. Available from: https://doi.org/10.1002/adbi.201900138.

[9] C. Dincer, R. Bruch, A. Kling, P.S. Dittrich, G.A. Urban, Multiplexed point-of-care testing – xPOCT, Trends Biotechnol. 35 (2017) 728–742. Available from: https://doi.org/10.1016/j.tibtech.2017.03.013.

[10] A.P.F. Turner, Biosensors: sense and sensibility, Chem. Soc. Rev. 42 (2013) 3184. Available from: https://doi.org/10.1039/c3cs35528d.

[11] A.D. Warren, G.A. Kwong, D.K. Wood, K.Y. Lin, S.N. Bhatia, Point-of-care diagnostics for noncommunicable diseases using synthetic urinary biomarkers and paper microfluidics, Proc. Natl. Acad. Sci. 111 (2014) 3671–3676. Available from: https://doi.org/10.1073/pnas.1314651111.

[12] J.F.C. Loo, A.H.P. Ho, A.P.F. Turner, W.C. Mak, Integrated printed microfluidic biosensors, Trends Biotechnol. (2019). Available from: https://doi.org/10.1016/j.tibtech.2019.03.009.

[13] A. Kurdekar, L.A.A. Chunduri, E.P. Bulagonda, M.K. Haleyurgirisetty, V. Kamisetti, I.K. Hewlett, Comparative performance evaluation of carbon dot-based paper immunoassay on Whatman filter paper and nitrocellulose paper in the detection of HIV infection, Microfluid. Nanofluidics 20 (2016) 99. Available from: https://doi.org/10.1007/s10404-016-1763-9.

[14] D. Jemmeli, E. Marcoccio, D. Moscone, C. Dridi, F. Arduini, Highly sensitive paper-based electrochemical sensor for reagent free detection of bisphenol A, Talanta 216 (2020) 120924. Available from: https://doi.org/10.1016/j.talanta.2020.120924.

[15] G.C. Ilacas, A. Basa, K.J. Nelms, J.D. Sosa, Y. Liu, F.A. Gomez, Paper-based microfluidic devices for glucose assays employing a metal-organic framework (MOF), Anal. Chim. Acta 1055 (2019) 74–80. Available from: https://doi.org/10.1016/j.aca.2019.01.009.

[16] L. Ge, S. Wang, X. Song, S. Ge, J. Yu, 3D Origami-based multifunction-integrated immunodevice: low-cost and multiplexed sandwich chemiluminescence immunoassay on microfluidic paper-based analytical device, Lab Chip 12 (2012) 3150. Available from: https://doi.org/10.1039/c2lc40325k.

[17] G. Xu, D. Nolder, J. Reboud, M.C. Oguike, D.A. van Schalkwyk, C.J. Sutherland, et al., Paper-origami-based multiplexed malaria diagnostics from whole blood, Angew. Chem. Int. Ed. 55 (2016) 15250–15253. Available from: https://doi.org/10.1002/anie.201606060.

[18] M. Zhang, L. Ge, S. Ge, M. Yan, J. Yu, J. Huang, et al., Three-dimensional paper-based electrochemiluminescence device for simultaneous detection of Pb^{2+} and Hg^{2+} based on potential-control technique, Biosens. Bioelectron. 41 (2013) 544–550. Available from: https://doi.org/10.1016/j.bios.2012.09.022.

[19] W. Lee, D. Kwon, W. Choi, G.Y. Jung, A.K. Au, A. Folch, et al., 3D-printed microfluidic device for the detection of pathogenic bacteria using size-based separation in helical channel with trapezoid cross-section, Sci. Rep. 5 (2015) 7717. Available from: https://doi.org/10.1038/srep07717.

[20] I.G. Siller, J.-A. Preuss, K. Urmann, M.R. Hoffmann, T. Scheper, J. Bahnemann, 3D-printed flow cells for aptamer-based impedimetric detection of *E. coli* crooks strain, Sensors 20 (2020) 4421. Available from: https://doi.org/10.3390/s20164421.

[21] C. Park, J. Lee, Y. Kim, J. Kim, J. Lee, S. Park, 3D-printed microfluidic magnetic preconcentrator for the detection of bacterial pathogen using an ATP luminometer and antibody-conjugated magnetic nanoparticles, J. Microbiol. Methods 132 (2017) 128–133. Available from: https://doi.org/10.1016/j.mimet.2016.12.001.

[22] H.Y. Jeong, B.H. Woo, N. Kim, Y.C. Jun, Multicolor 4D printing of shape-memory polymers for light-induced selective heating and remote actuation, Sci. Rep. 10 (2020) 6258. Available from: https://doi.org/10.1038/s41598-020-63020-9.

[23] J. Casals-Terré, J. Farré-Lladós, J.A. López, T. Vidal, M.B. Roncero, Enhanced fully cellulose based forward and reverse blood typing assay, J. Biomed. Mater. Res., B: Appl. Biomater. 108 (2020) 439–450. Available from: https://doi.org/10.1002/jbm.b.34400.

[24] C.-K. Chiang, A. Kurniawan, C.-Y. Kao, M.-J. Wang, Single step and mask-free 3D wax printing of microfluidic paper-based analytical devices for glucose and nitrite assays, Talanta 194 (2019) 837–845. Available from: https://doi.org/10.1016/j.talanta.2018.10.104.

[25] H. Towbin, T. Staehelin, J. Gordon, Electrophoretic transfer of proteins from polyacrylamide gels to nitrocellulose sheets: procedure and some applications, Proc. Natl. Acad. Sci. U.S.A. 76 (1979) 4350–4354. Available from: https://doi.org/10.1073/pnas.76.9.4350.

[26] J.-H. Wu, C.-H. Wang, Y.-D. Ma, G.-B. Lee, A nitrocellulose membrane-based integrated microfluidic system for bacterial detection utilizing magnetic-composite membrane microdevices and bacteria-specific aptamers, Lab Chip 18 (2018) 1633–1640. Available from: https://doi.org/10.1039/C8LC00251G.

[27] E. Eriksson, J. Lysell, H. Larsson, K.Y. Cheung, D. Filippini, W.C. Mak, Geometric flow control lateral flow immunoassay devices (GFC-LFIDs): a new dimension to enhance analytical performance, Research 2019 (2019) 1–8. Available from: https://doi.org/10.34133/2019/8079561.

[28] B.R. Lutz, P. Trinh, C. Ball, E. Fu, P. Yager, Two-dimensional paper networks: programmable fluidic disconnects for multi-step processes in shaped paper, Lab Chip 11 (2011) 4274–4278. Available from: https://doi.org/10.1039/C1LC20758J.

[29] Y. Lu, W. Shi, J. Qin, B. Lin, Fabrication and characterization of paper-based microfluidics prepared in nitrocellulose membrane by wax printing, Anal. Chem. 82 (2010) 329–335. Available from: https://doi.org/10.1021/ac9020193.

[30] E.A. Phillips, R. Shen, S. Zhao, J.C. Linnes, Thermally actuated wax valves for paper-fluidic diagnostics, Lab Chip 16 (2016) 4230–4236. Available from: https://doi.org/10.1039/C6LC00945J.

[31] J. Potter, P. Brisk, W.H. Grover, Using printer ink color to control the behavior of paper microfluidics, Lab Chip 19 (2019) 2000–2008. Available from: https://doi.org/10.1039/C9LC00083F.

[32] S. Kim, D. Kim, S. Kim, Simultaneous quantification of multiple biomarkers on a self-calibrating microfluidic paper-based analytic device, Anal. Chim. Acta 1097 (2020) 120–126. Available from: https://doi.org/10.1016/j.aca.2019.10.068.

[33] A.M. Yehia, M.A. Farag, M.A. Tantawy, A novel trimodal system on a paper-based microfluidic device for on-site detection of the date rape drug "ketamine", Anal. Chim. Acta 1104 (2020) 95–104. Available from: https://doi.org/10.1016/j.aca.2020.01.002.

[34] F.M. Gharaghani, M. Akhond, B. Hemmateenejad, A three-dimensional origami microfluidic device for paper chromatography: application to quantification of Tartrazine and Indigo carmine in food samples, J. Chromatogr. A 1621 (2020) 461049. Available from: https://doi.org/10.1016/j.chroma.2020.461049.

[35] J. Yan, L. Ge, X. Song, M. Yan, S. Ge, J. Yu, Paper-based electrochemiluminescent 3D immunodevice for lab-on-paper, specific, and sensitive point-of-care testing, Chem. Eur. J. 18 (2012) 4938–4945. Available from: https://doi.org/10.1002/chem.201102855.

[36] M. Pollard, E. Hunsicker, M. Platt, A tunable three-dimensional printed microfluidic resistive pulse sensor for the characterization of algae and microplastics, ACS Sens. 5 (2020) 2578–2586. Available from: https://doi.org/10.1021/acssensors.0c00987.

[37] M.A. Raoufi, S.R. Bazaz, H. Niazmand, O. Rouhi, M. Asadnia, A. Razmjou, et al., Fabrication of unconventional inertial microfluidic channels using wax 3D printing, Soft Matter 16 (2020) 2448–2459. Available from: https://doi.org/10.1039/C9SM02067E.

[38] N. Bhattacharjee, C. Parra-Cabrera, Y.T. Kim, A.P. Kuo, A. Folch, Desktop-stereolithography 3D-printing of a poly(dimethylsiloxane)-based material with sylgard-184 properties, Adv. Mater. 30 (2018) 1–7. Available from: https://doi.org/10.1002/adma.201800001.

[39] A.V. Bickham, C. Pang, B.Q. George, D.J. Topham, J.B. Nielsen, G.P. Nordin, et al., 3D printed microfluidic devices for solid-phase extraction and on-chip fluorescent labeling of preterm birth risk biomarkers, Anal. Chem. 92 (2020) 12322–12329. Available from: https://doi.org/10.1021/acs.analchem.0c01970.

[40] W. Su, B.S. Cook, Y. Fang, M.M. Tentzeris, Fully inkjet-printed microfluidics: a solution to low-cost rapid three-dimensional microfluidics fabrication with numerous electrical and sensing applications, Sci. Rep. 6 (2016) 1–12. Available from: https://doi.org/10.1038/srep35111.

[41] I.C. Samper, S.A.N. Gowers, M.L. Rogers, D.-S.R.K. Murray, S.L. Jewell, C. Pahl, et al., 3D printed microfluidic device for online detection of neurochemical changes with high temporal resolution in human brain microdialysate, Lab Chip 19 (2019) 2038–2048. Available from: https://doi.org/10.1039/C9LC00044E.

[42] J. Sun, Y. Li, G. Liu, F. Chu, C. Chen, Y. Zhang, et al., Patterning a superhydrophobic area on a facile fabricated superhydrophilic layer based on an inkjet-printed water-soluble polymer template, Langmuir 36 (2020) 9952–9959. Available from: https://doi.org/10.1021/acs.langmuir.0c01769.

[43] U.N. Lee, X. Su, D.J. Guckenberger, A.M. Dostie, T. Zhang, E. Berthier, et al., Fundamentals of rapid injection molding for microfluidic cell-based assays, Lab Chip 18 (2018) 496–504. Available from: https://doi.org/10.1039/C7LC01052D.

[44] L. Zhou, J. Ye, J. Fu, Q. Gao, Y. He, 4D printing of high-performance thermal-responsive liquid metal elastomers driven by embedded microliquid chambers, ACS Appl Mater Interfaces 12 (2020) 12068–12074. Available from: https://doi.org/10.1021/acsami.9b22433.

[45] J. ter Schiphorst, S. Coleman, J.E. Stumpel, A. Ben Azouz, D. Diamond, A.P.H.J. Schenning, Molecular design of light-responsive hydrogels, for in situ generation of fast and reversible valves for microfluidic applications, Chem. Mater. 27 (2015) 5925–5931. Available from: https://doi.org/10.1021/acs.chemmater.5b01860.

[46] L. D'Eramo, B. Chollet, M. Leman, E. Martwong, M. Li, H. Geisler, et al., Microfluidic actuators based on temperature-responsive hydrogels, Microsyst. Nanoeng. 4 (2018) 17069. Available from: https://doi.org/10.1038/micronano.2017.69.

[47] J.W. Seo, S.R. Shin, Y.J. Park, H. Bae, Hydrogel production platform with dynamic movement using photo-crosslinkable/temperature reversible chitosan polymer and stereolithography 4D printing technology, Tissue Eng. Regen. Med. 17 (2020) 423–431. Available from: https://doi.org/10.1007/s13770-020-00264-6.

[48] F. Peng, J. He, J.F.C. Loo, S.K. Kong, B. Li, D. Gu, Identification of serum MicroRNAs as diagnostic biomarkers for influenza H7N9 infection, Virol. Rep. (2017) 7. Available from: https://doi.org/10.1016/j.virep.2016.11.001.

[49] H.P. Ho, F.C. Loo, S.Y. Wu, D. Gu, K.-T. Yong, S.K. Kong, MicroRNA biosensing with two-dimensional surface plasmon resonance imaging, in: A. Rasooly, B. Prickril (Eds.), Biosensors and Biodetection: Methods and Protocols Volume 1: Optical-Based Detectors, New York, NY, Springer, New York, NY, 2017, pp. 117–127. Available from: https://doi.org/10.1007/978-1-4939-6848-0_8.

[50] M. Ji, Y. Xia, J.F.-C. Loo, L. Li, H.-P. Ho, J. He, et al., Automated multiplex nucleic acid tests for rapid detection of SARS-CoV-2, influenza A and B infection with direct reverse-transcription quantitative PCR (dirRT-qPCR) assay in a centrifugal microfluidic platform, RSC Adv. 10 (2020) 34088–34098.

[51] B. Lopez-Jimena, S. Wehner, G. Harold, M. Bakheit, S. Frischmann, M. Bekaert, et al., Development of a single-tube one-step RT-LAMP assay to detect the Chikungunya virus genome, PLoS Negl. Trop. Dis. 12 (2018) e0006448. Available from: https://doi.org/10.1371/journal.pntd.0006448.

[52] P.J. Asiello, A.J. Baeumner, Miniaturized isothermal nucleic acid amplification, a review, Lab Chip 11 (2011) 1420–1430. Available from: https://doi.org/10.1039/C0LC00666A.

[53] M.G. Mohsen, E.T. Kool, The discovery of rolling circle amplification and rolling circle transcription, Acc. Chem. Res. 49 (2016) 2540–2550. Available from: https://doi.org/10.1021/acs.accounts.6b00417.

[54] O. Piepenburg, C.H. Williams, D.L. Stemple, N.A. Armes, DNA detection using recombination proteins, Water 4 (2006) 2–8. Available from: https://doi.org/10.1371/journal.pbio.0040204.

[55] J.M. Peters, B.-M. Koo, R. Patino, G.E. Heussler, C.C. Hearne, J. Qu, et al., Enabling genetic analysis of diverse bacteria with Mobile-CRISPRi, Nat. Microbiol. 4 (2019) 244–250. Available from: https://doi.org/10.1038/s41564-018-0327-z.

[56] J.S. Gootenberg, O.O. Abudayyeh, M.J. Kellner, J. Joung, J.J. Collins, F. Zhang, Multiplexed and portable nucleic acid detection platform with Cas13, Cas12a, and Csm6, Science 360 (2018) 439–444. Available from: https://doi.org/10.1126/science.aaq0179.

[57] S. Lutz, P. Weber, M. Focke, B. Faltin, J. Hoffmann, C. Müller, et al., Microfluidic lab-on-a-foil for nucleic acid analysis based on isothermal recombinase polymerase amplification (RPA), Lab Chip 10 (2010) 887–893. Available from: https://doi.org/10.1039/B921140C.

[58] I.L.G. Law, J.F.C. Loo, H.C. Kwok, H.Y. Yeung, C.C.H. Leung, M. Hui, et al., Automated real-time detection of drug-resistant *Mycobacterium tuberculosis* on a lab-on-a-disc by Recombinase Polymerase Amplification, Anal. Biochem. 544 (2018) 98–107. Available from: https://doi.org/10.1016/j.ab.2017.12.031.

[59] C. Myhrvold, C.A. Freije, J.S. Gootenberg, O.O. Abudayyeh, H.C. Metsky, A.F. Durbin, et al., Field-deployable viral diagnostics using CRISPR-Cas13, Science 360 (2018) 444–448. Available from: https://doi.org/10.1126/science.aas8836.

[60] M.M. Ali, C.L. Brown, S. Jahanshahi-Anbuhi, B. Kannan, Y. Li, C.D.M. Filipe, et al., A printed multicomponent paper sensor for bacterial detection, Sci. Rep. 7 (2017) 12335. Available from: https://doi.org/10.1038/s41598-017-12549-3.

[61] S. Muyldermans, Nanobodies: natural single-domain antibodies, Annu. Rev. Biochem. 82 (2013) 775–797. Available from: https://doi.org/10.1146/annurev-biochem-063011-092449.

[62] Z.A. Ahmad, S.K. Yeap, A.M. Ali, W.Y. Ho, N.B.M. Alitheen, M. Hamid, scFv antibody: principles and clinical application, Clin. Dev. Immunol. 2012 (2012) 980250. Available from: https://doi.org/10.1155/2012/980250.

[63] E.A. Miller, M.W. Traxlmayr, J. Shen, H.D. Sikes, Activity-based assessment of an engineered hyperthermophilic protein as a capture agent in paper-based diagnostic tests, Mol. Syst. Des. Eng. 1 (2016) 377–381. Available from: https://doi.org/10.1039/C6ME00032K.

[64] J. Loo, P.-M. Lau, S.-K. Kong, H.-P. Ho, An assay using localized surface plasmon resonance and gold nanorods functionalized with aptamers to sense the cytochrome-C released from apoptotic cancer cells for anti-cancer drug effect determination, Micromachines 8 (2017) 338. Available from: https://doi.org/10.3390/mi8110338.

[65] J.F. Loo, P. Lau, H. Ho, S. Kong, An aptamer-based bio-barcode assay with isothermal recombinase polymerase amplification for cytochrome-c detection and anti-cancer drug screening, Talanta 115 (2013) 159–165. Available from: https://doi.org/10.1016/j.talanta.2013.04.051.

[66] V.C.F. dos Santos, N.B.F. Almeida, T.A.S.L. de Sousa, E.N.D. Araujo, A.S.R. de Andrade, F. Plentz, Real-time PCR for direct aptamer quantification on functionalized graphene surfaces, Sci. Rep. 9 (2019) 19311. Available from: https://doi.org/10.1038/s41598-019-55892-3.

[67] A.M. Liao, W. Pan, J.C. Benson, A.D. Wong, B.J. Rose, G.T. Caltagirone, A simple colorimetric system for detecting target antigens by a three-stage signal transformation–amplification strategy, Biochemistry 57 (2018) 5117–5126. Available from: https://doi.org/10.1021/acs.biochem.8b00523.

[68] J.N. Anker, W.P. Hall, O. Lyandres, N.C. Shah, J. Zhao, R.P. Van Duyne, Biosensing with plasmonic nanosensors, Nat. Mater. 7 (2008) 442–453. Available from: https://doi.org/10.1038/nmat2162.

[69] K.M. Mayer, J.H. Hafner, Localized surface plasmon resonance sensors, Chem. Rev. 111 (2011) 3828–3857. Available from: https://doi.org/10.1021/cr100313v.

[70] J.F.-C. Loo, Y.-H. Chien, F. Yin, S.-K. Kong, H.-P. Ho, K.-T. Yong, Upconversion and downconversion nanoparticles for biophotonics and nanomedicine, Coord. Chem. Rev. 400 (2019) 213042. Available from: https://doi.org/10.1016/j.ccr.2019.213042.

[71] J. Owen, L. Brus, Chemical synthesis and luminescence applications of colloidal semiconductor quantum dots, J. Am. Chem. Soc. 139 (2017) 10939–10943. Available from: https://doi.org/10.1021/jacs.7b05267.

[72] Y. Wang, R. Hu, G. Lin, I. Roy, K.-T. Yong, Functionalized quantum dots for biosensing and bioimaging and concerns on toxicity, ACS Appl. Mater. Interfaces 5 (2013) 2786–2799. Available from: https://doi.org/10.1021/am302030a.

[73] M. Trotter, D. Juric, Z. Bagherian, N. Borst, K. Gläser, T. Meissner, et al., Inkjet-printing of nanoparticle gold and silver ink on cyclic olefin copolymer for DNA-sensing applications, Sensors 20 (2020) 1333. Available from: https://doi.org/10.3390/s20051333.

[74] C.A. Proença, T.A. Baldo, T.A. Freitas, E.M. Materón, A. Wong, A.A. Durán, et al., Novel enzyme-free immunomagnetic microfluidic device based on Co0.25Zn0.75Fe2O4 for cancer biomarker detection, Anal. Chim. Acta 1071 (2019) 59–69. Available from: https://doi.org/10.1016/j.aca.2019.04.047.

[75] H. de Puig, I. Bosch, L. Gehrke, K. Hamad-Schifferli, Challenges of the nano–bio interface in lateral flow and dipstick immunoassays, Trends Biotechnol. 35 (2017) 1169–1180. Available from: https://doi.org/10.1016/j.tibtech.2017.09.001.

[76] C.-K. Su, J.-C. Chen, One-step three-dimensional printing of enzyme/substrate–incorporated devices for glucose testing, Anal. Chim. Acta 1036 (2018) 133–140. Available from: https://doi.org/10.1016/j.aca.2018.06.073.

[77] E.A. Miller, Y. Jabbour Al Maalouf, H.D. Sikes, Design principles for enhancing sensitivity in paper-based diagnostics via large-volume processing, Anal. Chem. 90 (2018) 9472–9479. Available from: https://doi.org/10.1021/acs.analchem.8b02113.

[78] D.J. Kinahan, S.M. Kearney, O.P. Faneuil, M.T. Glynn, N. Dimov, J. Ducrée, Paper imbibition for timing of multi-step liquid handling protocols on event-triggered centrifugal microfluidic lab-on-a-disc platforms, RSC Adv. 5 (2015) 1818–1826. Available from: https://doi.org/10.1039/C4RA14887H.

[79] N. Godino, E. Vereshchagina, R. Gorkin, J. Ducrée, Centrifugal automation of a triglyceride bioassay on a low-cost hybrid paper-polymer device, Microfluid. Nanofluidics 16 (2014) 895–905. Available from: https://doi.org/10.1007/s10404-013-1283-9.

[80] H. Hwang, S.-H. Kim, T.-H. Kim, J.-K. Park, Y.-K. Cho, Paper on a disc: balancing the capillary-driven flow with a centrifugal force, Lab Chip 11 (2011) 3404–3406. Available from: https://doi.org/10.1039/C1LC20445A.

[81] L. Joswig, M.J. Vellekoop, F. Lucklum, Miniature 3D-printed centrifugal pump with non-contact electromagnetic actuation, Micromachines 10 (2019) 631. Available from: https://doi.org/10.3390/mi10100631.

[82] M.R. Behrens, H.C. Fuller, E.R. Swist, J. Wu, M.M. Islam, Z. Long, et al., Open-source, 3D-printed peristaltic pumps for small volume point-of-care liquid handling, Sci. Rep. 10 (2020) 1543. Available from: https://doi.org/10.1038/s41598-020-58246-6.

[83] J.F.C. Loo, H.C. Kwok, C.C.H. Leung, S.Y. Wu, I.L.G. Law, Y.K. Cheung, et al., Sample-to-answer on molecular diagnosis of bacterial infection using integrated lab-on-a-disc, Biosens. Bioelectron. 93 (2017) 212–219. Available from: https://doi.org/10.1016/j.bios.2016.09.001.

[84] Y. Wang, Y. Zhang, Z. Qiao, W. Wang, A 3D printed jet mixer for centrifugal microfluidic platforms, Micromachines 11 (2020) 695. Available from: https://doi.org/10.3390/mi11070695.

[85] S.J. Oh, T.S. Seo, Combination of a centrifugal microfluidic device with a solution-loading cartridge for fully automatic molecular diagnostics, Analyst 144 (2019) 5766–5774. Available from: https://doi.org/10.1039/C9AN00900K.

[86] E. Sweet, R. Mehta, Y. Xu, R. Jew, R. Lin, L. Lin, Finger-powered fluidic actuation and mixing via MultiJet 3D printing, Lab Chip 20 (2020) 3375–3385. Available from: https://doi.org/10.1039/D0LC00488J.

[87] M. Sharafeldin, K. Kadimisetty, K.R. Bhalerao, I. Bist, A. Jones, T. Chen, et al., Accessible telemedicine diagnostics with ELISA in a 3D printed pipette tip, Anal. Chem. 91 (2019) 7394–7402. Available from: https://doi.org/10.1021/acs.analchem.9b01284.

[88] J.L. Garcia-Cordero, D. Kurzbuch, F. Benito-Lopez, D. Diamond, L.P. Lee, A.J. Ricco, Optically addressable single-use microfluidic valves by laser printer lithography, Lab Chip 10 (2010) 2680–2687. Available from: https://doi.org/10.1039/C004980H.

[89] M.F. Santangelo, S. Libertino, A.P.F. Turner, D. Filippini, W.C. Mak. Integrating printed microfluidics with silicon photomultipliers for miniaturised and highly sensitive ATP bioluminescence detection. Biosens Bioelectron. 99 (2018) 464–470. Available from: https://doi.org/10.1016/j.bios.2017.07.055.

[90] C. Xiao, J. Eriksson, A. Suska, D. Filippini, W.C. Mak. Print-and-stick unibody microfluidics coupled surface plasmon resonance (SPR) chip for smartphone imaging SPR (Smart-iSRP). Anal Chim Acta. 1201 (2022) 339606. Available from: https://doi.org/10.1016/j.aca.2022.339606.

[91] S. Lee, S.K. Thio, S.-Y. Park, S. Bae, An automated 3D-printed smartphone platform integrated with optoelectrowetting (OEW) microfluidic chip for on-site monitoring of viable algae in water, Harmful Algae 88 (2019) 101638. Available from: https://doi.org/10.1016/j.hal.2019.101638.

[92] S. Kecili, H.C. Tekin, Adhesive bonding strategies to fabricate high-strength and transparent 3D printed microfluidic device, Biomicrofluidics 14 (2020) 024113. Available from: https://doi.org/10.1063/5.0003302.

[93] A. Bertoncini, S.P. Laptenok, L. Genchi, V.P. Rajamanickam, C. Liberale, 3D-Printed high-NA catadioptric thin lens for suppression of XPM background in stimulated Raman scattering microscopy, J. Biophotonics (2020) e202000219. Available from: https://doi.org/10.1002/jbio.202000219. n/a.

[94] A.I. Shallan, P. Smejkal, M. Corban, R.M. Guijt, M.C. Breadmore, Cost-effective three-dimensional printing of visibly transparent microchips within minutes, Anal. Chem. 86 (2014) 3124–3130. Available from: https://doi.org/10.1021/ac4041857.

[95] L.H. Duong, P.-C. Chen, Simple and low-cost production of hybrid 3D-printed microfluidic devices, Biomicrofluidics (2019) 13. Available from: https://doi.org/10.1063/1.5092529.

[96] D. Wolfe, K.W. Goossen, Evaluation of 3D printed optofluidic smart glass prototypes, Opt. Express 26 (2018) A85–A98. Available from: https://doi.org/10.1364/OE.26.000A85.

[97] V. Romanov, R. Samuel, M. Chaharlang, A.R. Jafek, A. Frost, B.K. Gale, FDM 3D printing of high-pressure, heat-resistant, transparent microfluidic devices, Anal. Chem. 90 (2018) 10450–10456. Available from: https://doi.org/10.1021/acs.analchem.8b02356.

[98] S.S. Hinman, K.S. McKeating, Q. Cheng, Plasmonic sensing with 3D printed optics, Anal. Chem. 89 (2017) 12626–12630. Available from: https://doi.org/10.1021/acs.analchem.7b03967.

[99] H. Gong, M. Beauchamp, S. Perry, A.T. Woolley, G.P. Nordin, Optical approach to resin formulation for 3D printed microfluidics, RSC Adv. 5 (2015) 106621–106632. Available from: https://doi.org/10.1039/C5RA23855B.

[100] N. Vaidya, O. Solgaard, 3D printed optics with nanometer scale surface roughness, Microsyst. Nanoeng. 4 (2018) 18. Available from: https://doi.org/10.1038/s41378-018-0015-4.

[101] J.L. Delaney, C.F. Hogan, J. Tian, W. Shen, Electrogenerated chemiluminescence detection in paper-based microfluidic sensors, Anal. Chem. 83 (2011) 1300–1306. Available from: https://doi.org/10.1021/ac102392t.

[102] J.L. Delaney, E.H. Doeven, A.J. Harsant, C.F. Hogan, Reprint of: Use of a mobile phone for potentiostatic control with low cost paper-based microfluidic sensors, Anal. Chim. Acta 803 (2013) 123–127. Available from: https://doi.org/10.1016/j.aca.2013.08.014.

[103] S. Smith, P. Madzivhandila, L. Ntuli, P. Bezuidenhout, H. Zheng, K. Land, Printed paper–based electrochemical sensors for low-cost point-of-need applications, Electrocatalysis 10 (2019) 342–351. Available from: https://doi.org/10.1007/s12678-019-0512-8.

[104] T.A. Freitas, C.A. Proença, T.A. Baldo, E.M. Materón, A. Wong, R.F. Magnani, et al., Ultrasensitive immunoassay for detection of *Citrus tristeza* virus in citrus sample using disposable microfluidic electrochemical device, Talanta 205 (2019) 120110. Available from: https://doi.org/10.1016/j.talanta.2019.07.005.

[105] T. Pardy, H. Sink, A. Koel, T. Rang, Development of a low-cost, wireless smart thermostat for isothermal DNA amplification in lab-on-a-chip devices, Micromachines 10 (2019) 437. Available from: https://doi.org/10.3390/mi10070437.

[106] X. Li, X. Pan, J. Lu, Y. Zhou, J. Gong, Dual-modal visual/photoelectrochemical all-in-one bioassay for rapid detection of AFP using 3D printed microreactor device, Biosens. Bioelectron. 158 (2020) 112158. Available from: https://doi.org/10.1016/j.bios.2020.112158.

[107] Y. Chu, A.-P. Deng, W. Wang, J.-J. Zhu, Concatenated catalytic hairpin assembly/hyperbranched hybridization chain reaction based enzyme-free signal amplification for the sensitive photoelectrochemical detection of human telomerase RNA, Anal. Chem. 91 (2019) 3619–3627. Available from: https://doi.org/10.1021/acs.analchem.8b05610.

[108] H. Cheng, C. Liu, J. Jiang, G. Luo, Y. Lu, K. Jin, et al., Analysis of ctDNA to predict prognosis and monitor treatment responses in metastatic pancreatic cancer patients, Int. J. Cancer 140 (2017) 2344–2350. Available from: https://doi.org/10.1002/ijc.30650.

[109] Z. An, O. Aksoy, T. Zheng, Q.-W. Fan, W.A. Weiss, Epidermal growth factor receptor and EGFRvIII in glioblastoma: signaling pathways and targeted therapies, Oncogene 37 (2018) 1561–1575. Available from: https://doi.org/10.1038/s41388-017-0045-7.

[110] Y. Liu, T. Lei, Z. Liu, Y. Kuang, J. Lyu, Q. Wang, A novel technique to detect EGFR mutations in lung cancer, Int. J. Mol. Sci. (2016) 17. Available from: https://doi.org/10.3390/ijms17050792.

[111] Y. Wang, J. Luo, J. Liu, S. Sun, Y. Xiong, Y. Ma, et al., Label-free microfluidic paper-based electrochemical aptasensor for ultrasensitive and simultaneous multiplexed detection of cancer biomarkers, Biosens. Bioelectron. 136 (2019) 84–90. Available from: https://doi.org/10.1016/j.bios.2019.04.032.

[112] K. Wang, J. Yang, H. Xu, B. Cao, Q. Qin, X. Liao, et al., Smartphone-imaged multilayered paper-based analytical device for colorimetric analysis of carcinoembryonic antigen, Anal. Bioanal. Chem. 412 (2020) 2517–2528. Available from: https://doi.org/10.1007/s00216-020-02475-1.

[113] F.F. Tao, X. Xiao, K.F. Lei, I.-C. Lee, Paper-based cell culture microfluidic system, BioChip J. 9 (2015) 97–104. Available from: https://doi.org/10.1007/s13206-015-9202-7.

[114] L. Zhu, X. Liu, J. Yang, Y. He, Y. Li, Application of multiplex microfluidic electrochemical sensors in monitoring hematological tumor biomarkers, Anal. Chem. 92 (2020) 11981–11986. Available from: https://doi.org/10.1021/acs.analchem.0c02430.

[115] K. Pantel, C. Alix-Panabières, Circulating tumour cells in cancer patients: challenges and perspectives, Trends Mol. Med. 16 (2010) 398–406. Available from: https://doi.org/10.1016/j.molmed.2010.07.001.

[116] V. Plaks, C.D. Koopman, Z. Werb, Circulating tumor cells, Science 341 (2013) 1186–1188. Available from: https://doi.org/10.1126/science.1235226.

[117] C. Alix-Panabieres, K. Pantel, Circulating tumor cells: liquid biopsy of cancer, Clin. Chem. 59 (2013) 110–118. Available from: https://doi.org/10.1373/clinchem.2012.194258.

[118] M. Ignatiadis, M. Lee, S.S. Jeffrey, Circulating tumor cells and circulating tumor DNA: challenges and opportunities on the path to clinical utility, Clin. Cancer Res. 21 (2015) 4786. Available from: https://doi.org/10.1158/1078-0432.CCR-14-1190.

[119] Y. Song, Z. Zhu, Y. An, W. Zhang, H. Zhang, D. Liu, et al., Selection of DNA Aptamers against epithelial cell adhesion molecule for cancer cell imaging and circulating tumor cell capture, Anal. Chem. 85 (2013) 4141–4149. Available from: https://doi.org/10.1021/ac400366b.

[120] L. Chen, M. Peng, N. Li, Q. Song, Y. Yao, B. Xu, et al., Combined use of EpCAM and FRα enables the high-efficiency capture of circulating tumor cells in non-small cell lung cancer, Sci. Rep. 8 (2018) 1188. Available from: https://doi.org/10.1038/s41598-018-19391-1.

[121] C.-H. Chu, R. Liu, T. Ozkaya-Ahmadov, M. Boya, B.E. Swain, J.M. Owens, et al., Hybrid negative enrichment of circulating tumor cells from whole blood in a 3D-printed monolithic device, Lab Chip 19 (2019) 3427–3437. Available from: https://doi.org/10.1039/C9LC00575G.

[122] J. Chen, C.-Y. Liu, X. Wang, E. Sweet, N. Liu, X. Gong, et al., 3D printed microfluidic devices for circulating tumor cells (CTCs) isolation, Biosens. Bioelectron. 150 (2020) 111900. Available from: https://doi.org/10.1016/j.bios.2019.111900.

[123] Q. Tang, X. Li, C. Lai, L. Li, H. Wu, Y. Wang, et al., Fabrication of a hydroxyapatite-PDMS microfluidic chip for bone-related cell culture and drug screening, Bioactive Mater. 6 (2021) 169–178. Available from: https://doi.org/10.1016/j.bioactmat.2020.07.016.

[124] P. Liu, B. Li, L. Fu, Y. Huang, M. Man, J. Qi, et al., Hybrid three dimensionally printed paper-based microfluidic platform for investigating a cell's apoptosis and intracellular cross-talk, ACS Sens. 5 (2020) 464–473. Available from: https://doi.org/10.1021/acssensors.9b02205.

[125] T.-H. Ulep, R. Zenhausern, A. Gonzales, D.S. Knoff, P.A. Lengerke Diaz, J.E. Castro, et al., Smartphone based on-chip fluorescence imaging and capillary flow velocity measurement for detecting ROR1 + cancer cells from buffy coat blood samples on dual-layer paper microfluidic chip, Biosens. Bioelectron. 153 (2020) 112042. Available from: https://doi.org/10.1016/j.bios.2020.112042.

[126] World Health Organization, Fact Sheets on Food Safety (retrieved 17th February, 2021). Available from: https://www.who.int/news-room/fact-sheets/detail/food-safety.

[127] Y. Zhao, D. Zeng, C. Yan, W. Chen, J. Ren, Y. Jiang, et al., Rapid and accurate detection of *Escherichia coli* O157:H7 in beef using microfluidic wax-printed paper-based ELISA, Analyst 145 (2020) 3106–3115. Available from: https://doi.org/10.1039/D0AN00224K.

[128] A. Nilghaz, X. Lu, Detection of antibiotic residues in pork using paper-based microfluidic device coupled with filtration and concentration, Anal. Chim. Acta 1046 (2019) 163–169. Available from: https://doi.org/10.1016/j.aca.2018.09.041.

[129] Z.T. Al-Aqbi, Y.C. Yap, F. Li, M.C. Breadmore, Integrated microfluidic devices fabricated in poly (methyl methacrylate) (PMMA) for on-site therapeutic drug monitoring of aminoglycosides in whole blood, Biosensors (Basel) 9 (2019) 19. Available from: https://doi.org/10.3390/bios9010019.

[130] Nitrate and nitrite in Drinking-water, WHO (2003) 1-16 (retrieved 17th February, 2021) http://apps.who.int/iris/bitstream/handle/10665/75380/WHO_SDE_WSH_04.03_56_eng.pdf

[131] E. Trofimchuk, A. Nilghaz, S. Sun, X. Lu, Determination of norfloxacin residues in foods by exploiting the coffee-ring effect and paper-based microfluidics device coupling with smartphone-based detection, J. Food Sci. 85 (2020) 736–743. Available from: https://doi.org/10.1111/1750-3841.15039.

[132] E. Trofimchuk, Y. Hu, A. Nilghaz, M.Z. Hua, S. Sun, X. Lu, Development of paper-based microfluidic device for the determination of nitrite in meat, Food Chem. 316 (2020) 126396. Available from: https://doi.org/10.1016/j.foodchem.2020.126396.

[133] Z. Gu, M.-L. Wu, B.-Y. Yan, H.-F. Wang, C. Kong, Integrated digital microfluidic platform for colorimetric sensing of nitrite, ACS Omega 5 (2020) 11196–11201. Available from: https://doi.org/10.1021/acsomega.0c01274.

[134] J. Yukird, V. Soum, O.-S. Kwon, K. Shin, O. Chailapakul, N. Rodthongkum, 3D paper-based microfluidic device: a novel dual-detection platform of bisphenol A, Analyst 145 (2020) 1491–1498. Available from: https://doi.org/10.1039/C9AN01738K.

[135] P.B. Tchounwou, C.G. Yedjou, A.K. Patlolla, D.J. Sutton, Heavy metal toxicity and the environment, in: A. Luch (Ed.), Molecular, Clinical and Environmental Toxicology: Volume 3: Environmental Toxicology, Springer Basel, Basel, 2012, pp. 133–164. Available from: https://doi.org/10.1007/978-3-7643-8340-4_6.

[136] M.F. Santangelo, I. Shtepliuk, D. Filippini, D. Puglisi, M. Vagin, R. Yakimova, et al., Epitaxial graphene sensors combined with 3D-printed microfluidic chip for heavy metals detection, Sensors 19 (2019) 2393. Available from: https://doi.org/10.3390/s19102393.

[137] C. Richard, A. Neild, V.J. Cadarso, The emerging role of microfluidics in multi-material 3D bioprinting, Lab Chip 20 (2020) 2044–2056. Available from: https://doi.org/10.1039/C9LC01184F.

[138] L. Meng, A.P.F. Turner, W.C. Mak, Soft and flexible material-based affinity sensors, Biotechnol. Adv. 39 (2019) 107398. Available from: https://doi.org/10.1016/j.biotechadv.2019.05.004.

[139] Q. Shang, Y. Su, Y. Liang, W. Lai, J. Jiang, H. Wu, et al., Ultrasensitive cloth-based microfluidic chemiluminescence detection of *Listeria monocytogenes* hlyA gene by hemin/G-quadruplex DNAzyme and hybridization chain reaction signal amplification, Anal. Bioanal. Chem. 412 (2020) 3787–3797. Available from: https://doi.org/10.1007/s00216-020-02633-5.

[140] C. Warr, J.C. Valdoz, B.P. Bickham, C.J. Knight, N.A. Franks, N. Chartrand, et al., Biocompatible PEGDA resin for 3D printing, ACS Appl Bio Mater 3 (2020) 2239–2244. Available from: https://doi.org/10.1021/acsabm.0c00055.

[141] M. Goyal, T. Knackstedt, S. Yan, S. Hassanpour, Artificial intelligence-based image classification methods for diagnosis of skin cancer: challenges and opportunities, Comput. Biol. Med. 127 (2020) 104065. Available from: https://doi.org/10.1016/j.compbiomed.2020.104065.

CHAPTER

Design and fabrication technologies for microfluidic sensors

2

Anis Nurashikin Nordin[1] and Asrulnizam Abd Manaf[2]

[1]*Department of Electrical and Computer Engineering, Faculty of Engineering, International Islamic University Malaysia, Kuala Lumpur, Malaysia*

[2]*Collaborative Microelectronic Design Excellence Center, Universiti Sains Malaysia, Sains@USM, Pulau Pinang, Malaysia*

2.1 Introduction

Microfluidics technology, which emerged in the early 1980s, introduced devices that could manipulate and analyze fluids at microscale. This lab-on-chip technology has revolutionized conventional laboratory-based tests, allowing rapid and efficient experiments to be conducted on small sample volumes. Using these miniature devices, integration of different equipment is done seamlessly on a single platform. Sample testing that requires centrifugation, such as separation of plasma from blood cells, can be automated by allowing the sample to go through different layers in the microfluidic platform. Mixing of the samples and reagents, as well as preparation of samples, can be done by changing the design of the fluidics channel to merge flow from different inlets. These devices, also known as micro-total analysis systems, are widely popular since they reduce the risk of contamination, cost less per analysis, allow automation and reduction of tedious operations, provide enhanced sensitivity and specificity, and have better reliability than conventional physical laboratory-based tests. Microfluidic devices can now be found in a wide range of sensor applications such as clinical diagnostics, environmental testing, and food quality testing.

In this chapter, the design and modeling techniques for popular microfluidic devices such as micromixers and microdroplet generators are described. Fundamental design principles such as laminar and electrokinetic flow are explained first, followed by different modeling techniques for these devices. Fabrication of these devices traditionally comprise polydimethylsiloxane (PDMS) substrates using soft lithography techniques that require cleanroom environments and expensive equipment such as mask aligners and evaporators for lithography. Taking into consideration that affordability and accessibility of such equipment are limited in poorly resourced and remote neighborhoods, in this chapter, we will also review and highlight various low-cost fabrication methods that can be applied in such settings. Paper- and plastic-based fluidics have become extremely popular in recent years because of their cost-

Microfluidic Biosensors. DOI: https://doi.org/10.1016/B978-0-12-823846-2.00004-3

effective methods and ease of fabrication. Finally, the chapter concludes with future outlooks for microfluidic platforms.

2.2 Design and modeling of microfluidic platforms for sensing

A microfluidic system is usually involved in the transportation and handling of fluids using micrometer-sized channels and components. Effective designs of microfluidic systems require an understanding of basic fluid mechanics concepts and terms such as laminar flow and its Reynolds number. These concepts will be illustrated in popular microfluidic device designs such as micromixers and droplet generators.

2.2.1 Micromixers

Applications of microfluidic devices in chemical, biological, and analytical detection processes typically involve the mixing mechanism of samples and reagents [1]. Micromixers can control the efficiency and sensitivity of microfluidic devices, and are considered to be one of the most important components of microfluidics. Unlike macro-scale fluidic devices where the mixing of fluids often relies on convection effects, mixing in the microscale is often achieved using diffusion or special microstructures that can induce chaotic advection and large surface-to-volume ratio [2]. Micromixers can be broadly classified as active and passive mixers. Passive micromixers require no external energy input and rely on diffusion to mix fluids. Active mixers, on the other hand, need external energy sources such as ultrasound, time-dependent electric and magnetic fields, etc. to disturb the fluids and enhance the mixing effect [3]. Flow rates of the fluids are generally very low in the microfluidic devices and the regime of the fluids in the microchannels is usually laminar flow with the Reynolds number of <1. Reynolds number Re is a dimensionless parameter that characterizes the flow regime in channels and represents the ratio of inertial forces to viscous force for given flow conditions. The Reynolds number defines the relative significance of inertial to viscous forces. It is given as follows:

$$Re = \frac{\rho v_s D_H}{\mu} \tag{2.1}$$

where ρ is the fluid density (kg/m^3), μ is the absolute dynamic fluid viscosity (Pa s), v_s is the mean fluid velocity (m/s), and D_H is the hydraulic diameter of the mixing channel (m) which is defined in Eq. (2.1) as follows:

$$D_H = \frac{4A}{P} = \frac{4wd_m}{2w + 2d_m} \tag{2.2}$$

The hydraulic diameter is commonly used when the fluid flows in a noncircular cross-sectioned channel. Hydraulic diameter uses the perimeter and the area of the channel to provide the diameter of the area engraved within the wetted perimeter [4].

It has been reported that the intensity of mixing is a nearly linear function when *Re* is in the range of 2000–10,000 and is independent of the width of the mixing channel [3]. Under typical operating conditions, the flow in passive microstructured devices is laminar where the fluid flows in parallel layers with no disruption between the layers. *Re* in case of laminar flow is usually less than 2300. In this condition, due to the low rate of fluid flow, viscous forces are dominant [5]. Molecular diffusion across the channels is also slow, making efficient mixing in microfluidic devices rather difficult. Mixing efficiency is a key parameter for a micromixer [2]. Mixing efficiencies can be evaluated based on the intensity of segregation. The standard deviation of pixel intensity or point concentration is often used as the mixing index (MI) to evaluate the mixing efficiency and is expressed by [6]:

$$\mathrm{MI} = \sqrt{\frac{1}{N}\sum_{i=1}^{N}(c_i - \overline{c})^2} \tag{2.3}$$

where c_i is the point concentration/pixel intensity, $\overline{c}$ is the mean concentration/intensity, and N is the number of sampling points.

Numerous researches have been done to discover new ways to improve laminar mixing by modification of the channel walls, geometry, and surface properties. Reports have shown that different geometries such as "T-," "O-," and "H"-shaped micromixers can be introduced to improve mixing efficiencies. Amongst the three, experimental data show that the H-micromixer is more efficient at all flow rates [7]. This is due to the repeated effect of splitting and recombining the two species to be mixed. When implemented, this process is more effective than molecular diffusion at low Reynolds numbers. Experimental results have shown that the H-micromixer's efficiency can be as high as 98% at $Re = 0.083$.

The configuration of inlets also play an important role in the efficiency of mixing since the geometry at the junction of the microchannel influences the transport and mixing characteristics of the fluids. The velocity profile during laminar flow in straight-line microchannels is usually parabolic where the highest velocity occurs at the center of the channel. When the fluid passes through an intersection, the velocity distribution changes considerably depending on the angle of the inlet junction. The velocity distributions at the intersection for a mixing angle of 300° are shown in Fig. 2.1A [3].

For all inlet angles, Poiseuille (laminar) flow is formed after the inlet streams merge at the junction of inlet channels. However, the velocity distribution at the intersection varies noticeably. When the streams meet at the junction, one side of the boundary layer separates from the channel wall and is mixed with the boundary layer of the other stream under the inertia and shear forces of the fluids. The shear effect is related to the mixing angle at the intersection. As the mixing angle increases, the advection effect is enhanced and the diffusion distance between the

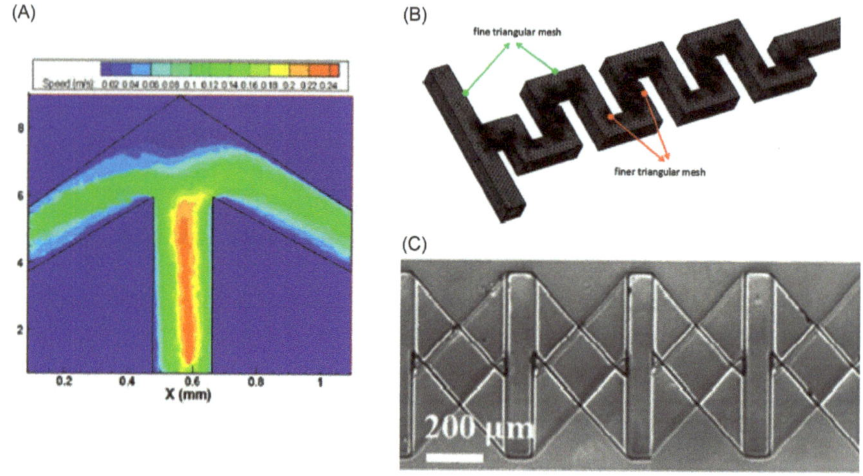

FIGURE 2.1

Passive micromixing. (A) Using angled fluidic channels that create different velocity distributions. In this image, fluidic channels are angled at 300° [3]. (B) Numerical simulation and mesh of a micromixer [8]. (C) Two-layer serpentine crossing microchannels [9].

molecules of two fluids is reduced. To balance mixing efficiency and energy consumption, the optimal mixing angle is kept between 90° and 180° [3].

Numerical simulation-based micromixers are designed mostly by trial and error. Even though qualitative and quantitative predictions of mixing can be made, the uncertainty from modeling accuracy, discretization, and iteration are difficult to control [10]. Therefore, researching the fabrication of micromixers and experiment of mixing are crucial to confirm the numerical simulation accuracy.

A numerical study on mixing of 3D serpentine micromixer was performed by Raza et al. [11]. Li and Chen [8] reported the simulation of 3D serpentine micromixer with added obstacles in the microchannel. Ruijin et al. [12] proposed an efficient mixing system using splitting-merging passive micromixer. A computational fluid dynamic investigation of modified zigzag micromixer with enhanced chaotic advection was executed by Chen and Li [13]. The et al. [14] analyzed the geometric effects of a trapezoidal-zigzag micromixer that allow numerous mixing mechanisms to improve the mixing performance. 3D serpentine and square-wave micomixer with cubic grooves have been designed and simulated by Lin [15]. Gidde et al. [16] have conducted numerical simulation for mixing in chamber based of circular and square planar micromixer. Selected micromixers from the abovementioned works are illustrated in Fig. 2.1B and C.

Hossain et al. [9] fabricated a serpentine micromixer with two-layer crossing channel by soft lithography with PDMS. The SU8 mold is prepared by photolithography technique on a silicon substrate. The same fabrication method is employed by Tran-Minh et al. [17] to demonstrate a passive micromixer with B-spline curve channel with ellipse-like obstacles [18] and to fabricate chamber-based micromixer with four different shape obstacles. Passive micromixers with 3D channel geometry are difficult to fabricate by photolithography technique [19,20]. The obstacles in the microchannel to improve mixing are smaller than the channel height. All fabricated micromixers usually have obstacles span as tall as the full microchannel height as shown. Therefore, a better fabrication method is needed to realize passive micromixers with 3D channel geometry and obstacles. An alternative method using computer numerical control (CNC) engraving and milling machine has been proposed to fabricate chain and serpentine micromixers from plexiglass [21,22]. However, this method is limited by the fabrication resolution of the equipment.

2.2.2 Droplet generator

Other than mixing, another important application of microfluidics is in microdroplet generation. This microfluidic design has widespread applications in the food, cosmetics, and pharmaceuticals, and it is commonly used in international companies such as Emultech (The Netherlands), Capsum (France), and Microcaps (Switzerland) [23]. Droplet-based microfluidics have several advantages including small volume of reagents consumed, large production of monodisperse droplets, fast reactions due to its high surface-area-to-volume ratio, and its ability to independently control each droplet [24].

Droplet generation originates from fluid instabilities. In general, there are two types of techniques in microfluidic droplet formation: passive and active. Passive microfluidic devices squeeze, drip, jet, tip-stream, or tip-multi-break droplets as shown in Fig. 2.2A. In contrast, active microfluidic devices combine external forces, such as electric, magnetic, and centrifugal, with intrinsic forces in the channels to form droplets. In addition to the Reynolds number, droplet generator design requires the usage of other dimensionless parameters such as capillary number, Weber number, Bond number, viscosity ratio, and flow rate ratio [24].

In this section, we will discuss methods to generate droplets using passive, pressure-driven flow methods. Droplet formation in microfluidic devices involves the deformation and breaking of the liquid–liquid interface. For passive microfluidics, there are four types of forces that influence droplet generation: inertial, viscous, gravity, and capillary forces. The effect of the forces are characterized by the stress (forces per unit area) that they generate in the channel. The different values of stress generated by each type of force can be calculated as follows: inertial stress $f_i \sim \rho_s v_s^2$, viscous stress $f_v \sim \eta_s v_s / L$, gravity $f_g \sim \rho_s g L$, and capillary pressure $f_\gamma \sim \gamma / L$, assuming that the volume of fluid is flowing at velocity v_s in a microfluidic device with characteristic length L [24]. Similar to the design of microfluidic mixers, the Reynolds number in the droplet can be calculated using Eq. (2.1) with the hydraulic diameter replaced by the characteristic length L.

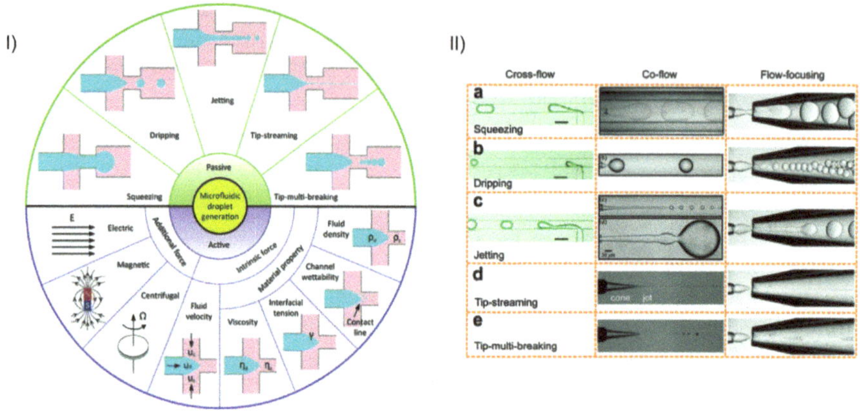

FIGURE 2.2

(A) Schematic of droplet generation by passive and active methods. In the passive method, droplets can be produced in squeezing, dripping, jetting, tip-streaming, and tip-multi-breaking modes, depending on the competition of capillary, viscous, and inertial forces. In active control, droplet generation can be manipulated by either applying additional forces from electrical, magnetic, and centrifugal controls, or modifying intrinsic forces via tuning fluid velocity and material properties including viscosity, interfacial tension, channel wettability, and fluid density. (B) Images of droplet generation with different modes in cross-flow, co-flow, and microcapillary flow-focusing geometries. (a) Squeezing mode, (b) dripping mode, (c) jetting mode. The upper image in co-flow is a narrowing jet, while the lower one is a widening jet. (d) Tip-streaming mode. (e) Tip-multi-breaking mode. Neither tip-streaming nor tip-multi-breaking modes has been reported in cross-flow geometry yet [24].

For droplet generation, the other dimensionless numbers such as capillary, Weber, Bond, viscosity, and flow rate play a more important role compared to the Reynolds number. During the process of droplet generation, three major steps are normally involved:

1. Dispersion of continuous phase fluid that meets at the junction to form an immiscible interface
2. Creation of an unstable state at the junction due to large deformation in the fluids
3. Spontaneous fragmentation of the interface to create discrete droplets.

These three steps involve the calculation of interfacial tension in the fluids. Characterization of interfacial tension can be described using the capillary number *Ca,* which is the ratio of viscous stress to capillary pressure $fv/f\gamma$. The inertial stress f_i is not directly dependent on the device length-scale L. However, when L is reduced, viscous stress fv and capillary pressure $f\gamma$ increase and the gravitational effect f_g is weakened. Devices with very small L will experience dominant viscous and capillary forces. In microflows, Ca is usually in the range of 10^{-3} to 10 [24]. Another important parameter for microdroplet generation is the Bond number Bo, which is defined as the ratio of gravity to capillary pressure $f_g/f\gamma$.

$$B_o = \frac{\Delta\rho L^2}{\gamma} \tag{2.4}$$

where $\mathrm{Dr} = \Delta\rho$ is the density difference between dispersed and continuous phases. Typically, $\Delta\rho$ and L are small for liquid–liquid microflows, thereby $B_o \ll 1$ in microfluidic droplet generation.

Other two dimensionless numbers are also relevant: the viscosity ratio λ and the flow rate ratio φ of dispersed to continuous phase fluids. These are defined as follows [24]:

$$\lambda = \frac{\eta_d}{\eta_c}, \tag{2.5}$$

and

$$\phi = \frac{Q_d}{Q_c}. \tag{2.6}$$

where Q_d is the dispersed flow rate and Q_c is the continuous flow rate.

For passive droplet generators, Q_d and Q_c are usually controlled by either syringe pumps that supply constant flow rates or pressure regulator and gravity-based pressure units [25] that set stable source pressures P_d and P_c without additional energy input. The energy introduced from the syringe pumps or pressure controllers is partially converted into interfacial energy that facilitates the destabilization of the liquid–liquid interface allowing the formation of droplets.

The key design aspect of droplet microfluidics is the geometry of the junction, which controls the local fluid fields and determines the droplet formation and breakup. There are three categories of channel geometries for droplet formation namely: co-flow, cross-flow, and flow-focusing categories.

Fig. 2.2B illustrates the different fundamental modes in shear-based droplet generation. The squeezing, dripping, and jetting modes can be generated in cross-flow, co-flow, and flow-focusing geometries, while tip-streaming and tip-multi-breaking modes are used in single-flow channels. The squeezing mode generates larger droplets compared to the dripping mode. Both squeezing (Fig. 2.2B(a)) and dripping modes (Fig. 2.2B(b)) generate monodisperse droplets while the jetting mode can generate polydisperse droplets (Fig. 2.2B(c)). For single-flow channels, tip-streaming can produce monodisperse droplets (Fig. 2.2B(d)) and tip-multi-breaking can be used to generate polydiverse and geometric size progression droplets (Fig. 2.2B(e)) [24].

The cross-flow junction is most commonly designed as a T-junction. In this category, the dispersed and continuous fluids flow independently in separate channels and later meet at the cross-junction [26]. When the dispersed phase is partially blocked by the continuous phase, a shear gradient is established. The dispersed phase then elongates and eventually breaks into droplets. The size of the droplets depends on the shear stress.

Co-flow fluidics use parallel microchannels of different diameters which create co-flow streams of dispersed and continuos fluids. The dispersed fluid phase is introduced into an inner channel and the continuous phase flows into an outer concentric channel in the same direction, as shown in Fig. 2.5. The size of droplets is determined by the flow rate where higher continuous flow rates produce smaller droplets due the higher viscous shear stress [27]. On the other hand, higher dispersed flow rates produce larger droplets. Interfacial tension also affects the size of the droplets where higher interfacial tensions produce smaller droplets due to the higher resistance to droplet pinchoff.

In the flow-focused geometries, both dispersed and continuous phases flow in the same channel. The channel has a pinchoff that creates shear stress that focuses the liquid into uniform droplets. The droplet formation is strongly affected by the channel geometry, flow rate, fluid viscosity, and addition of surfactants [26]. The size of the droplets can be controlled by changing the size of the small orifice.

Nonhomogeneous droplets with specific shapes and additional features can be prepared using more complex microfluidic channels. These multicomponent, multi-emulsion droplets can be used as template for fabricating microcapsules that can be used in a variety of applications such as drug delivery, tissue engineering [26], food delivery, single-cell analysis, and synthesis of nanomaterials [28]. Parallel, biphase co-flow channels can be nested to form Janus or multicomponent droplets as shown in Fig. 2.3A. In this type of geometry, the flow rate of the fluid affects the component's proportion, size, and frequency [29]. Double emulsion droplets can be precisely controlled using a flow-focused microfluidic device as shown in Fig. 2.3B that can be used to create solid–oil–water droplets. The combination of vicious shear and hydrodynamic flow-focusing from the outer phase allows the single core to be completely encapsulated [28]. The microchannel can be modified further to allow microencapsulation of multiple cores within a single droplet. Fig. 2.3C illustrates how two injection microchannels are combined in a Y-channel that has a thin neck to allow formation of droplets [28]. Adjusting the flow rate of the fluids allows different combinations of the solid cores within the emulsion droplets.

2.3 Fabrication methods

Microfluidic devices are conventionally fabricated using PDMS molding and soft lithography techniques. These techniques, while precise, require a cleanroom environment combined with expensive lithographic and plasma bonding equipment. These techniques are unaccessible for research and development labs that are in countries with limited infrastructure and resources. Another disadvantage of PDMS-based devices is that they are unsuitable for high-throughput mass

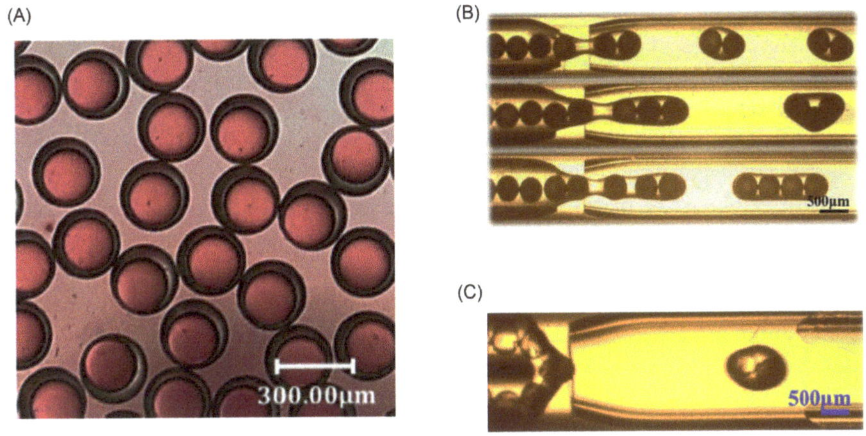

FIGURE 2.3

(A) Optical image of single-emulsion droplet prepared using nested co-flow microchannels [29]. (B) Multicore microencapsulation of single-component droplet using microchannels [28]. (C) Multicore microencapsulation of multicomponent droplets using parallel microchannels [28].

fabrication. Most of the PDMS molding techniques used to fabricate microfluidic channels from molding to bonding are done manually in research labs, and automation of these processes for commercial devices has proven to be very challenging. Consequently, most commercial microfluidic devices are fabricated using low-cost transparent plastics using injection molding. In this section, we will first describe traditional method of fabricating microfluidic channels using PDMS followed by three different low-cost techniques to fabricate microfluidic devices using paper and plastic as comparison.

2.3.1 Soft lithography

The common fabrication techniques of the microfluidic devices are based on soft lithography using PDMS which was first introduced in 1998 [30]. The basic principle of soft lithography method is based on realization of microfluidic devices by PDMS replication. The photoresist mold fabrication relies on photolithography, where in later process the PDMS is poured and cured on the mold. The cured PDMS is then lifted up from the mold and it contains replication of the mold pattern. Finally, the replicated PDMS is bonded with glass to form a closed channel.

In microfluidic fabrication, SU8 is the common negative photoresist used due to its high resolution, mold durability, and capacity for high aspect ratio structures [31]. Other than photolithography, mold preparations were also demonstrated by grayscale photolithography and multidirectional exposure photolithography. Grayscale photolithography utilizes the conventional photolithography setup to fabricate 3D mold structures for replication of PDMS microfluidic devices by using a grayscale photomask. In contrast to a binary photomask that offers the all-or-none binary exposure scheme, a grayscale photomask offers a gradient exposure dose to create multilevel 3D mold structures in a single lithographic exposure. Grayscale lithography is an interesting method to realize 3D microstructures in negative and positive photoresist [32,33].

Various grayscale masks have been prepared in the literature, and are classified into binary halftone or analog continuous tone grayscale masks [34]. Chrome halftone grayscale photomask is a binary halftone grayscale mask that has various thickness of chromium to represent the grayscale levels. The various thicknesses of chromium are created by using electron beam lithography. Examples of analog continuous grayscale masks are high-energy-beam-sensitive (HEBS) glass, laser direct write (LDW) glass, and thin-film metallic mask. Electron beam lithography is used to pattern write on the HEBS glass mask [35]. Meanwhile, direct pattern writing using lasers can be used for both LDW glass and thin-film metallic masks [36–38]. Grayscale photolithography is an effective approach with single masks and exposure process to create 3D mold structures. However, the method of preparing grayscale photomask using laser or electron beam is costly and requires numerous calibration and optimization steps.

The following section shows how grayscale fabrication technique can be used to fabricate passive micromixers. Planar, multistep, B-spline, and 3D serpentine micromixers can all be fabricated using this technique. These various designs were developed based on previous work of other researchers [17]. All micromixers were designed with 40 mm mixing channel length and 600 μm width, with two inlets and one outlet. The depth of mixing channel is 500 μm, which is represented by 64% grayscale concentration in the software mask design. The width of the mixing channel is maintained in microscale since width or diameter of the channel dominates the Reynolds number equation as described by Eq. (2.1). The flow in passive micromixer must be kept laminar and this is provided by the designed width in microscale that leads to a low Reynolds number flow [39]. The depth of inlet and outlet wells is designed as 1 mm with 92% grayscale concentration. These wells are designed with 2 mm diameter to match the tygon tube diameter and to prevent sample fluid leaking during micromixing test.

The dimensions of each mixing units for multistep micromixer are described in Fig. 2.4A–C, which shows the PET mask design for a multistep micromixer. The grayscale masks were used to produce microfluidic structures. Multistep micromixer was designed with various channel depths, represented by the grayscale concentration of 62%, 72%, 82%, and 92%. The microchannels were fabricated by replicating the PDMS on an SU8–10 mold. There are four main processes including emulsion mask fabrication, UV exposure, development, and PDMS replication.

The following steps are used for grayscale fabrication technique:

1. The PET mask is first used as a master to prepare the emulsion mask.

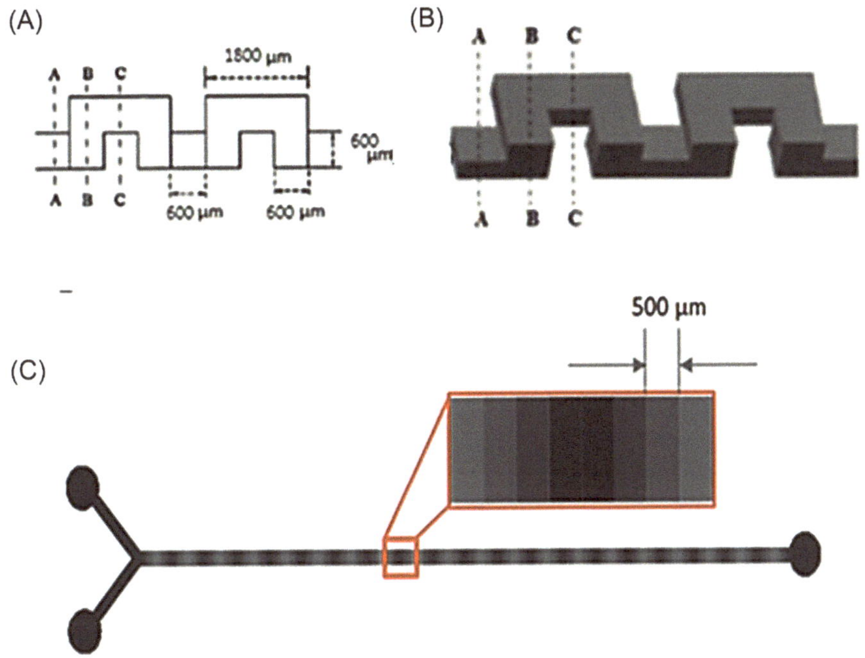

FIGURE 2.4

(A) Dimensions of a mixing unit for multistep micromixer. (B) 3D view of multistep micromixer [8,9]. (C) PET mask design for multistep micromixer using grayscale technique.

2. Next, lithography techniques are used to develop the SU8−10 mold using 60 seconds of UV exposure. Fig. 2.8 shows the images of the fabricated micromixer molds.
3. PDMS replication is performed by pouring PDMS on the SU8−10 micromixer mold. The device is then cured and peeled off from the mold.
4. After PDMS replication, the PDMS layer with microchannel is sealed with a flat PDMS layer by using a thin layer of spin-coated PDMS as adhesive glue to form a closed channel (Fig. 2.8B).
5. The thicknesses of the SU8−10 mold are measured using scanning electron microscope (SEM) where the samples are tilted by 90°.

In addition to planar and multistep micromixers, the B-spline 3D serpentine micromixer structure can also be fabricated using the grayscale fabrication technique in PDMS. The step-by-step process for fabrication of B-spline structures is similar to planar and multistep micromixer. The SEM image of the B-spline micromixer mold is shown in Fig. 2.5B and the fabricated micromixer in PDMS is shown in Fig. 2.5C. The B-spline micromixer had a thickness of 506 μm.

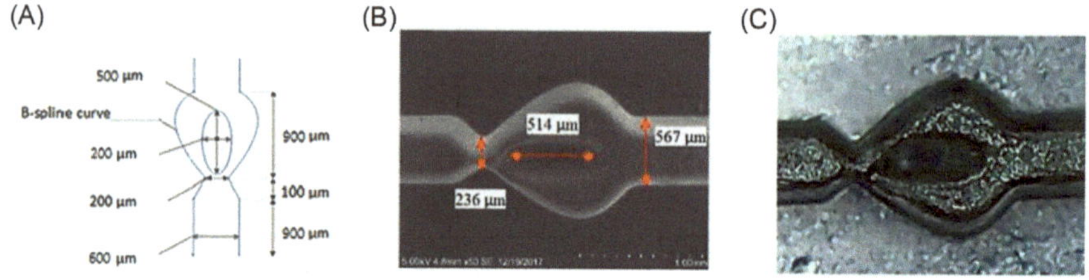

FIGURE 2.5

(A) Dimensions of a mixing unit for a B-spline micromixer [17]. Microscopic images of the B-spline micromixer. (B) Top view of SU8–10 mold. (C) Bottom view of the fabricated B-spline micromixer in PDMS. *PDMS*, Polydimethylsiloxane.

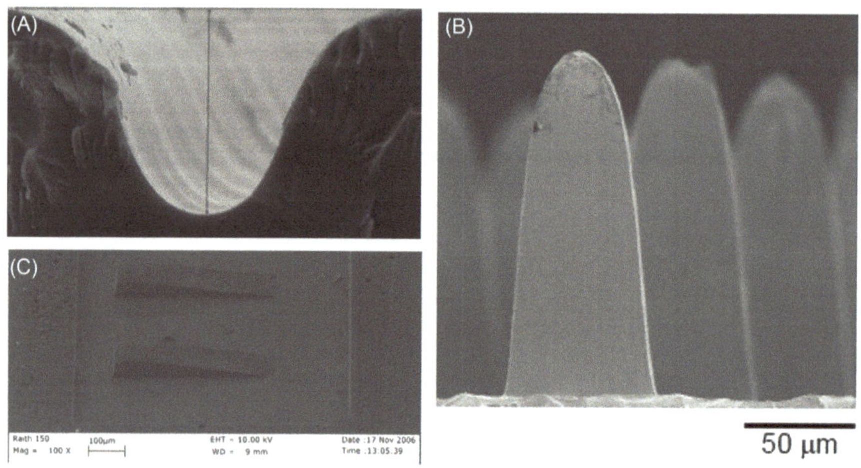

FIGURE 2.6

3D microstructures obtained by grayscale lithography, (A) micro-curved channel [42], (B) pyramid-shaped PMMA structure [43], and (C) tapered photoresist structure [35].

An alternative to grayscale lithography is the microfluidic photomask [40]. The microfluidic photomask can be created by varying the concentration of dye solutions in PDMS channel to characterize the grayscale levels. Luo and Zhang [41] reported a grayscale digital maskless lithography technique to fabricate a curved microlens array. Fig. 2.6A–C illustrates the micro-curved channel, pyramid-shaped poly(methyl methacrylate) (PMMA) structure, and tapered photoresist structure, respectively, that are fabricated using grayscale photolithography.

Another method to fabricate 3D mold structures is by using multidirectional exposure photolithography. Multidirectional exposure includes inclined UV exposure, rotated UV exposure, and reverse-side/back-side exposure as shown in Fig. 2.6C. These techniques are capable of producing microstructures with resolution of hundreds of nanometers to microns [44,45]. In multidirectional lithography method, the precise control of the angle is crucial, and this can be controlled by either a tilting/moving UV source or rotating the substrate stage [44].

Soft lithography process produces high-resolution replicas even though there is potential for defects in PDMS replication that can occur from pattern deformation of the mold or when removing the cast from the mold [46]. However, the versatility in design, biocompatibility, and integration potential of soft lithography ensures to continue as a promising technique in research requiring custom microfluidic design.

2.3.2 3D printing fabrication method

Despite the extensive research carried out using PDMS microfluidics, automation of the manual PDMS molding process is very challenging and unsuitable for high-throughput mass fabrication. Most commercial microfluidic manufacturers prefer instead to fabricate using low-cost transparent plastics that can utilize high-throughput injection molding. This results in a plastic molding process that is totally different from the soft lithographic method, especially if the devices consist of multiple layers. Injection molding has its own disadvantages though, as it requires an expensive metal mold for each device design.

A possible alternative is 3D printing technology, which can print in clear biocompatible polymers at high resolutions. 3D printing offers a middle level solution in which it allows automated and cost-efficient fabrication that is extremely suitable for incremental development for research [47]. There are two different techniques in 3D printing that can be applied to micro- and millifluidics namely stereolithography (SL) printing and fused filament fabrication [48]. SL printing involves layer by layer photocuring of liquid polymer resin using focused laser or LED light source. Microchannels can be formed by polymerizing the walls of channels and draining uncured resin [49]. Alignment and plasma bonding is not required for these type of devices, making the fabrication process much simpler compared to traditional PDMS devices. Fig. 2.7 illustrates the comparison of fabrication steps between PDMS and SL microfluidic devices.

In 3D printing process, the basic principle is using a 3D digital model that is carved by layers. The layers are created by the printer using layer-by-layer manufacturing technology. The new layer of material is added on top of the previous layer to build multilevel and multilayer microfluidic devices directly or indirectly via molding process [50–52]. In direct 3D printing, multiple copies of microfluidic devices are produced by a rapid prototyping process. On the other hand, 3D-printed transfer molding offers faster and more reliable process where the 3D-printed mold is used as a master for polymer replication. Fused deposition modeling is the most common 3D printing technology for fabrication of polymer composites with resolution of 50–200 μm [53]. In this method, a nozzle is used to print multiple layers on top of each other where the material is melted in the nozzle and then it is extruded onto the previous layer and left for binding before cooling. This method was first demonstrated for fabrication of master template for soft

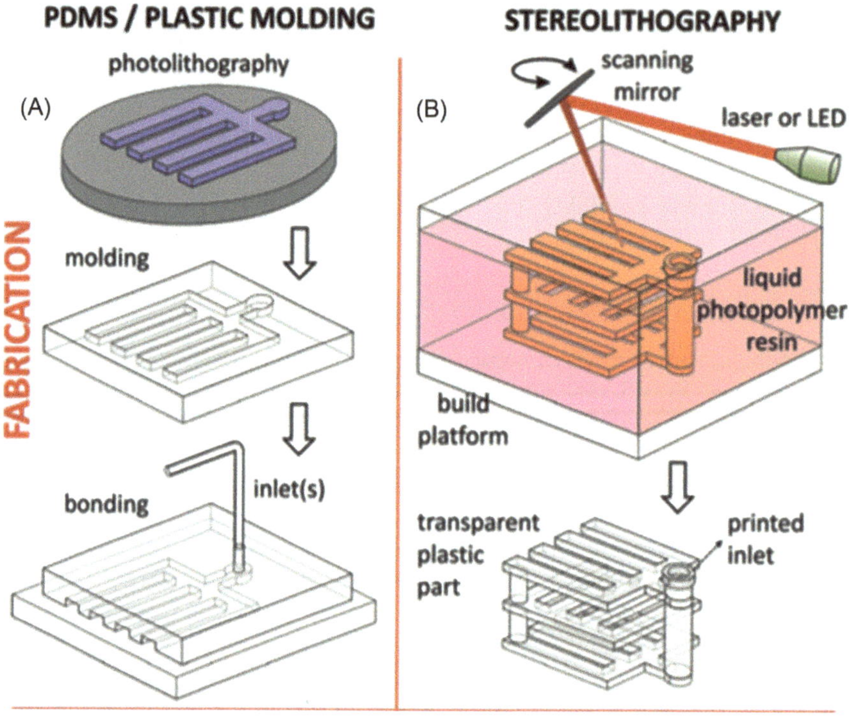

FIGURE 2.7

Comparison between (A) PDMS/plastic molding and (B) stereolithography printing [49]. *PDMS*, Polydimethylsiloxane.

lithography of microfluidic devices in 2002 [54]. Since then, this method has been used with SL printing, multijet printing, wax printers, and office quality laser printers [55–57]. In summary, the 3D printing process offers an attractive approach for the rapid fabrication of multilayered microfluidic devices, but the limitations in terms of speed, the available resolution of the printer, surface roughness, and material types must be resolved first [58]. The abovementioned microfluidic fabrication processes in this and previous section are summarized in Table 2.1.

Another significant method for layer-by-layer fabrication is microstereolithography. The principle of microstereolithography is based on the layer-by-layer surface scanning by a focused laser beam on photoresist layers which will produced a polymerized solid in order to create 3D microstructures. Focused beam of light used in the laser-based method provides a precise control of exposure with resolution depending on the laser wavelength. However, this method suffers from the issue of low throughput where the structure is slowly built up by volume elements and requires sophisticated control of the laser beam movement [64–66]. After the laser exposure on the first 2D layer is performed, the second layer of photoresist is spread on top of the 2D microstructure. The next layers are created with the same principle until the desired 3D microstructures are created with

Table 2.1 3D microfluidic fabrication techniques.

Item	References	Lithographic techniques	Height dimensions	Materials
1	[35]	Grayscale photolithography (PMMA grayscale mask prepared by electron beam lithography)	22 μm	AZ4562
2	[37]	Direct write lithography (using software grayscale mask file)	3 μm 4 μm	Shipley 1827, AZ4620
3	[41]	Digital micromirror device-based grayscale digital maskless lithography	3 μm	GP-28
4	[40]	Grayscale photolithography (PDMS grayscale photomask)	16 μm	SU8
5	[59]	Multilayer photolithography	25 μm	SU8
6	[60]	Multilayer photolithography	6 μm	AZ5214-E
7	[61]	Electron beam lithography	1 μm	PMMA
8	[62]	Electron beam lithography	100 nm	PMMA
9	[63]	Electron beam lithography	100 nm	PMMA
10	[64]	Laser direct write	17 μm	Hydrogel
11	[65]	Laser direct write	30 μm	SZ2080
12	[44]	Rotated photolithography	1 mm	SU8
13	[42]	Incline photolithography	500 μm	SU8
14	[43]	X-ray lithography	27 μm	PMMA
15	[51]	3D printing	50 μm	Resin
16	[52]	3D printing	8 mm	PEGDMA

PDMS, Polydimethylsiloxane; *PMMA*, poly(methyl methacrylate).

micrometer precision [67]. The structures are directly created by the computer-aided design (CAD) pattern which eliminates the need of photomasks [68].

2.3.3 Laminate fabrication method

Laminate microfluidic devices are produced by stacking and bonding several independently cut layers to form microfluidic features and microchannels [69]. Laminate process starts with the material selection. The laminate device creation includes three most common materials which are adhesive transfer tapes, polymer layers, and glass slides [70]. The adhesive transfer tapes offer inherent bonding in the material. Meanwhile, polymers and glass layers are selected for their optical clarity and sample compatibility.

The next process in the laminate method is cutting the desired features for each layer. Cutting methods are chosen based on the types of laminate materials, thickness and dimensional accuracy of the layers [71]. Two main methods for layers cutting are by knife plotter and laser cutter. Both methods offer speed and simplicity of the cutting process setup. A general process of cutting involves designing of a device using CAD software, cutting the device geometry, and bonding the layers of the device to form closed channels. The device geometry is created by precisely cutting the material with a blade for knife plotter or by using a focused beam for laser cutter. Knife plotter offers less resolution and it is very challenging to cut features smaller than 500 μm. On the other hand, laser cutter offers better resolution with accuracy of features size as small as 25 μm [71].

The final step in laminate process is bonding of the independent layers to form a functioning device. Two most common bonding techniques are use of adhesives and thermal bonding. Adhesive bonding employs directly applying the cut double-sided tape between layers of different materials. Adhesive bonding suffers from an uneven bonding and channel heights. In addition, it is difficult for the bonded layers to maintain high pressures beyond 5 bar [72]. Thermal bonding is carried out by raising the temperature of the individual layers to a temperature near the glass transition temperature of one or both of the materials, then a force is applied across the surface of the layers. However, thermal bonding is not possible for all materials, with the main disadvantages of features warping from heating and cool down process and sometimes leaves bubbles between layers [73].

The limitations of laminate process include the difficulties of aligning the individual layers and poor bonding that may lead to channel leakage. In contrast, laminate devices offer a simple approach to provide a lab-on-a-chip platform with multiple flow layers to incorporate inlets, storing areas, functional and sensing layers [74].

2.3.4 Computer numerical control micromilling

CNC milling is a subtractive fabrication process that uses rotating cutting tools to remove material from a starting stock piece, commonly referred to as the workpiece. The CNC milling system typically consists of (1) a worktable for positioning the workpiece, (2) a cutting tool (most commonly an endmill), and (3) an overhead spindle for securing and rotating the cutting tool. CNC helps to automate the fabrication process, thus improving repeatability and precision, reducing human error, and adding advanced capabilities (e.g., the direct conversion of CAD models to finished parts). Recent technological advances have improved precision and resolution down to the micron scale, allowing the CNC to be coined as *micromilling* to describe fabrication of parts with microscale resolution. Modern CNC mills are versatile and capable of fabricating devices with features ranging in size from several microns to several meters. The wide availability of cutting tool shapes, materials, and sizes makes the mill amenable to fabricating many types of features in many different materials. It can fabricate a part directly from a 3D CAD model, making it easier and faster to convert design concepts into working prototypes. Micromilling can be useful in microfluidics applications for two main functions:

- Machining the mold to be used in subsequent fabrication steps.
- Machining microchannels and features directly as the final device. In this case, micromilling offers a key advantage as a plastic workpiece can be milled into a device in less than 30 minutes for simple designs which significantly reduces the turnaround time from design to prototype.

Three main aspects that need to be considered when fabricating microfluidic devices are the compatibility of the method with the material of choice, the ability of the fabrication technique to achieve the desired features, and the finishing quality of the fabricated product. Most micromilled microfluidic devices use elastomer material such as PDMS and polycarbonate. Milling of elastomers, although possible, can be difficult due to large elastic deformations that prevent effective material removal. Other microfluidic substrates such as glass and ceramics, which are generally difficult to fabricate, are also difficult to mill due to their brittleness which leads to chipping, cracking, and other defects. Micromilling can achieve complex features such as undercuts, rounded surfaces, and sharp internal corners provided that appropriate tooling and extended process times are available. CNC micromilling has been successfully used to fabricate microfluidic devices on glass [75] and complex molds [76].

Quality of the finished parts may be assessed by its surface roughness, replication fidelity of features, and optical characteristics after fabrication. For micromilling, surface roughness depends on the cutting tool (e.g., tool features, profile, and wear) and the operational parameters chosen for the tool. These parameters can be optimized to reduce surface roughness. Besides surface roughness, feature transfer fidelity is another important metric of quality. To ensure that the final part has dimensions that match the original design, one must account for mold shrinkage, a known effect where parts thermally "shrink" as a result of cooling. During milling, friction from the cutting tool can cause increase in temperatures, and this affects the surface quality. When milling in plastics, these temperature fluctuations are generally localized around the cutting tool and can be mitigated through proper selection of cutting parameters (i.e., feeds and speeds) and adequate use of coolant. Thus, thermal effects when milling plastics are not as significant to feature fidelity and tolerances as tool alignment and workholding.

Other than the profile and material, important considerations for endmills are its flute length and cutting diameter. These features directly determine the maximum cutting depth and minimum width of the microchannel. Though it is possible, it is generally not advisable to mill deeper than the flute length of a given endmill (often three times the diameter). Special extended-reach endmills can be used if deeper channels or taller features are necessary, but they are generally more expensive. This type of endmills also tend to flex more due to their length, increasing their likelihood of breaking. Another important consideration is that the shank diameter of the endmill must match the size of the tool holder or collet. The helix angle facilitates removal of chips. Insufficient chip removal will lead to clogging of the flutes, which will create heat and ultimately damage the device, endmill, or both. When milling plastics and other soft materials, heat build-up can be particularly problematic, so lower helix angles are preferred (30° is the industry standard) because they provide more space between flutes, thereby improving chip removal.

Fig. 2.8A–D summarizes the standard characterization methods, its process flow, fabrication, and parameters of interest for micromilling [77]. Micromilling can also be combined with 3D printing to produce more complex structures as

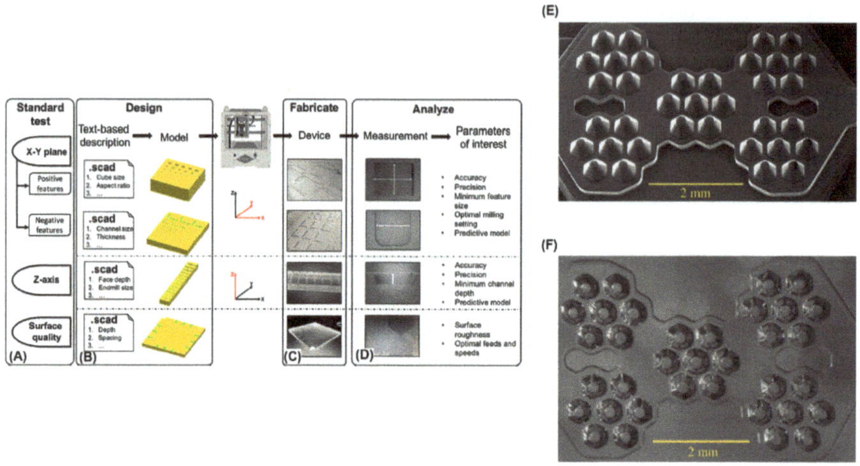

FIGURE 2.8

Set of standard tests proposed to characterize CNC micromills in terms of spatial accuracy and the surface quality of microchannels. (A) Three categories of standard tests include spatial accuracy in $X-Y$ plane both for positive and negative features, spatial accuracy in Z axis, and surface roughness. (B) Researchers can obtain CAD models of the standard tests using the text-based description of each test in OpenSCAD. (C) Models are fabricated using any CNC mill. (D) Parameter of interest will be measured and compared to existing benchmarks to characterize the performance of the CNC micromill [77]. Combined 3D printing and CNC micromilling for microfluidic device. (E) SEM image of the side view of the mold with constant microwell heights. (F) Optical micrograph of the fabricated PDMS layers with the variable and constant depth microwells [76]. *CAD*, Computer-aided design; *CNC*, computer numerical control; *PDMS*, polydimethylsiloxane; *SEM*, scanning electron microscope.

shown in Fig. 2.8E and F [76]. Two molds were used namely one 3D-printed template and two CNC micromilled parts. The CNC micromilling parts are bonded to each other by chloroform via the solvent bonding method and used as the mold for the soft lithography process. Fig. 2.8E shows the SEM image of the mold, while Fig. 2.8F illustrates the optical micrograph of PDMS layers with the variable and constant depth (300 μm) microwells.

It can be concluded that micromilling offers advantages in terms of (1) versatility across various materials as it can be used to create both molds and devices and (2) rapid design-to-prototype turnaround times. These features make it ideal for prototyping and research purposes as it allows several design iterations at lower costs. For large volume production, micromilling cannot compete with injection molding and hot embossing. Fig. 2.9 provides a summary of the comparison of micromilling with three most commonly used microfabrication methods for

(A) Technical Capabilities

Categories	Milling	Embossing	Stereolithography	Injection Molding
Material Capabilities				
Thermoplastics	●●●	●●●	●●○*	●●●
Thermosets	●●○*	●●○*	●●○*	●●●
Elastomers	●○○	●●○†	●●○*	●●●
Metals	●●●	●○○‡	●○○†	●○○*
Glass/Ceramics	●○○†	●○○§	●○○†	○○○*
Wax	●●●	●●●	○○○	●●●
Feature Capabilities				
Additional Heights	No added complexity	Additional layer per height	No added complexity	No added complexity
Aspect Ratio	8:1	2:1	Method dependent	8:1
Contoured 3D Features	Continuous	Layered	Layered	Continuous
Sharp Corners	External Only	Internal / External	Internal / External	Internal / External
Undercuts	Special tooling	Impractical	Yes	Special tooling
Results				
Surface Roughness	0.4 - 2 μm	Replicates mold roughness	0.4 - 6 μm	Replications mold roughness
Autofluorescence	Not affected	Increased by processing	Material dependent	Not affected
Legend ●●● Excellent ●●○ Most conditions ●○○ Specific conditions ○○○ Impractical	* *Only cured thermosets* † *Poor consistency and characterization*	* *Only uncured thermosets* † *Only thermoplastic elastomers* ‡ *Limited to specific features and thin sheets* § *Layered mix with polymer*	* *Uses resins, that when cured have similar properties to desired polymer* † *Requires polymer/wax addative*	* *Requires polymer/wax addative*

(B) Cost Comparison

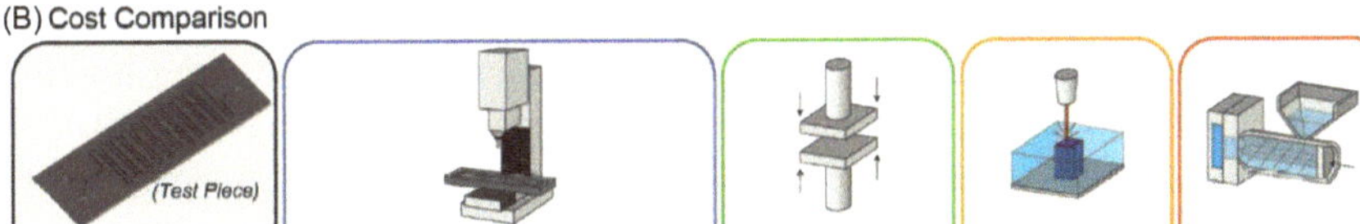

	Milling (On-site)		Milling (Outsourced)		Embossing (On-site)		Stereolithography (Outsourced)		Injection Molding (Outsourced)	
Setup Costs										
Equipment	$15k <		N/A		$15k <		N/A		N/A	
Tooling / Supplies	$500		N/A		$50		N/A		N/A	
Process Times and Costs										
Outsourced Expenses	Time	Cost	Time	Cost	Time	Cost	Time	Cost	Time	Cost
Mold / Tooling	N/A	N/A	N/A	N/A	4 - 15 d	$ 55 - 321	N/A	N/A	N/A	$ 2255
Device Fabrication	N/A	N/A	11 - 15 d	$ 137	N/A	N/A	4 - 6 d	$ 33	11 - 15 d	$ 2
On-site Expenses										
Machine Setup	10 m	N/A	N/A	N/A	5 m	N/A	N/A	N/A	N/A	N/A
Material Setup	< 5 m	$ 1	N/A	N/A	< 5 m	$ 1	N/A	N/A	N/A	N/A
Device Fabrication	10 m	N/A	N/A	N/A	30 m	N/A	N/A	N/A	N/A	N/A
Subtotal:	25 m	$ 1	N/A	N/A	40 m	$ 1	N/A	N/A	N/A	N/A
Expenses (per device)										
1 Devices	< 1 h	$ 1	11 - 15 d	$ 137	4 - 15 d	$ 56 - 322	4 - 6 d	$ 33	11 - 15 d	$ 2257
25 Devices	1 d	$ 1	11 - 15 d	$ 137	6 - 17 d	$ 3 - 14	4 - 6 d	$ 33	11 - 15 d	$ 92
50 Devices	3 d	$ 1	11 - 15 d	$ 137	8 - 19 d	$ 2 - 7	4 - 6 d	$ 33	11 - 15 d	$ 47

FIGURE 2.9

A comparison between milling and other microfabrication methods for plastics, in terms of: (A) material compatibility, feature capability, and quality; and (B) cost. In (A), three filled circles = "excellent," three open circles = "impractical" or "inadequate"; see legend (bottom left of A). (B) For process times, "Time" represents the time of fabrication for one device for both on-site and outsourced devices. "Cost" (in USD) is an estimate, where onsite fabrication is calculated based on cost of goods (not including labor) used (estimated from the labs of the authors), and outsourced fabrication is based on the lowest quoted price we obtained for the different quantities [78]. *N/A*, Not applicable.

plastics, that is, injection molding, hot embossing, and SL printing [78]. This comparison is based on (1) complexity of the fabrication procedure; (2) fabrication material of the microchannel or mold; (3) application of microfluidic device; (4) cost of fabrication; (5) resolution of microchannel; and (6) surface roughness. These technical features are tabulated in Fig. 2.9A. Hot embossing, injection molding, and micromilling can be used for polymeric materials. Main strength of micromilling is however for machining metals and it is unsuitable for elastomers, glass, and ceramics. In terms of feature capabilities, both micromilling and SL techniques allow fabrication of complex devices without much added process complexity.

Overall cost of fabrication is also an important factor that needs to be considered when choosing a particular fabrication method. The key costs involved during fabrication are summarized in Fig. 2.9B which include costs of equipment, cost of operation, and cost of production. For academia, the chosen fabrication method should fit well with the existing infrastructure in the university such as availability of central cleanroom facilities and equipment to conduct lithography. For industries, return on investment is a key factor where the cost of equipment should be able to be offset by the profits obtained when selling the finished products. Injection molding has the highest start-up cost based on infrastructure, equipment setup, and technical know-how for producing quality parts [78]. Due to this, injection molding is usually outsourced to companies such as Microfluidic ChipShop, Micronit, SimTech, thinXXS, Symbient.

2.4 Characterization and measurement techniques for microfluidic platforms

2.4.1 Leakage assessments

A key factor of any microfluidic device performance evaluation is the bonding strength between layers. It is important to have no leakage during fluid flow. Leakage tests are usually done by inserting colored fluids in the microfluidic channels. Microchannels that have low bonding strengths between the top and bottom layers will have leakage along its sidewalls.

To illustrate a leakage assessment technique, an example of a PDMS-based refractive index sensor is shown in Fig. 2.10. This sensor was fabricated using grayscale photolithography technique. The dimensions of the refractive index sensor are 7.5 cm in length and 2 cm in width. The sensor consists of a microchannel with three different depths and two wells for fluid inlet and outlet. The tapered fiber optic is embedded in the microchannel with the sensing region located at the middle of microchannel as shown as SEM images in Fig. 2.10A.

During fluid injection, the core region will be surrounded by fluid and the refractive index of the fluid can be detected by the shift of the transmission spectrum measured by optical spectrum analyzer. Green dye is injected in the PDMS refractive sensor as shown in Fig. 2.10B. Leakage can be observed at the ends

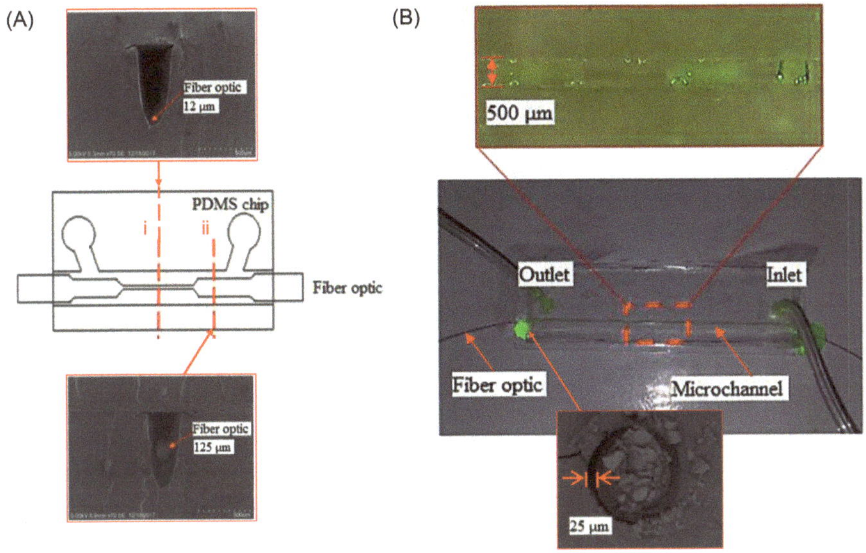

FIGURE 2.10

(A) Tapered fiber optic refractive index sensor within a PDMS microfluidic channel and its cross sections at core and cladding region. (B) Leakage that occurs at the ends of the microchannels can be observed when injected with green dye. The inset (top) shows the presence of trapped air bubbles within the microchannel. The bottom inset shows the gap between the fiber optic and the microchannel resulting in incomplete sealing [79]. *PDMS*, Polydimethylsiloxane.

where there is dripping fluid. This occurs after multiple fluid injections indicating that there is low adhesion between the PDMS layers. The fluid in the microchannel also has trapped air bubbles.

To study the cause of the dripping fluid, the cross-section image at the end of the microchannel was analyzed using an SEM. The SEM micrograph is shown in Fig. 2.10B that indicates that a 25 μm gap between the fiber optic and PDMS was found at the end of the microchannel. This gap causes the leakage problem and is a result of poor PDMS sealing. This condition will introduce the air bubbles in the microchannel.

To solve the leakage problem, modifications were made to the sealing process by using a thin layer of PDMS mixture as adhesive glue to seal the microchannel and curing at room temperature for 24 hours. After 24 hours, the microchannel was completely sealed. One difficulty of using PDMS mixture as adhesive glue is determining the appropriate amount of PDMS mixture. A thick layer of spin-coated PDMS mixture will increase the possibility of PDMS reflows inside the microchannel. This will lead to clogging in microchannel. Alternatively, silicon adhesive can be used to seal the microchannel.

The finish quality of the microchannel may also create the leakage problems. Channels with high surface roughness will create uneven surfaces. The characteristics of the surface observation can be evaluated using the surface profiler such as Atomic Force Microscopy (AFM) and SEM as shown in Fig. 2.10B bottom inset.

2.4.2 Mixing assessment

Other than leakage assessment, another important parameter that needs to be measured for microfluidic mixers is its mixing efficiency. The performances of the mixers can be measured through a combination of flowing dye/plain water into the channel and capturing images of the mixing fluids using visualization experiments. In this example, red dye solution and distilled water were introduced simultaneously via the inlets into four different designs of the micromixers at a flow rate of 1.5 mL/min. Mixing was observed visually using an inverted microscope (Nikon eclipse TS100) and was displayed at the computer monitor. Once mixing was done, 3 μL samples were taken from the outlet and placed onto a microdrop plate. The microdrop plate is then placed into a microplate reader (SkanIt Go by Thermo Scientific) and the absorbance readings are recorded. The measurement setup is as shown in Fig. 2.11A.

Fig. 2.12 shows the images of the mixed fluids at three different points of the mixer namely at the Y-junction, middle of the channel, and at the end of the channel. This procedure was repeated for all the four different types of micromixers, namely the planar, multistep, B-spline, and 3D serpentine micromixers. The images were captured using an inverted microscope which is normally used for microfluidic observations.

All the fabricated micromixers, planar, multistep, B-spline, and 3D serpentine are passive micromixers. Passive micromixers have low Reynolds number because there are no external forces involved in fluid flow. Flow in passive micromixers is laminar and mixing mainly relies on diffusion. In Fig. 2.12A for the simple planar Y-junction mixer, it can be seen that fluid streams of the red dye solution and distilled water remain separated until the end of mixing channel. This is due to the fact that all walls inside the mixing channel are flat and thus no chaotic advection occurs in the fluid flow. Mixing in the planar micromixer is inefficient because it relies only on diffusion.

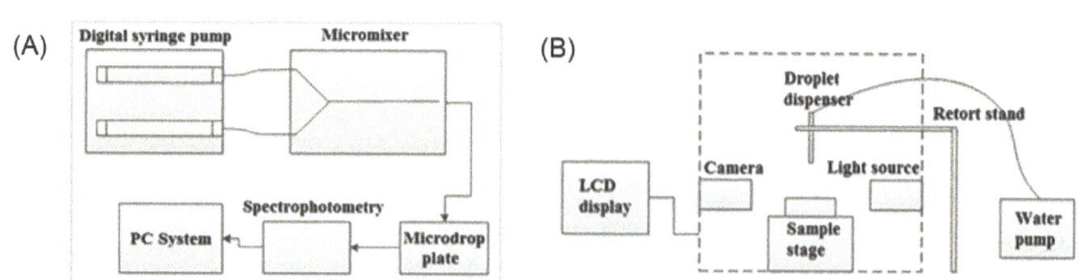

FIGURE 2.11

(A) Mixing assessment setup and (B) contact angle measurement setup [80].

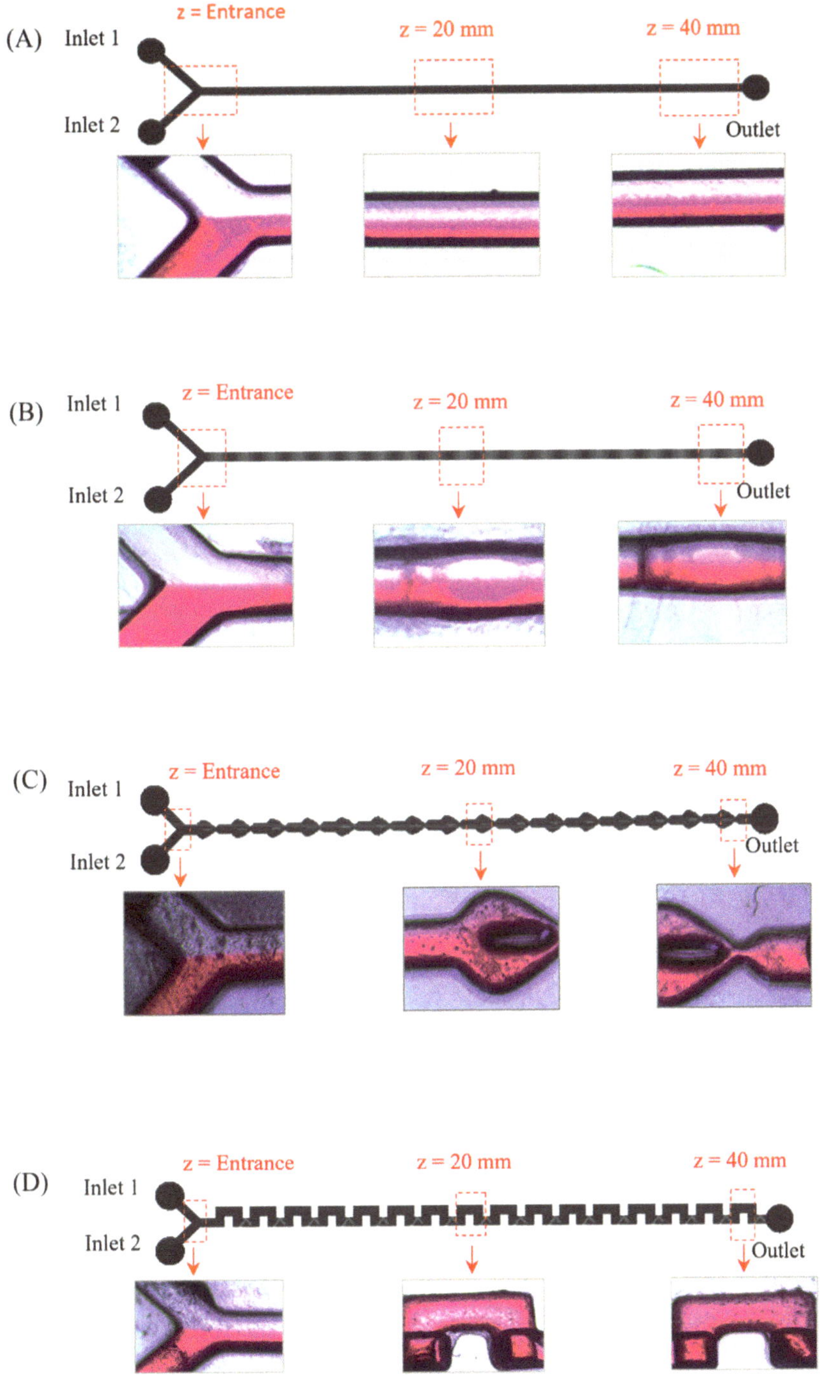

FIGURE 2.12

Characterization of mixed fluids in different microchannel patterns: (A) simple Y-shaped planar micromixer, (B) multistep micromixer (C) B-spline and (D) serpentine micromixer [80].

Fig. 2.12B shows mixing in a multistep micromixer. This multistep micromixer has a vertical wall which has steps of four different heights. This device has improved mixing as compared to the planar micromixer. The design of multistep micromixer enables continuous stretching and folding in the fluid streams, which significantly promotes mixing. However, complete mixing between the two fluids was still not achieved in this micromixer within the mixing channel length of 40 mm, as seen at the end of mixing channel.

Fig. 2.12C shows the mixing in the B-spline micromixer. It can be observed that complete mixing occurs at the end of the channel. This mixer design includes B-spline curves with obstacles placed at each B-spline portion. This causes the fluid flows to split and recombine at each B-spline portion. Having this design allows increased contact between the two solutions resulting in better mixing via diffusion.

The 3D serpentine micromixer achieved the best mixing as shown in Fig. 2.12D. Result shows that mixing occurs from the first serpentine cycle. This is due to the fluid manipulation in the two layers crossing channel of 3D serpentine micromixer. Chaotic advection was generated by the stretching and folding in fluid stream. Splitting and recombining flows also increased the mixing process due to the increase of the contact area at the fluid flows interface. As seen at the end of mixing channel, the red color dominates with less visible white color. This denotes that a good mixing has been achieved at the end of the mixing channel.

2.4.3 Characterization of mixed fluids

The concentration of the mixed solutions at the outlet will be different from the concentration of the two initial solutions. The concentration of red dye solution is 1 mg/mL while that of distilled water is 0 mg/mL. For an ideal case of complete mixing, the concentration at the outlet will be reduced but the volume will be doubled. It can be calculated by Eq. (2.7) as follows:

$$\text{Concentration at outlet} = \frac{(\text{concentration 1})(\text{volume 1}) + (\text{concentration 2})(\text{volume 2})}{(\text{volume 1} + \text{volume 2})} \tag{2.7}$$

Using this equation, the calculated concentration at the outlet is 0.5 mg/mL. This indicates that concentration has reduced by 50% of its initial concentration. According to the linear relationship of concentration and absorbance in Beer–Lambert equation, two solutions are completely mixed when the absorbance at the outlet achieves 50% of the absorbance value in initial solutions. Eq. (2.8) is used to calculate the percentage of mixing in all types of micromixers.

$$\text{Mixing percentage} = \frac{\text{Absorbance at outlet}}{\text{Absorbance at inlet}} \times 100\% \tag{2.8}$$

The mixing percentage is tabulated in Table 2.2. The table shows that the 3D serpentine micromixer achieves the best mixing of 55.2%, the nearest to the ideal 50%. In this condition, almost all red dye solution has mixed with distilled water.

Table 2.2 Mixing percentage of the micromixers.

Type of micromixer	Absorbance at outlet	Absorbance at inlet	Mixing percentage
3D serpentine	0.74	1.44	55.2
B-spline	0.78		58.2
Multistep	0.85		63.4
Planar	1.09		81.3

Multistep and B-spline micromixer designs also achieve good mixing with mixing percentages of 63.4% and 58.2%, respectively. Although the planar micromixer was observed to have nearly no mixing from the images, the photometric analysis has verified that mixing occurs with the reduction of absorbance and concentration values with the mixing percentage of 81.3% at the planar micromixer outlet.

2.4.4 Droplet characterization

In general, droplets can be characterized based on contact angle measurement, droplet size, and stability of the droplet during flow in the microchannel.

2.4.4.1 Contact angle measurement technique

Contact angles are measured between the polymer surface and a liquid droplet to determine the wetting behavior of a polymer surface. The liquid is said to be "wetting" the surface when the contact angle value is smaller than 90° and the surface is not "wetted" when the contact angle is greater than 90°. Surfaces are classified depending on the contact angle value made with a droplet of distilled water. The surface is "hydrophilic" if the contact angle is smaller than 90° and "hydrophobic" when the contact angle is greater than 90° [81]. Superhydrophobic surfaces have a great impact in microfluidic applications. To evaluate the droplet size, the contact angle measurement was carried out for the fabricated micropillar array samples. The measurement was done at room temperature by using the goniometer (Rame-hart Instrument Co., USA). The sample was placed on the sample stage and pure water droplet with 3 μL volume was dispensed repeatedly at three different areas of the surface. The contact angle value was calculated by the software provided with the instrument and displayed at the computer monitor. Fig. 2.11B shows the schematic diagram and image of contact angle measurement setup.

2.4.4.2 Optical imaging of droplets

One of the criterion for evaluating droplet is its stability and size. These can be quantified by utilizing optical imaging to detect droplets in microfluidic channels and then analyzing the images in real time or offline as shown in Fig. 2.13A [82].

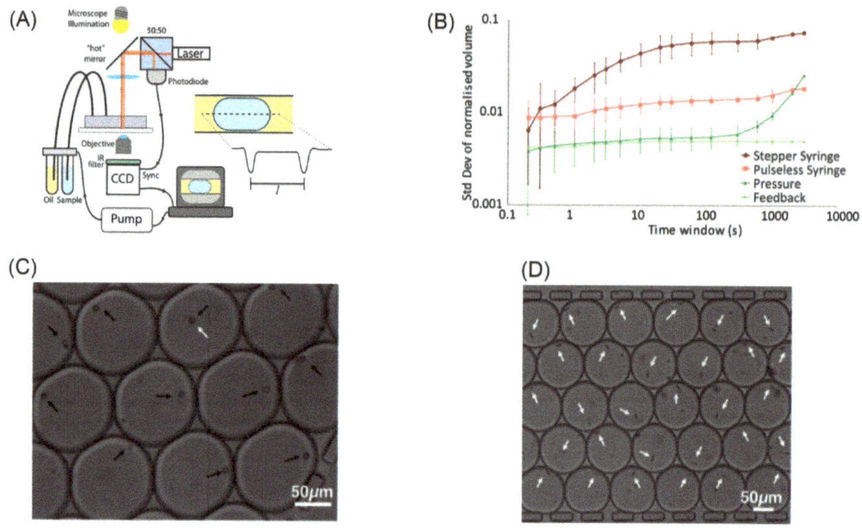

FIGURE 2.13

(A) Measurement setup for optical imaging of droplets (B) and droplet uniformity over a long period of time using the feedback composed of optical imaging and pressure pump system to regulate the droplet size in comparison with the systems without feedback loop [82]. (C) Droplets sorted using optical imaging system, by their content. Droplets containing *Dunaliella tertiolecta* in droplets and *Phaeodactylum tricornutum* encapsulated in droplets. (D) Black and white arrows show the location of *P. tricornutum* and *D. tertiolecta* in the droplets, respectively. The white arrow in panel (D) shows that two different species were used in a single run [83].

The capacity to control or stabilize droplet size during the experiment is the most significant benefit of adopting real-time processing. However, real-time analysis is very limited in terms of how many images per second the system can process as it needs a lot of processing power. Nevertheless, with the very fast increase of computer power and improved algorithms of image processing, the speed could increase in the near future. When utilizing a line camera, real-time image processing has been shown to work at up to 1000 Hz rates [82]. However, the real-time analysis of full images and a droplet control using a feedback loop can work at up to 250 Hz when each and every droplet is analyzed. Despite this slow processing speed, the speeds of pressure pumps are substantially slower. This problem can be rectified by employing fast response pressure pumps as they have faster response time compared to syringe pumps. The use of flow controller [84] can increase uniformity of the formed droplets as shown in Fig. 2.13B.

Apart from droplet size measurement, optical imaging in microfluidics can also be used for sorting droplets that contain single cells [83] or microorganisms

[86]. Droplets may be sorted at a rate of up to 10 Hz with an incredible accuracy of 90% without the use of cell labeling (Fig. 2.13C and D). In addition, it was shown that the system can separate empty droplets from occupied ones and also distinguish between different cell types that are contained inside the droplets. This method also allows study of droplet formation, precision control, and forces acting on them [87].

Optical imaging principle can also be used for measurements of droplet velocity [88] in a microfluidic channel. In this example, the droplet is exposed to three different color flashes in a very short time frame and the image taken is still in a single frame. Thus, in one frame instead of one real image of droplet, one would get three images of the same droplet just illuminated with different light sources at different time frames. The detector then distinguishes the different colors, detects the droplet, and calculates the droplet speed. In another similar optical imaging system, a flow cell with two light paths is used to measure the size and velocity of droplets that are flowing through the microchannel [89]. In both cases, offline post-processing of images or videos containing droplets is required.

2.4.4.3 Electrical techniques for microfluidic droplet detection

Label free techniques that utilize electrical measurement methods can also be used for droplet detection. Electrochemical, impedance, and microwave techniques are most commonly utilized in this approach. Multiple sensor arrays can be integrated into the microfluidic chip. When a droplet or microparticle passes over the contacts, an electric signal is produced which is then recorded and evaluated [90–92]. The benefit of this technique is the possibility of sensing the content of individual droplets without any chemical or physical intrusion.

An example of a device using technique is shown in Fig. 2.14A where the schematic of the microfluidic chip with T-junction droplet generator and coplanar electrodes placed beneath the microchannel forming the capacitive sensors is shown. A single microfluidic channel travels over a variety of electrode

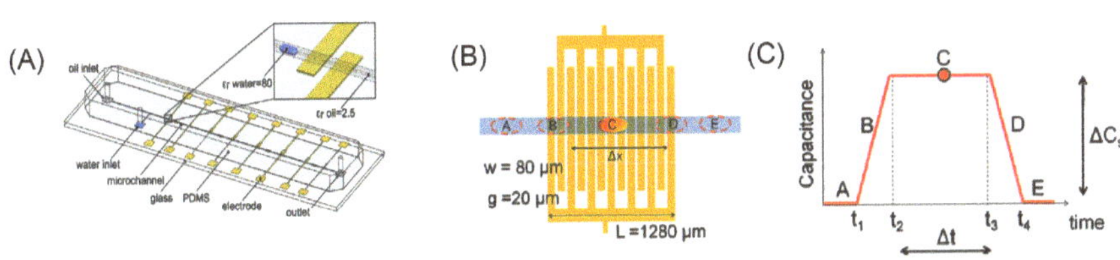

FIGURE 2.14

(A) Schematic of T-junction droplet generator with integrated capacitive droplet detectors. (B) Closeup schematic of the interdigitated capacitive droplet sensor. (C) Measured capacitive signals at certain points on the sensor. Capacitive detector integrated into a microfluidic chip [85].

configurations with varying electrode gaps and widths. The challenge is the difficulty of integrating electrical contacts onto the chip, which makes the structure complex and difficult to fabricate. Integrated detectors within the microfluidic channels are however superior to designs with external detectors which are less sensitive. Embedded detectors also allow the droplet's presence, speed, and size to be measured without any cross contamination [85].

2.5 Point-of-care microfluidic sensors using hybrid fabrication techniques

Resource limited and remote areas require innovative low-cost technologies for screening and prevention. In developing countries, lack of proper diagnosis and treatment as well as access to healthcare centers are the reasons for more than 95% of deaths due to infectious diseases [94]. Even in developed countries, remote areas with limited access to health care require innovative cost-effective technologies in health screening and disease prevention. Affordability and accessibility of diagnostic tools are major problems in resource poor and remote neighborhoods. Portable, simple, and low-cost stand-alone devices can advance much needed healthcare in these areas. The World Health Organization has provided guidelines for a successful diagnostic technology in the developing world, which has the "ASSURED" acronym, meaning affordable, sensitive, specific, user-friendly, robust and rapid, equipment free, and deliverable. Low-cost microfluidic techniques have demonstrated great potential for frugal diagnostics where resources and infrastructure are limited. In this section, we illustrate examples of innovative, low-cost paper and plastic fluidic devices for diseases such as whooping cough, malaria, and others. These devices are highly suitable to be used in areas where there is either no or limited access to electricity, laboratories, or primary healthcare.

2.5.1 PDMS–paper hybrid microfluidic device for detection of whooping cough

The whooping cough is a respiratory disease that can be a life threat to children and is prevalent in developing countries. It was estimated that about 24.1 million whooping cough cases and with 160,700 deaths occurred in children globally in 2014 [95]. Cases are often underdiagnosed due to the similarity of its symptoms with common cold at the beginning of infection [96]. Having a low-cost, rapid diagnostic tool would be extremely beneficial as it can quickly detect, treat, and prevent the spread of the infection. Current diagnostic methods include bacterial culture, enzyme-linked immunosorbent assay (ELISA), and real-time polymerase chain reaction (RT-PCR) which are either time-consuming or require expensive instrumentation.

Recent research have used microfluidic devices to implement a DNA amplification method, known as the loop-mediated isothermal amplification (LAMP) for pathogen detection [97]. The LAMP technique is attractive as it can amplify target DNA at constant temperatures, and it has shown high specificity in disease diagnosis and can be miniaturized into portable diagnostics using microfluidics. The work by Dou illustrates a microfluidic device that uses the LAMP technique for instrument-free point of care detection of the whooping cough bacteria, *Bordetella pertussis* [95]. The sensor is a hybrid paper and PDMS device in which the paper substrate is used for storage and protection of DNA primers and the transparent PDMS substrate is used for robust liquid manipulation.

To use the sensor, LAMP primers and PC DNA with its primer mix are preloaded into separate inlets. The reaction mixtures and positive control were introduced next into the microfluidic device using a pipette and the flow is driven by capillary action. Once complete, the inlet and outlets reservoirs are sealed with epoxy glue to prevent the reagent evaporation during on-chip LAMP reactions. Next the microfluidic device was heated at 63°C for 45 minutes and to 95°C for 2 minutes to terminate the LAMP reactions. The hybrid microfluidic platform is housed in 3D-printed casing and is equipped with a battery-powered bioanalyzer and heater to allow portability and field detection. This hybrid PDMS–paper fluidic device enables better primer distribution and more stable diagnostic performance over long periods of time compared to paper-free non-hybrid devices. This testing method is low-cost where the cost of the ready-to-use paper/polymer hybrid microfluidic device was only ~30 cents, and the cost per assay was ~3 dollars. Another advantage of this device is that it can perform rapid tests, where the whole assay process takes <1 hour and it does not require any additional expensive equipment. Test results can be easily detected by the naked eye under a portable UV light pen. The limit of detection (LOD) was found to be as low as five DNA copies per LAMP zone. These significant features make this microfluidic approach suitable for POC detection of *B. pertussis* and has good potential for applications in low-resource settings [95].

2.5.2 Point-of-care detection of protein biomarkers using enhanced centrifugation-assisted lateral flow immunoassay

Lateral flow immunoassay (LFIA) is one of the most popular tools for POC detection due to its low cost, simplicity, and portability. Compared to ELISA techniques which usually take several hours to complete, lateral flow devices allow detection to be completed within 10–20 minutes. Sensitivities of these rapid lateral flow devices, however, are limited and signal-enhancing materials as well as amplification methods are often introduced to improve its performance. Signal amplification, sample preparation, and handling often adds complexity to LFIAs making them less suitable for POC applications especially when body fluids are involved. When using clinical samples or samples using body fluids,

ELISA-derived methods are often preferred as they can produce sensitive, specific, and precise quantification of protein biomarkers. Unfortunately, ELISA requires the use of cumbersome instruments and experienced personnel to perform sample incubation with washing cycles in-between and final read-out, making them only suitable for nonemergency detections in centralized laboratories.

To solve these problems, a group of researchers have integrated centrifugal microfluidics in lateral flow devices in an effort to allow automation for sample preparation, lateral flow actuation, washing, and signal amplification in a single device [93]. Centrifugal microfluidics uses centrifugal forces and valves for liquid manipulation. This centrifugation-assisted LFIA was used to detect protein biomarkers in whole blood for human prostate specific antigen (PSA). PSA detection allows long-term monitoring of the prostate condition and also provides information and prognosis for prostate cancer [93]. This system, named as the ECLFIA disk is shown in Fig. 2.15 consists of a PMMA microfluidic layer, an arc nitrocellulose (NC) membrane, three pillar valves, and a layer of single-sided adhesive tape. The NC membrane has the capture antibody and anti-mouse IgG as the test and control lines respectively (Fig. 2.15A). The microfluidic channels were patterned on the PMMA microfluidics using a CNC milling machine as shown in Fig. 2.15B. Metal pillars were added in certain parts of the channels and act as valves that can be opened using electromagnetic force. Adhesive tape is used to seal the chambers and to keep both the metal pillars and the NC membrane in place.

To perform testing, undiluted blood samples are loaded into the disk, and the disk is centrifuged at different speeds to first separate the plasma and the blood cells,

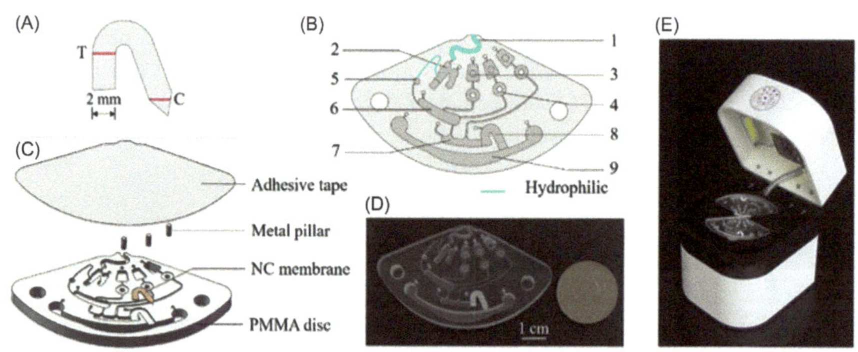

FIGURE 2.15

Schematic of the ECLFIA system. (A) Patterned NC membrane (T, test line; C, control line). (B) Illustration of the patterned PMMA layer (1, inlet channel; 2, blood cell separator; 3, reagent chamber; 4, pillar valve; 5, lyophilizing chamber; 6, cushion chamber; 7, lateral flow chamber; 8, NC chamber; and 9, waste chamber). (C) Configuration of the ECLFIA disc. (D) Image of an assembled ECLFIA disc. (E) Image 459 of the supporting device [93].

to transfer the supernatant into the siphon channel, to allow antibody–antigen conjugation to occur in the lateral flow membrane, and for the final washing steps. An optimized rotation speed is used to allow lateral flow reaction to occur in the NC membrane. These rotation speeds and valve control can be preprogrammed into the system to allow for automation. This system has advantages such as integrated sample preparation, enhanced lateral flow detection and, signal amplification. Only 20 μL of blood sample is required for detection of PSA and the entire process can be completed within 15 minutes. This system also has better sensitivities with its LOD = 0.028 ng/mL which has a 21.4-fold improvement compared to LFIAs.

2.5.3 Point-of-care pathogen diagnostics using an integrated rotary microfluidic system

Recent months has seen the emergence of infectious diseases due to increasing world population and increased human contact with unfamiliar microbes that promotes dissemination. To prevent and contain disease spread, timely and precise diagnostics are of great importance since it helps to reduce human fatalities, economic losses and disruption of day-to-day activities. Nucleic acid based test (NAT) has been traditionally the "gold standard" of diagnostics since it allows highly sensitive and specific pathogen detection [103]. Unfortunately, NAT requires skilled personnel, specialized laboratories, and expensive equipment making it unsuitable for applications in remote settings.

To overcome this, researchers have developed an integrated rotary microfluidic system that can perform several functions including DNA extraction, LAMP, lateral flow detection using microfluidics, and rotational speed control as shown in Fig. 2.16A. Microfluidics technology was used to allow small reagent volume consumption, rapid analysis, high throughput automation of the whole NAT procedure. A single unit of the microdevice contains a solid phase DNA extraction using glass microbeads (Fig. 2.16B). The use of glass bead-based centrifugal microfluidics allows automation of solid phase nucleic acid extraction. The DNA absorbs on the surface of the microbeads. Keeping the strategy of cost effective microfluidics, LAMP reaction was used instead of PCR to remove the necessity of having an expensive thermocycler, while the specificity and sensitivity was enhanced using selective six primer sets [98]. The LAMP reaction chamber (green box in Fig. 2.16B) mixes the eluted DNA with a LAMP cocktail. The sample then flows into the lateral flow strip which contains a buffer loading pad, a conjugate pad and a detection zone Fig. 2.16C. When combined with a colorimetric based lateral flow strip, this system was able to identify foodborne bacterial pathogen in water and milk. The LOD for monoplex *Salmonella* Typhimurium, multiplex *Salmonella* Typhimurium, and *Vibrio parahaemolyticus* was 50 CFU in 80 minutes. This results show promising applications in the fields of clinical diagnostics, food safety, and environment monitoring.

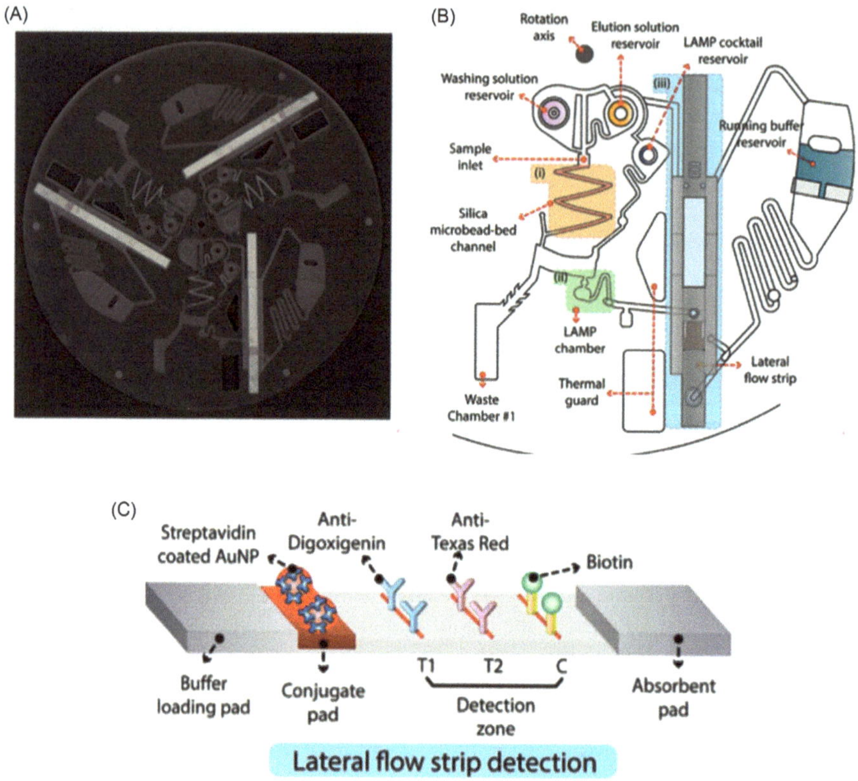

FIGURE 2.16

(A) A digital image of the integrated rotary microdevice. (B) Schematic illustration of the integrated rotary microdevice for the DNA extraction, the LAMP reaction, and the lateral flow strip detection. (C) Schematics of a lateral flow strip which consists of a buffer loading pad, a conjugate pad that contains streptavidin-coated AuNPs, a detection zone where an anti-Digoxigenin, an anti-Texas red, a biotin are immobilized in the test line 1, the test line 2 and the control line, respectively, and an absorbent pad [98]. *LAMP*, Loop-mediated isothermal amplification.

2.5.4 Automated microfluidics for rapid CRISPR-based COVID-19 detection

Rapid, early-stage screening is crucial during pandemics to effectively identify infected patients and contain the disease. During the COVID-19 pandemic, RT-PCR has been the most commonly used method to detect patients due to its reliability, specificity, and sensitivity. Unfortunately RT-PCR is time-consuming, requires specialized equipment and laboratories, making it unsuitable for point-of-care applications in the field. An alternative to this is to use CRISPR diagnostic methods which can be automated for rapid and accurate pathogen detection

[102]. In this work, the researchers combined automation, microfluidics, on-chip electric field control, with CRISPR to allow rapid detection of SARS-CoV-2 RNA in clinical samples. Automation of the CRISPR method reduces testing time significantly where this method takes about 35 minutes from raw sample to result—a significant improvement over RT-PCR tests which takes several hours to complete.

The key component of this system is an electrokinetic microfluidic technique named isotachophoresis (ITP) which is used in two stages during testing. ITP uses electric field to selectively focus ions at a specific zone in the chamber allowing the samples and reagents to be preconcentrated, purified, mixed and have accelerated reactions. The overall flow of the SARS-CoV-2 RNA detection system is shown in Fig. 2.17A. ITP is first used to automatically extract nucleic acids from raw biological samples or from nasopharyngeal swab samples from COVID-19 patients and healthy controls. After extraction, RT-LAMP preamplification is performed. Next, ITP-CRISPR-based fluorescent detection of N, E, and RNase P genes is performed on chip. The detailed steps show that a microfluidic ITP chip (NS-12A) is used to extract total nucleic acids from the lysed samples (Fig. 2.17B; mode 1) and ITP-CRISPR is used for target detection using Cas12-based enzymatic reactions (Fig. 2.17B; mode 2). On a different chip, ITP can also be used with electric field gradients to preconcentrate, cofocus Cas12-gRNA, target DNA and reporter ssDNA into a 100 pL reaction volume. Next, CRISPR-based fluorescent detection can be used to detect N, E, and RNase P genes as shown in Fig. 2.17C.

The performance of the ITP-CRISPR was tested on a total of 64 patient specimens. This ITP–CRISPR system could correctly detect 30 out of 32 positive COVID samples with no false positives observed. In comparison, the RT-PCR assay detected 32 positive samples and 32 were negative samples for SARS-CoV-2. The ITP-CRISPR method correctly detects positive samples when the Ct value is less than 33, which matches with the estimated LOD of 10 copies per microliter. The ITP–CRISPR assay showed 96.9% overall agreement with a kappa value of 0.94 (95% CI: 0.85–1.0).

2.5.5 Array of microfluidic button valves for high throughput COVID-19 serology assays

During a pandemic such as COVID-19, where mass and periodic testing is required to control the spread of the disease, serological assays are often preferred for screening purposes as they are low in cost and produce almost instantaneous results. Low-cost serosurveys can be used to determine prevalence of the infection in a population, monitor the disease's spread rate, and identify which public health policies should be used to contain the pandemic. While still lacking the sensitivities of RT-PCR, serological assays can be used instead to differentiate between early and late infections by measuring the IgM and IgG isotypes,

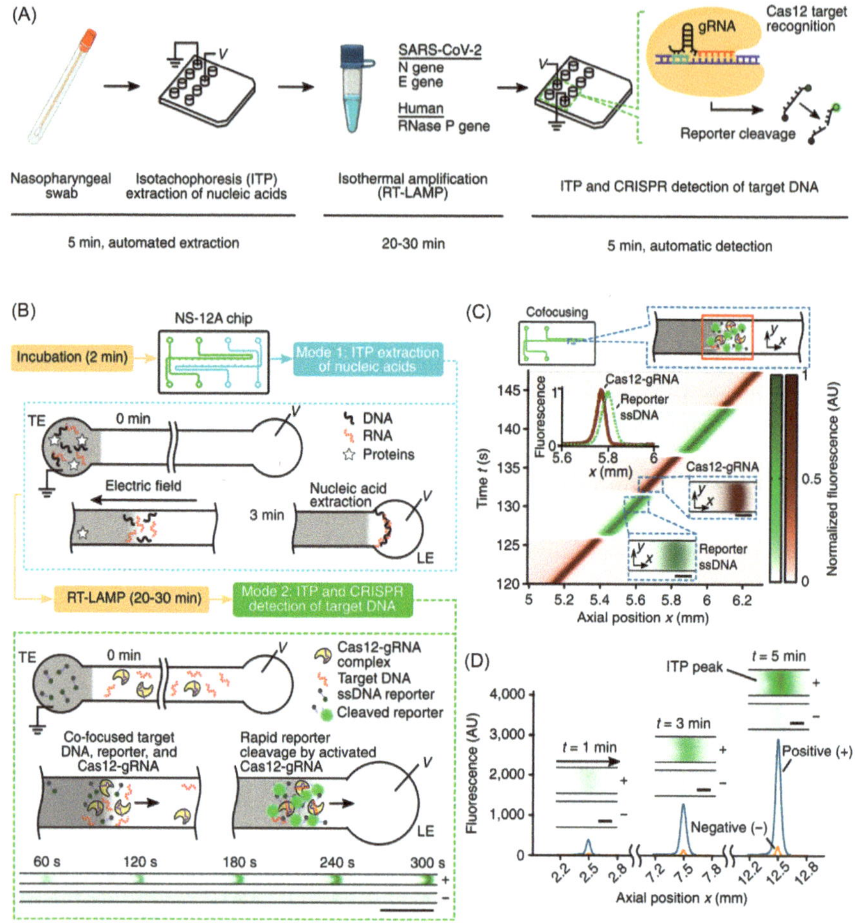

FIGURE 2.17

An electric field-mediated microfluidic assay for SARS-CoV-2 RNA detection using ITP and CRISPR–Cas12. (A) Schematic of SARS-CoV-2 detection workflow from sample to result. Microfluidic ITP is used to extract nucleic acids from raw NP sample, followed by off-chip RT-LAMP preamplification and on-chip ITP-CRISPR – based fluorescent detection of N, E, and RNase P genes. CRISPR–Cas12 activation by the presence of target cDNA of SARS-CoV-2 results in nonspecific ssDNA cleavage and unquenching of a reporter ssDNA labeled with a fluorophore and quencher [99]. (B) Assay working principle. A single microfluidic chip with two channels is used for ITP extraction of nucleic acids (mode 1) and ITP–CRISPR detection (mode 2). In mode 1 (within *dotted blue rectangle*), on application of an electric field, nucleic acids with electrophoretic mobility bracketed by the leading (LE) and trailing (TE) electrolyte ions selectively focus within the electromigrating LE–TE interface, leaving behind impurities [100,101]. Following off-chip

(*Continued*)

respectively. In this work, researchers from Mexico demonstrated a microfluidic platform for semiautomatic detection of IgG and IgM for four different SARSCoV-2 antigens [104].

Four proteins of the virus SARS-CoV-2, that is, spike (S), the S1 subunit (S1), the receptor-binding domain, and nucleocapsid (N) can be detected using this microfluidic array. The array consists of 4 rows and 50 columns, with a total of 200 microchambers, and each chamber is equipped with a button valve. The device is fabricated using multilayer soft lithography techniques to form chambers with 6 μL volumes. Each chamber acts as a fluorescence biosensor using the mechanically induced trapping of molecular interactions technique. Reagents are first loaded into the chambers via the inlets, followed by antigen immobilization of the four antigens. Next, 50 serum samples are pipetted into each well followed by washing steps. Anti-IgG or anti-IgM (fluorescent labeled antibodies) are injected next into the chambers and left to incubate for 30 minutes. An inverted fluorescence microscope is used next to measure the signal intensities of the antibodies. All these steps can be automated and controlled from a computer to ensure repeatability and minimal user intervention.

This device was tested with 100 serum samples from patients of a hospital in Mexico City. Sixty-six samples were from individuals who tested positive for SARS-CoV-2 through PCR with reverse transcription [104]. The device has a sensitivity of 95% and specificity of 91% and has the potential of being a low-cost screening test that can be used to determine the prevalence of infection in a population and monitor the spread of the disease. In this work, researchers from Mexico demonstrated a microfluidic platform for semiautomatic detection of IgG and IgM for four different SARS-CoV-2 antigens that is shown in Fig. 2.18 [104].

◀ RT-LAMP of ITP-extracted nucleic acids, in mode 2 (within the green dashed rectangle), ITP is used to effect target DNA detection using a CRISPR–Cas12 enzyme assay. A positive sample shows a strong fluorescent signal compared to the negative control. (scale bar, 0.5 mm.) (C) Electric field control of Cas12–gRNA and nucleic acids. Experimental visualization of the moving ITP interface in mode 2 using a fluorescently tagged gRNA (red) and ssDNA reporter (green). Spatiotemporal intensity plots of the green and red fluorescence emission show that Cas12–gRNA and nucleic acids electromigrate and co-focus in a ~100-pL ITP interface volume. Top Inset shows ssDNA fluorescence intensity profile (green) at 135 s and comparison with the Cas12–gRNA profile (red). Bottom Insets show instantaneous fluorescence images of the ITP peak which includes labeled Cas12–gRNA (red) and reporter ssDNA (green). (scale bar, bottom insets, 50 μm.) (D) Example of quantitative measurements of on-chip fluorescence detection from cleaving of quencher/fluorophore ssDNA reporters by ITP-focused and activated Cas12–gRNA complex. Raw on-chip fluorescence signal versus axial location for ITP-CRISPR detection of E gene of SARS-CoV-2 positive and negative controls in mode 2. Insets show instantaneous epifluorescence microscopy images of the moving ITP interface. (scale bar, 50 μm) [102].

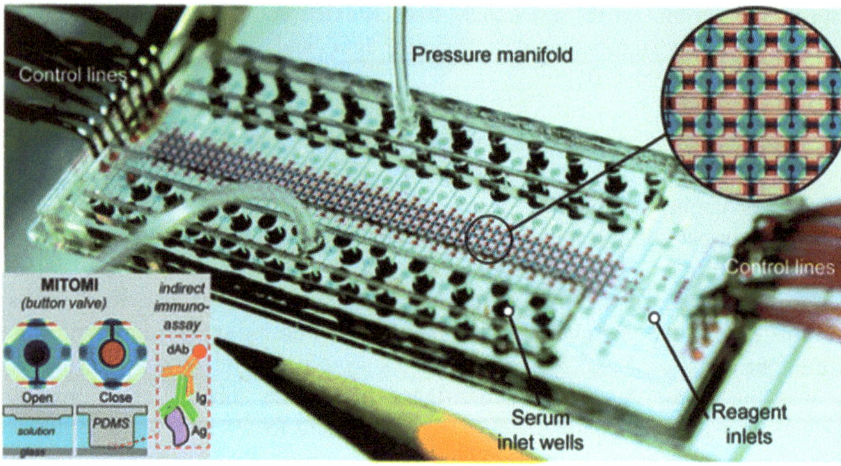

FIGURE 2.18

Microfluidic device for COVID-19 antibody detection. The tip of a pencil is shown for reference. Closeup shows an array of microchambers (blue) surrounded by valves (red) and a MITOMI button valve (red circle) in the center. Bottom inset shows top-view photographs and a cross-sectional view of the actuation of the MITOMI button. An indirect immunoassay is performed on the surface of a glass substrate [104]. *Ag*, antigen; *dAb*, detection antibody; *Ig*, serum antibodies.

2.5.6 Optofluidic sensor for coagulation risk monitoring of COVID-19 patients

COVID-19 patients may experience different levels of illness severity ranging from asymptomatic to multiple organ failures. Severely ill COVID-19 patients may develop coagulation disorders that could lead to clot formation, such as pulmonary embolism, that affects the patients' survival [105]. To improve the recovery rates, doctors may choose to use anticoagulation drugs to treat patients at the intensive care unit. In this research, a point-of-care device that monitors the coagulation status of the patient to allow the doctors to prescribe evidence-based treatment was developed by a team of researchers at Massachusetts Institute of Technology [106]. This device combines microfluidics with a photonic sensor to form an integrated optofluidic device. Microfluidics was used to perform blood pre-processing, and a photonic sensor was used to measure the blood coagulation level. The device is shown in Fig. 2.19A. Whole blood sample is injected into Inlet-1 and passes through a curvature which branches into two channels of different diameters (Fig. 2.19B) to allow blood and plasma separation. Due to the large channel dimension, combined with the centrifugal force, the rich blood component flows into the collector channel, while the plasma component flows into the smaller serpentine channel. Coagulating agent is injected in Inlet-2 and merges with the plasma component into the sensor unit chamber (Fig. 2.19C). The optofluidic sensor detects the coagulation level of

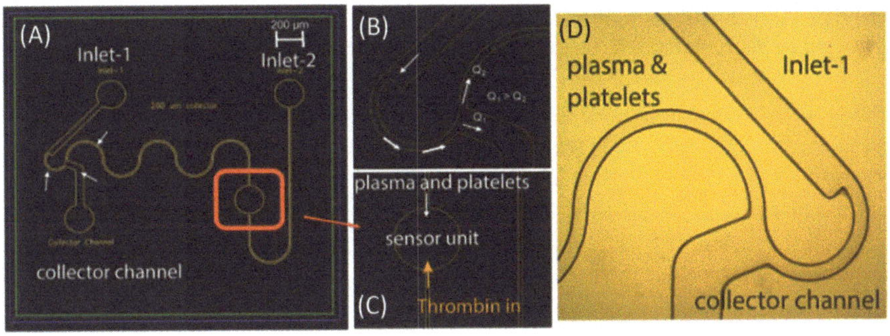

FIGURE 2.19

A microfluidic device for on-chip plasma separation. (A) CAD design of the microfluidic device. (B) Bifurcation channel for blood–plasma separation. Channels of different width combined with centrifugation allow RBC to be collected in a chamber and plasma to flow through. (C) Sensor unit measures the refractive index of blood plasma and coagulation level. (D) Microscopic image of the bifurcation channel [106]. *CAD*, Computer-aided design; *RBC*, rich blood component.

blood by measuring the refactive index of blood. The optofluidic sensor is an optical ring resonator which produces a specific resonant peak for a specific material under test. When anticoagulation agents are introduced into the sensor chamber, the refractive index (and dielectric constant) changes and this can be measured by observing the changes in the resonant peaks of the optical ring resonator. Due to the completeness of the system that combines centrifugation, blood–plasma separation, sensing chamber, and sensor all in one system, it is highly portable and suitable as a point-of-care device for monitoring of COVID-19 patients.

2.6 Conclusions and outlooks

Microfluidic devices have made tremendous progress since their inception in the 1980s, in which miniature devices were first created to manipulate small volumes of fluids. It can be said that these four fields, namely molecular biology, molecular analysis, national security, and microelectronics have helped to develop microfluidics technology into what we have today, where complex lab experiments can be performed in microscale in a single fluidic device. Microfluidics are now embedded in point-of-care platforms that can be used in a wide range of applications such as pregnancy test kits, detection of tropical diseases, detections of toxins in food, detection of chemical contaminants, and detection of biological infectious agents.

Microfluidic design theory and equations have also evolved from Hagen–Poiseuille, Navier–Stokes equations, which were developed in the early 1800s, into a complete field of its own or computational fluid dynamics. There are currently several software tools such as COMSOL, Fluent and Autodesk CFD which can solve these fluidic equations through finite element modeling. These tools can be used to model fluid flow in complex structures such as mixers and droplet generators.

Microfluidics have traditionally capitalized on the advances in integrated circuit technology. Initially, fabrication of microfluidic devices used methods and materials from microelectronics such as photolithography, silicon and glass to manufacture microchannels. Slowly, these methods have been adapted into more specific methods and materials such as soft lithography to incorporate channels on "soft" substrates such as PDMS. Mass production of microfluidic devices that are based on glass or PDMS is still difficult, where challenges arise in terms of production efficiency, integration with electronics. and obtaining regulatory approval. Due to this, microfluidic devices have slowly moved away from academic research to more commercially viable substrates such as paper and plastics. Commercially, lateral flow devices have been the most successful format of microfluidics and they have been utilized in a wide range of applications such as pregnancy test kits, detection of tropical diseases, detections of toxins in food, detection of chemical contaminants, and detection of biological infectious agents. While simple lateral flow devices have easily reached to the consumers in the form of disposable test kits, there is still a bottleneck in the implementation of more complex microfluidic devices in in the commercial world.

In order to have a wider adoption of microfluidic devices commercially, better fabrication methods need to be used. Current challenges that hinder commercialization of microfluidic devices include production efficiency, integration difficulties, cost of equipment, and acquiring regulatory approval. A possible solution to these challenges is 3D printing, which can cater to smaller volumes, has lower upfront costs, and can be customized to a specific application. Compared to plastic/PDMS molding. 3D printing can utilize different materials, depending on the properties needed, at acceptable resolution for many microfluidics applications. This method is especially suitable during the prototyping stage when using the print-on-demand method. There is a low upfront cost and numerous product iterations can be made until the design is finalized. In terms of scaling, 3D-printed devices can be used when required in low volumes or in the range of hundreds of devices. For mass production where thousands of devices are needed, injection molding provides a more efficient and reliable method for production. With the increasing demand for microfluidic devices in biotechnology applications, there has also been an increase in the number of companies that provide services to mass produce microfluidic devices. ThinXXS, Germany and Micronit, Holland, Schott Minifab provide OEM services to mass produce custom designs of microfluidics for clients.

Testing and characterization is also an important aspect in the development and commercialization of microfluidic devices. Burst pressure and leakage tests are among important tests that can be done to check the reliability of the fluidic devices. These tests can be done manually or can be automated using flow control systems which usually include syringe pumps, valves, flow switches, microfluidic pumps, flow matrix switches, and controllers. Presence of leakage, bubbles, and burst pressures can be detected using flow sensors, pressure sensors, and bubble detectors. These systems are available at numerous microfluidic companies such as Elveflow, Fluigent, Darwin Microfluidics, and others.

2.7 Future outlook

Recent event of COVID-19 outbreak has placed a great demand on public health. There are several challenges that need to be addressed in the light of the pandemic. The first challenge is the need for diagnostic tools that are portable and can provide rapid screening while maintaining high sensitivities. Once patients have been identified, more sensors are needed to constantly monitor their disease progression. The second challenge is the necessity of big data, both in terms of data collection from patients, their medical histories, current conditions as well as their locations in order to contain the outbreak. Analytics on these data need to be performed quickly; infected individuals need to be traced and their close contacts be identified. Public health needs to rely on more sophisticated sensors that can not only detect diseases quickly, but they should also be able to be scaled-up rapidly in order to serve the entire population.

Microfluidic point-of-care platforms can be seen as an emerging technology that can help to provide the tools to improve public health as well as provide solutions for personalized and evidence-based treatment. When coupled with ehealth and artificial intelligence, microfluidic platforms can be a crucial tool to provide point-of-care diagnostics, screening, and monitoring of the masses. The number of microfluidics startups has also increased in recent years, with the majority of companies based in the United States. These companies offer a plethora of devices with applications ranging from water quality analysis, single-cell analysis, plasma extraction, ELISA assays, and wearable systems.

For applications in low-resource settings, small sample preparation techniques coupled with paper- or plastic-based channels and low-cost optical or electrical (e.g., Arduino-based) detection systems can be used either at home or by community health workers. These systems can also be programmed to either being operated as a stand-alone or wirelessly connected to smartphones together with apps for data visualizations. Implementations in low-resource settings have several challenges, namely the sensitivities of these low-cost devices and providing proper training to the local communities. These challenges can be overcome by performing adequate number of trials and iterations in improving the products to achieve the desired sensitivities. Educational awareness activities and technology transfer may also be done to overcome the lack of expertise in local communities. The advantages of using rapid, low-cost, point-of-care microfluidic devices in low-resource settings include having a reduced budget for monitoring and containment of diseases, bringing affordable clinical testing to the masses, increasing health awareness to the poor, and local empowerment in healthcare. With all these advantages, it can be seen that microfluidic devices have the potential to be a key player in the field of personalized diagnostics both in developing and developed countries.

References

[1] G.S. Jeong, S. Chung, C. Kim, S. Lee, Applications of micromixing technology, Analyst 135 (2010) 460–473.

[2] G. Cai, L. Xue, H. Zhang, J. Lin, A review on micromixers, Micromachines 8 (2017) 274.

[3] W. Zhang, X. Wang, X. Feng, C. Yang, Z. Mao, Investigation of mixing performance in passive micromixers, Ind. Eng. Chem. Res. 55 (2016) 10036–10043.

[4] K. Ward, Z.H. Fan, Mixing in microfluidic devices and enhancement methods, J. Micromech. Microeng. 25 (2015) 094001.

[5] N. Zaidon, Design and Modelling of a Microfluidic Concentration Gradient Generator Network Using Electric Circuit Analogy, International Islamic University Malaysia, 2017.

[6] R.H. Liu, M.A. Stremler, K.V. Sharp, M.G. Olsen, J.G. Santiago, R.J. Adrian, et al., Passive mixing in a three-dimensional serpentine microchannel, J. Microelectromech. Syst. 9 (2000) 190–197.

[7] M. Nimafar, V. Viktorov, M. Martinelli, Experimental comparative mixing performance of passive micromixers with H-shaped sub-channels, Chem. Eng. Sci. 76 (2012) 37–44.

[8] T. Li, X. Chen, Numerical investigation of 3D novel chaotic micromixers with obstacles, Int. J. Heat Mass. Transf. 115 (2017) 278–282.

[9] S. Hossain, I. Lee, S. Min, K. Kim, A micromixer with two-layer serpentine crossing channels having excellent mixing performance at low Reynolds numbers, Chem. Eng. J. 327 (2017) 268–277.

[10] S. Gambhire, N. Patel, G. Gambhire, S. Kale, A review on different micromixers and its micromixing within microchannel, Int. J. Curr. Eng. Technol. 4 (2011).

[11] W. Raza, S. Hossain, K. Kim, Effective mixing in a short serpentine split-and-recombination micromixer, Sens. Actuators B Chem. 258 (2018) 381–392.

[12] W. Ruijin, L. Beiqi, S. Dongdong, Z. Zefei, Investigation on the splitting-merging passive micromixer based on Baker's transformation, Sens. Actuators B Chem. 249 (2017) 395–404.

[13] X. Chen, T. Li, A novel passive micromixer designed by applying an optimization algorithm to the zigzag microchannel, Chem. Eng. J. 313 (2017) 1406–1414.

[14] H.L. The, B.Q. Ta, H.L. Thanh, T. Dong, T.N. Thoi, F. Karlsen, Geometric effects on mixing performance in a novel passive micromixer with trapezoidal-zigzag channels, J. Micromech. Microeng. 25 (2015).

[15] Y. Lin, Numerical characterization of simple three-dimensional chaotic micromixers, Chem. Eng. J. 277 (2015) 303–311.

[16] R.R. Gidde, P.M. Pawar, B.P. Ronge, N.D. Misal, R.B. Kapurkar, A.K. Parkhe, Evaluation of the mixing performance in a planar passive micromixer with circular and square mixing chambers, Microsyst. Technol. (2017).

[17] N. Tran-Minh, T. Dong, F. Karlsen, An efficient passive planar micromixer with ellipse-like micropillars for continuous mixing of human blood, Comput. Methods Prog. Biomed. 117 (2014) 20–29.

[18] M.S. Cheri, H. Latifi, M.S. Moghaddam, H. Shahraki, Simulation and experimental investigation of planar micromixers with short-mixing-length, Chem. Eng. J. 234 (2013) 247–255.

[19] A.A.S. Bhagat, E.T.K. Peterson, I. Papautsky, A passive planar micromixer with obstructions for mixing at low Reynolds numbers, J. Micromech. Miroeng. 17 (2007) 1017–1024.
[20] A. Rammohan, P.K. Dwivedi, R. Martinez-duarte, H. Katepalli, M.J. Madou, A. Sharma, One-step maskless grayscale lithography for the fabrication of 3-dimensional structures in SU-8, Sens. Actuators B Chem. 153 (2011) 125–134.
[21] X. Chen, T. Li, Z. Hu, A novel research on serpentine microchannels of passive micromixers, Microsyst. Technol. 23 (2017) 2649–2656.
[22] V. Viktorov, R. Mahmud, C. Visconte, Numerical study of fluid mixing at different inlet flow-rate ratios in Tear-drop and Chain micromixers compared to a new H-C passive micromixer, Eng. Appl. Comput. Fluid Mech. 10 (2016) 182–192.
[23] K. Doufène, C. Tourné-Péteilh, P. Etienne, A. Aubert-Pouëssel, Microfluidic systems for droplet generation in aqueous continuous phases: a focus review, Langmuir (2019).
[24] P. Zhu, L. Wang, Passive and active droplet generation with microfluidics: a review, Lab. Chip 17 (2017) 34–75.
[25] F. Dutka, A.S. Opalski, P. Garstecki, Nano-liter droplet libraries from a pipette: step emulsificator that stabilizes droplet volume against variation in flow rate, Lab. Chip 16 (2016) 2044–2049.
[26] L. Shang, Y. Cheng, Y. Zhao, Emerging droplet microfluidics, Chem. Rev. 117 (2017) 7964–8040.
[27] C. Cramer, P. Fischer, E.J. Windhab, Drop formation in a co-flowing ambient fluid, Chem. Eng. Sci. 59 (2004) 3045–3058.
[28] W. Gao, Y. Chen, Microencapsulation of solid cores to prepare double emulsion droplets by microfluidics, Int. J. Heat Mass. Transf. 135 (2019) 158–163.
[29] Y. Pang, Y. Du, J. Wang, Z. Liu, Generation of single/double Janus emulsion droplets in co-flowing microtube, Int. J. Multiph. Flow 113 (2019) 199–207.
[30] V. Faustino, S.O. Catarino, R. Lima, G. Minas, Biomedical microfluidic devices by using low-cost fabrication techniques: a review, J. Biomech. 49 (2016) 2280–2292.
[31] A. Campo, C. Greiner, SU-8: a photoresist for high-aspect-ratio and 3D submicron lithography, J. Micromech. Microeng. 17 (2007) R81–R95.
[32] D. Lai, J.M. Labuz, J. Kim, G.D. Luker, Simple multi-level microchannel fabrication by pseudo-grayscale backside diffused light lithography, RSC Adv. 3 (2013) 19467–19473.
[33] J. Schneider, H. Feick, D. Kaiser, M. Heller, D. Sarlette, 3D lithography for implant applications, in: IEEE, 2013, pp. 3–6.
[34] G.H. Chapman, Y. Tu, J. Dykes, M. Mio, J. Peng, Creating direct-write gray-scale photomasks with bimetallic thin film thermal resists, Photomask Technol. (2003) 400–411.
[35] V. Nock, R.J. Blaikie, Fabrication of optical grayscale masks for tapered microfluidic devices, Microelectron. Eng. 85 (2008) 1077–1082.
[36] T.A. Duncombe, K. Maurer, A.E. Herr, Rapid microfluidic prototyping of sophisticated protein analysis platforms using grayscale photopatterning, in: MEMS, IEEE, 2014, pp. 1027–1030.
[37] J. Loomis, K.M. Walsh, J. Loomis, D. Ratnayake, C. Mckenna, K.M. Walsh, Grayscale lithography—automated mask generation for complex three-dimensional topography, J. Micro/Nanolithography MEMS MOEMS 15 (2016).

[38] J.H. Teng, Q. Wang, G.H. Yuan, Reconfigurable phase-change photomask for gray-scale photolithography, Appl. Phys. Lett. 110 (2017) 201110.

[39] N.-T. Nguyen, Micromixers: Fundamentals, Design and Fabrication, William Andrew, 2011.

[40] Y. Nam, M. Kim, T. Kim, Fabricating a multi-level barrier-integrated microfluidic device using grey-scale photolithography, J. Micromech. Microeng. 23 (2013).

[41] N. Luo, Z. Zhang, Fabrication of a curved microlens array using double gray-scale digital maskless lithography, J. Micromech. Microeng. 27 (2017).

[42] K. Hung, F. Tseng, H. Chou, Application of 3D gray mask for the fabrication of curved SU-8 structures, Microsyst. Technol. 11 (2005) 365–369.

[43] H. Mekaru, Performance of SU-8 membrane suitable for deep X-ray grayscale lithography, Micromachines 6 (2015) 252–265.

[44] Y. Yoon, J. Park, M.G. Allen, S. Member, A. Various, Multidirectional UV lithography for complex 3-D MEMS structures, J. Microelectrom. Syst. 15 (2006) 1121–1130.

[45] J.B. Lee, K. Choi, K. Yoo, Innovative SU-8 lithography techniques and their applications, Micromachines 6 (2015) 1–18.

[46] P. Kim, K.W. Kwon, M.C. Park, S.H. Lee, S.M. Kim, Soft lithography for microfluidics: a review, BioChip J. 2 (2008) 1–11.

[47] R. Amin, A. Joshi, S. Tasoglu, Commercialization of 3D-printed microfluidic devices, J. 3D Print. Med. 1 (2017) 85–89.

[48] A.J.L. Morgan, L. Hidalgo San Jose, W.D. Jamieson, J.M. Wymant, B. Song, P. Stephens, et al., Simple and versatile 3D printed microfluidics using fused filament fabrication, PLoS One 11 (2016) e0152023.

[49] A.K. Au, W. Lee, A. Folch, Mail-order microfluidics: evaluation of stereolithography for the production of microfluidic devices, Lab. Chip 14 (2014) 1294–1301.

[50] C.C. Glick, M.T. Srimongkol, A.J. Schwartz, W.S. Zhuang, J.C. Lin, R.H. Warren, et al., Rapid assembly of multilayer microfluidic structures via 3D-printed transfer molding and bonding, Microsyst. Nanoeng. 2 (2016) 1–9.

[51] H. Nam, C. Yangfan, S. Yiwei, C. Yin, T. Qian, W. Hongkai, Direct, one-step molding of 3D-printed structures for convenient fabrication of truly 3D PDMS microfluidic chips, Microfluid. Nanofluidics 19 (2015) 9–18.

[52] A.R. Schultz, P.M. Lambert, N.A. Chartrain, D.M. Ruohoniemi, Z. Zhang, C. Jangu, et al., 3D printing phosphonium ionic liquid networks with mask projection microstereolithography, ACS Macro Lett. 3 (2014) 1205–1209.

[53] A.K. Au, W. Huynh, L.F. Horowitz, A. Folch, 3D-printed microfluidics, Angew. Chem. Int. Ed. 55 (2016) 3862–3881.

[54] S. Waheed, J.M. Cabot, N.P. Macdonald, T. Lewis, R.M. Guijt, B. Paull, et al., 3D printed microfluidic devices: enablers and barriers, Lab. Chip 16 (2016) 1993–2013.

[55] M. Heckele, W.K. Schomburg, Review on micro molding of thermoplastic polymers, J. Micromech. Microeng. 14 (2004) R1–R14.

[56] T.Y. Lin, T. Do, P. Kwon, P.B. Lillehoj, 3D printed metal molds for hot embossing plastic microfluidic devices, Lab. Chip 17 (2017) 241–247.

[57] K.M. Weerakoon-Ratnayake, C.E. O'Neil, F.I. Uba, S.A. Soper, Thermoplastic nanofluidic devices for biomedical applications, Lab. Chip 17 (2017) 362–381.

[58] J.U. Pucci, B.R. Christophe, J.A. Sisti Jr, E.S. Connolly, Three-dimensional printing: technologies, applications and limitations in neurosurgery, Biotechnol. Adv. 35 (2017) 521–529.

[59] A.J. Dy, A. Cosmanescu, J. Sluka, J.A. Glazier, D. Stupack, Fabricating microfluidic valve master molds in SU-8 photoresist, J. Micromech. Microeng. 24 (2014) 057001.

[60] P. Yao, G.J. Schneider, B. Miao, J. Murakowski, D.W. Prather, Multilayer three-dimensional photolithography with traditional planar method, Appl. Phys. Lett. 85 (2004) 3920.

[61] S.T.V. Grigali, A. Lazauskas, D. Jucius, D. Vir, B. Abakevi, S. Smetona, Microlens fabrication by 3D electron beam lithography combined with thermal reflow technique, Microelectron. Eng. 164 (2016) 23–29.

[62] M. Meunier, A.-P. Blanchard-Dionne, Electron beam lithography using a PMMA/P (MMA 8.5 MAA) bilayer for negative tone lift-off process, J. Vac. Sci. Technol. B 33 (2015) 061602.

[63] A. del Campo, E. Arzt, Fabrication approaches for generating complex micro- and nanopatterns on polymeric surfaces, Chem. Rev. 108 (3) (2008) 911–945.

[64] J. Xing, M. Zheng, X. Duan, Two-photon polymerization microfabrication of hydrogels: an advanced 3D printing technology for tissue engineering and drug delivery, Chem. Soc. Rev. 44 (2015) 5031–5039.

[65] S.J.M. Malinauskas, Ž. Albertas, S. Hasegawa, Y. Hayasaki, V. Mizeikis, Ultrafast laser processing of materials: From science to industry, Light Sci. Appl. 5 (2016) 3–5.

[66] H.G.T. Gissibl, S. Thiele, A. Herkommer, Two-photon direct laser writing of ultracompact multi-lens objectives, Nat. Photonics 10 (2016) 554.

[67] H. Becker, C. Gartner, Polymer microfabrication methods for microfluidic analytical applications, Electrophoresis 21 (2000) 12–26.

[68] H. Kim, S.K. Moon, M. Seo, Hybrid layering scanning-projection micro-stereolithography for fabrication of conical microlens array and hollow microneedle array, Microelectron. Eng. 15 (2015).

[69] A.R. Jafek, S. Harbertson, H. Brady, R. Samuel, B.K. Gale, Instrumentation for xPCR incorporating qPCR and HRMA, Anal. Chem. 90 (2018) 7190–7196.

[70] P. Nath, D. Fung, Y.A. Kunde, A. Zeytun, B. Branch, G. Goddard, Rapid prototyping of robust and versatile microfluidic components using adhesive transfer tapes, Lab. Chip 10 (2010) 2286–2291.

[71] D.I. Walsh, D.S. Kong, S.K. Murthy, P.A. Carr, Enabling microfluidics: from clean rooms to makerspaces, Trends Biotechnol. 35 (2017) 383–392.

[72] M. Serra, I. Pereiro, A. Yamada, J.L. Viovy, S. Descroix, D. Ferraro, A simple and low-cost chip bonding solution for high pressure, high temperature and biological applications, Lab. Chip 17 (2017) 629–634.

[73] D.J. Kinahan, L.A.N. Julius, C. Schoen, T. Dreo, J. Ducrée, Automated DNA purification and multiplexed LAMP assay preparation on a centrifugal microfluidic "lab-on-a-disc" platform, in: IEEE Micro Electro Mech. Syst, 2018, pp. 1134–1137.

[74] B.H. Weigl, R. Bardell, T. Schulte, F. Battrell, J. Hayenga, Design and rapid prototyping of thin-film laminate-based microfluidic devices, Circuit Des. 3 (2001) 267–274.

[75] X. Ku, Z. Zhang, X. Liu, L. Chen, G. Li, Low-cost rapid prototyping of glass microfluidic devices using a micromilling technique, Microfluid. Nanofluidics 22 (2018) 1–8. 2018 228.

[76] E. Behroodi, H. Latifi, Z. Bagheri, E. Ermis, S. Roshani, M. Salehi Moghaddam, A combined 3D printing/CNC micro-milling method to fabricate a large-scale

microfluidic device with the small size 3D architectures: an application for tumor spheroid production, Sci. Rep. 10 (2020) 1–14. 2020 101.
[77] A. Lashkaripour, R. Silva, D. Densmore, Desktop micromilled microfluidics, Microfluid. Nanofluidics 22 (2018) 1–13. 2018 223.
[78] D.J. Guckenberger, T.E. Groot, A.M.D. de Wan, D.J. Beebe, E.W.K. Young, Micromilling: a method for ultra-rapid prototyping of plastic microfluidic devices, Lab. Chip 15 (2015) 2364–2378.
[79] I.S.L. Abdul Hamid, Y. Mustapha Kamil, A. Abd Manaf, M.A. Mahdi, Fabrication and characterization of micro fluidic based fiber optic refractive index sensor, Sens. Bio-Sens. Res. 13 (2017) 70–74. Available from: https://doi.org/10.1016/J.SBSR.2017.03.003.
[80] I.S.L. Abdul Hamid, B. Khi Khim, S. Sal Hamid, M.F. Abd Rahman, A. Abd Manaf, Implementation of a single emulsion mask for three-dimensional (3D) microstructure fabrication of micromixers using the grayscale photolithography technique, Micromachines 11 (2020) 548.
[81] İ. Yilgor, E. Yilgor, S.C. Kosak, Superhydrophobic Polymer Surfaces: Preparation, Properties and Applications Superhydrophobic Polymer Surfaces: Preparation, Properties and Applications, Smithers Rapra Technology Ltd, 2016.
[82] D.F. Crawford, C.A. Smith, G. Whyte, Image-based closed-loop feedback for highly mono-dispersed microdroplet production, Sci. Rep. 7 (2017) 1–9. 2017 71.
[83] M. Girault, H. Kim, H. Arakawa, K. Matsuura, M. Odaka, A. Hattori, et al., An on-chip imaging droplet-sorting system: a real-time shape recognition method to screen target cells in droplets with single cell resolution, Sci. Rep. 7 (2017) 1–10. 2017 71.
[84] Elveflow, OB1 – 4 channel microfluidic flow controller – [WWW document]. <https://www.elveflow.com/microfluidic-products/microfluidics-flow-control-systems/ob1-pressure-controller/>, n.d. (accessed 08.17.21).
[85] C. Elbuken, T. Glawdel, D. Chan, C.L. Ren, Detection of microdroplet size and speed using capacitive sensors, Sens. Actuators A Phys. 171 (2011) 55–62.
[86] L. Dong, D.-W. Chen, S.-J. Liu, W. Du, Automated chemotactic sorting and single-cell cultivation of microbes using droplet microfluidics, Sci. Rep. 6 (2016) 1–8. 2016 61.
[87] T.M. Moyle, L.M. Walker, S.L. Anna, Controlling thread formation during tipstreaming through an active feedback control loop, Lab. Chip 13 (2013) 4534–4541.
[88] D.H. Tracy, A. Weber, P. Stokes, Low Cost Optical High Speed Discrete Measurement System, 2014.
[89] S. Hassan, A.M. Nightingale, X. Niu, Optical flow cell for measuring size, velocity and composition of flowing droplets, Micromachines 8 (2017) 58.
[90] T. Abadie, C. Sella, L. Thouin, Electrochemical detection of droplet content in microfluidic devices: evidence of internal recirculating convection within droplets, Electrochem. Commun. 80 (2017) 55–59.
[91] N.E. Yakdi, F. Huet, K. Ngo, Detection and sizing of single droplets flowing in a lab-on-a-chip device by measuring impedance fluctuations, Sens. Actuators B Chem. (2016) 794–804.
[92] G. Yesiloz, M.S. Boybay, C.L. Ren, Label-free high-throughput detection and content sensing of individual droplets in microfluidic systems, Lab. Chip 15 (2015) 4008–4019.

[93] M. Shen, N. Li, Y. Lu, J. Cheng, Y. Xu, An enhanced centrifugation-assisted lateral flow immunoassay for the point-of-care detection of protein biomarkers, Lab. Chip 20 (2020) 2626–2634.
[94] C.M. Michaud, Global burden of infectious diseases, Encyclopedia of Microbiology, Elsevier Inc, 2009, pp. 444–454.
[95] M. Dou, J. Sanchez, H. Tavakoli, J.E. Gonzalez, J. Sun, J. Dien Bard, et al., A low-cost microfluidic platform for rapid and instrument-free detection of whooping cough, Anal. Chim. Acta 1065 (2019) 71–78.
[96] V.M. Pierce, M. Elkan, M. Leet, K.L. McGowan, R.L. Hodinka, Comparison of the Idaho Technology FilmArray System to real-time PCR for detection of respiratory pathogens in children, J. Clin. Microbiol. 50 (2012) 364–371.
[97] T. Notomi, Y. Mori, N. Tomita, H. Kanda, Loop-mediated isothermal amplification (LAMP): principle, features, and future prospects, J. Microbiol. (2015).
[98] B.H. Park, S.J. Oh, J.H. Jung, G. Choi, J.H. Seo, D.H. Kim, et al., An integrated rotary microfluidic system with DNA extraction, loop-mediated isothermal amplification, and lateral flow strip based detection for point-of-care pathogen diagnostics, Biosens. Bioelectron. 91 (2017) 334–340.
[99] J.S. Chen, E. Ma, L.B. Harrington, M.D. Costa, X. Tian, J.M. Palefsky, et al., CRISPR-Cas12a target binding unleashes indiscriminate single-stranded DNase. activity, Science 360 (2018) 436–439.
[100] A. Persat, L.A. Marshall, J.G. Santiago, Purification of nucleic acids from whole blood using isotachophoresis, Anal. Chem. 81 (2009) 9507–9511.
[101] A. Rogacs, L.A. Marshall, J.G. Santiago, Purification of nucleic acids using isotachophoresis, J. Chromatogr. A 1335 (2014) 105–120.
[102] A. Ramachandran, D.A. Huyke, E. Sharma, M.K. Sahoo, C. Huang, N. Banaei, et al., Electric field-driven microfluidics for rapid CRISPR-based diagnostics and its application to detection of SARS-CoV-2, Proc. Natl. Acad. Sci. U.S.A. 117 (2020) 29518–29525.
[103] P. Craw, W. Balachandran, Isothermal nucleic acid amplification technologies for point-of-care diagnostics: a critical review, Lab. Chip (2012).
[104] R. Rodriguez-Moncayo, D.F. Cedillo-Alcantar, P.E. Guevara-Pantoja, O.G. Chavez-Pineda, J.A. Hernandez-Ortiz, J.U. Amador-Hernandez, et al., A high-throughput multiplexed microfluidic device for COVID-19 serology assays, Lab. Chip 21 (2021) 93–104.
[105] I. Biswas, G.A. Khan, Coagulation disorders in COVID-19: role of Toll-like receptors, J. Inflamm. Res. 13 (2020) 823–828.
[106] R. Singh, A.B.G. Frydman, L. Kimerling, A. Agarwal, B.W. Anthony, Integrated Optofluidic Sensor for Coagulation Risk Monitoring in COVID-19 Patients at Point-of-Care, 2020.

CHAPTER

Lab-in-a-fiber biosensors

3

Kaiwei Li[1], Jiajia Wang[2], Wenchao Zhou[3], Shuwen Zeng[4], Tuan Guo[5] and Lei Wei[6]

[1]*Key Laboratory of Bionic Engineering of Ministry of Education, Jilin University, Changchun, P.R. China*

[2]*College of Agricultural Equipment Engineering, Henan University of Science and Technology, Luoyang, P.R. China*

[3]*State Key Laboratory of Applied Optics, Changchun Institute of Optics, Fine Mechanics and Physics, Chinese Academy of Sciences, Changchun, P.R. China*

[4]*CNRS, XLIM Research Institute, UMR 7252, University of Limoges, Limoges, France*

[5]*Institute of Photonics Technology, Jinan University, Guangzhou, P.R. China*

[6]*School of Electrical and Electronic Engineering, Nanyang Technological University, Singapore, Singapore*

3.1 Introduction

Highly sensitive biomedical analysis technologies plays an important role in the fields of bioassay, disease diagnosis, food safety inspection, and environmental monitoring. Bioassay technologies such as surface plasmon resonance, enzyme-linked immunosorbent assay, absorbance spectroscopy, fluorescence spectroscopy, and Raman spectroscopy are well developed and established as the standard approaches in a variety of fields. These technologies commonly rely on bulky instruments and require tedious manual operations, which cannot meet the needs of rapid and portable detection ability in the case of heart infarction or outdoor environmental monitoring. In recent years, the industry has witnessed a growing need for small-sized, low-weight, and compact bio-analyzing instruments that can facilitate the point-of-care testing of disease, on-site testing of environmental samples, and on-site controlling of food qualities. Therefore, it is of great significance to develop portable biochemical sensor technologies with high sensitivity and specificity to fulfill the ever-growing demands mentioned above and thus enhance the quality of life of people.

The optical fiber biochemical sensing technology is a new type of detection technology that has emerged rapidly in recent years, and has a great prospect for applications in the development of portable and on-site bio-analyzing instruments [1]. The invention of optical fiber has played a dominant role in the development of the information society. Nowadays, the primary role of optical fiber is divided into two categories: optical communication and optical sensing. Compared with other optical bio-analyzing

Microfluidic Biosensors. DOI: https://doi.org/10.1016/B978-0-12-823846-2.00007-9

techniques, the optical fiber biochemical sensors have the advantages of a small footprint, antielectromagnetic interference, high sensitivity, fast response, and strong biological compatibility, etc. Benefiting from the well-established light source and demodulation module in the field of optical fiber communication, the optical fiber biochemical sensors can be easily integrated to enable multichannel portable measurement.

Generally, an optical fiber biochemical sensor utilizes the optical fiber as the fundamental transducer for biochemical signal detection and light guidance. In order to enable biochemical sensing, it also integrates multiple technologies, including advanced photonic technology, materials science, and molecular biology. The optical fiber biochemical sensor has flourished over the past few decades due to its unique merits and future prospects. Depending on the different sensing mechanisms, the optical fiber biochemical sensors can be classified into several categories: evanescent wave absorption sensors [2], label-free fiber-optic immunosensor [3], fiber-optic fluorescent sensors [4], and fiber-optic Raman probes [5], etc. Most of these sensors are directly made from telecommunication silica optical fibers consisting of a standard core-cladding structure. However, these fibers are designed to minimize the influence of the external medium on the light (hence information) propagating inside the optical fiber, which is unfavorable for detecting the biochemical substances or relevant reaction processes. Hence, to bring the light confined inside the core to come into contact with the outer medium, the standard optical fibers were usually tapered under a flame [6], etched with hydrofluoric acid [7], inscribed with gratings [8], and side polished [9] to leak the light out to the fiber surface. However, most of these fiber sensors suffer from low mechanical strength caused by the structural modifications and the poor manufacturing repeatability when faced with commercialization.

Recently, the lab-in-a-fiber biosensor has emerged as a new technique that has the potential to revolutionize optical bio-analyzing technology. The basic concept of the lab-in-a-fiber technology is to integrate multiple functionalities, including photonics sensing, microfluidics, and even sample pretreatment, which are conventionally performed in a large laboratory, into a single fiber device [10]. A conceptional schematic diagram of the lab-in-a-fiber biosensor technology is depicted in Fig. 3.1. Thanks to the rapid development of specialty fiber manufacturing technology, microstructured optical fibers with highly sophisticated cross-sections can be successfully fabricated for kilometers with high uniformity, and more importantly, it allows the incorporation of functional materials inside the fiber interactively. Leveraging on the microstructured optical fiber technology, various lab-in-a-fiber biosensing devices with user-designed geometry structures and functionalities have been proposed and explored. Depending on the different geometry structure and light guiding mechanisms, these fibers can be divided into the following: photonic crystal optical fibers [11], the hollow ring core fibers [12], and the suspended core fibers [13]. These lab-in-a-fiber sensors' sensing mechanisms also range from the fundamental evanescent wave absorption sensing [14,15] to fluorescence spectroscopy [16,17], Raman spectroscopy [18–22], and interferometer-based biosensing [23–26]. Such highly compact and sensitive biochemical sensing devices have found applications in cancer diagnosis, heart disease inspection, and uremic examination.

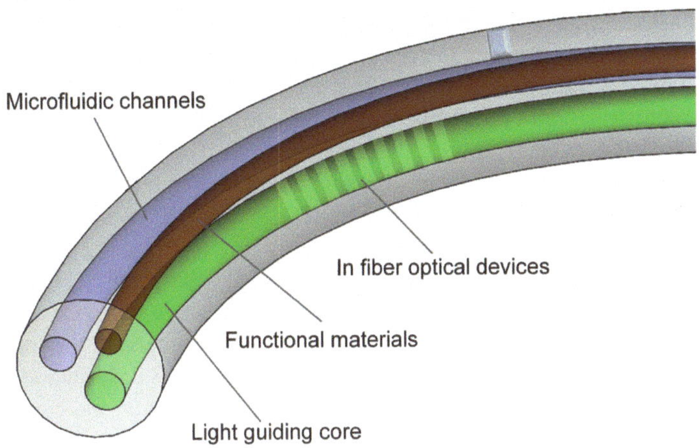

FIGURE 3.1

The conceptional schematics of the lab-in-a-fiber biosensor technology.

In this chapter, various kinds of lab-in-a-fiber biosensors developed in recent years are summarized and reviewed, focusing on those based on an optical sensing mechanism. Firstly, the basics of microstructured optical fibers and the fabrication technologies are introduced. Secondly, the recent developments in the lab-in-a-fiber biosensor technology are introduced. Finally, we describe the challenges and future development trends of the lab-in-a-fiber biosensor technology.

3.2 Basics of lab-in-a-fiber devices

Most biosensors involve fluid, which requires that the designed optical fiber itself contains microfluid channels and can accurately measure the fluid. Thus, the essential functions of a lab-in-a-fiber device are to provide a path for light propagation and a channel for liquid sample manipulation, and based on these providing an interface for the light–analyte interaction. When designing a lab-in-a-fiber device, it is specially taken care of that the liquid samples inside the channel can effectively influence the guiding light inside the fiber core.

3.2.1 Fiber structures and guiding mechanisms

The first microstructured optical fiber can be traced back to the 1970s when Kaiser demonstrated a low-loss single-material microstructured optical fiber with a pure silica core surrounding air holes [27]. However, the structural design was intended to achieve low transmission loss. In the1990s, Russell and coworkers demonstrated the first photonic crystal fiber that incorporates multiple periodically arranged air holes in the fiber to form a photonic bandgap structure and provide low-loss light guiding [28]. From then on, the photonic crystal optical fibers and microstructured optical fibers have experienced rapid development. The materials used for making the fibers progressed from pure silica to polymers [29] and chalcogenide glasses [30], and the single-material microstructured optical fibers were developed into multimaterial optical fibers [31]. Some new fiber designs include the suspended core fibers [13] and ring core fibers [12].

From the perspective of optical waveguiding, the probing light can be transmitted either in a solid fiber core with a high refractive index or inside the defect hollow core of a photonic crystal fiber. We have summarized the most representative designs for microstructured optical fibers in Fig. 3.2. It should be noted that for each design, the fiber structures can vary depending on the conditions of their application, and some advanced schemes that combine several designs are also available.

The photonic crystal optical fiber is the most extensively studied microstructured optical fiber. In principle, the photonic crystal optical fibers can be classified into two categories: the hollow-core photonic crystal fiber and the solid-core photonic crystal fiber. The hollow-core photonic crystal fiber, which has a core with a refractive index lower than the cladding, follows the photonic bandgap theory [11]. Light can be guided through coherent Bragg scattering from the periodic

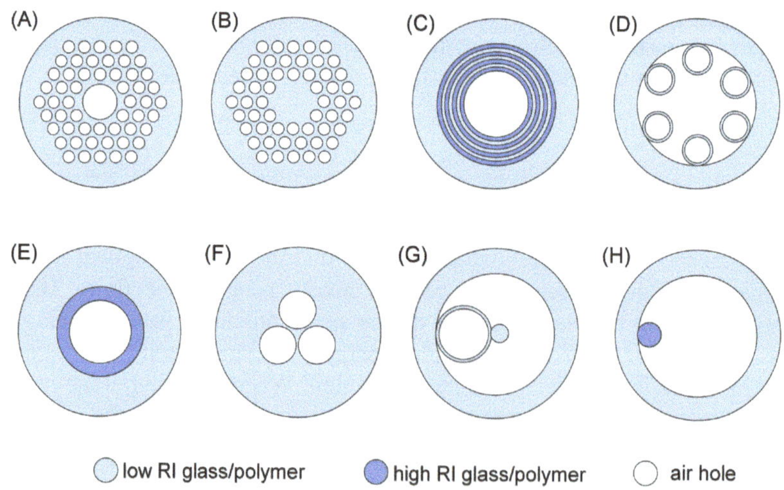

FIGURE 3.2

Representative designs for microstructured optical fibers. (A) Two-dimensional hollow-core photonic crystal fiber; (B) solid-core photonic crystal fiber; (C) one-dimensional hollow-core photonic bandgap fiber; (D) antiresonant hollow-core fiber; (E) ring core optical fiber; and (F–H) suspended core fiber.

photonic crystal structures. Fig. 3.2A and C shows the cross-section structure of the classical two-dimensional hollow-core photonic crystal fiber and one-dimensional hollow-core photonic bandgap fiber [32]. Recently, the antiresonant hollow-core fibers, which only have a single layer of ring tube elements surrounding the hollow core, have been invented [33], as shown in Fig. 3.2D. Light guidance in these fibers is based on antiresonance from the thin glass membranes constituted by the nontouching tubes surrounding the hollow core. As these hollow-core photonics optical fibers follow the photonic bandgap theory, only specific wavelengths that fall within the bandgap can be guided with low loss. The solid core photonic crystal fibers can be regarded as the index guiding fibers, and light can be guided with low loss without bandgaps. The air holes surrounding the solid core over the fiber length can serve as the microchannels for liquid samples.

In addition to the photonic crystal fibers, other microstructure optical fibers with unique designs have been reported. For example, the hollow ring core fiber [12], which introduces an air hole in the core region of conventional step-index fibers, can provide both light guidance in the circular solid core whose refractive index is slightly higher than that of the cladding and a fluid channel in the central hole (Fig. 3.2E). Alternatively, suspended core optical fibers surrounded by either multiple holes [13], thin-wall rings [34], or inner walls [17] attract a lot of attention (Fig. 3.2F–H). These fibers are relatively more straightforward in structure, and the fabrication can be much easier than the photonic crystal fibers. These solid core fibers follow the index guiding mechanism and do not show obvious transmission bandgaps.

3.2.2 Light–analyte interaction forms

The fundamental principle of an optical biosensor is to enable light–analyte interaction. Thus, the concentration of the analyte or the conformation information of a target molecule can be inferred through changes in frequency, phase, amplitude, and polarization states of the light signal. Since light can be guided inside optical fibers via different mechanisms, different forms of light–analyte interaction can be enabled. For light guided in a solid core, the light–analyte interaction always occurs via the evanescent wave, as shown in Fig. 3.3. The evanescent wave is a standing wave that has a penetration depth of half a wavelength [35]. It can penetrate into the surrounding liquids to interact with the substance to be examined near the fiber surface that falls within the penetration depth. Usually, the evanescent wave interaction is weak. However, by diminishing the size of the solid core [36], the interaction can be significantly enhanced. Also, using microstructured optical fibers with an extended length can facilitate the light–analyte interaction via the cumulative effect. The hollow channel can also serve as a fluid channel for light guided in a hollow-core photonic crystal fiber. Thus, the modal field of the guided light overlaps with the liquid sample under examination. In this situation, the light interacts with the analytes directly no matter whether the analytes

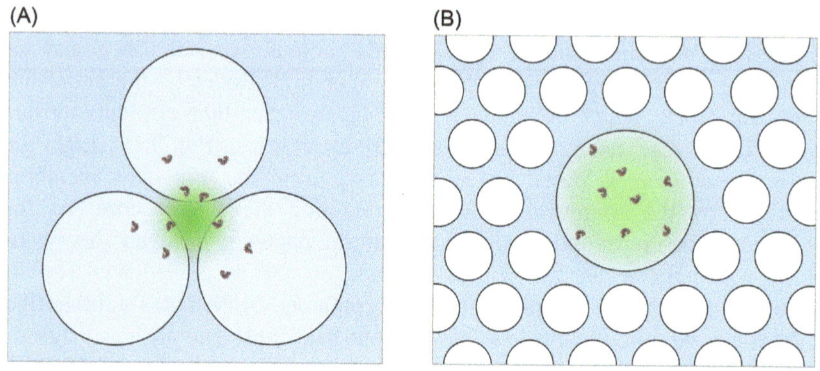

FIGURE 3.3

Forms of light–analyte interaction in microstructured fibers. (A) The evanescent wave interaction; and (B) direct light–analyte interaction.

are in a free state in the hollow core or captured by a receptor immobilized on the channel's inner wall.

3.2.3 Device fabrications

In order to manufacture microstructured optical fibers with well-defined cross-section geometries and optical characteristics, several methods have been developed. The most prevalent fabrication technique is the fiber-drawing technology conventionally used to fabricate telecommunication optical fibers [28]. The fiber-drawing process involves raw materials like tubes and rods and offers research with a high degree of freedom in designing and fabrication. This method can enable mass production with low costs. Recently, the traditional fiber tapering technology and the newly developed femtosecond laser micromachining technology have also been adopted for microstructured fiber device fabrications. In this section, we will give a brief introduction to these technologies.

3.2.3.1 Fiber drawing technique

The typical process of the fiber-drawing technique is displayed in Fig. 3.4, which mainly includes two steps: preform fabrication and fiber drawing. Firstly, a macro-sized fiber preform with the desired cross-section is fabricated from basic elements like tubes and rods. In order to obtain a preform with well-defined geometry structures, a variety of methods have been developed. Among these, the stacking methods shown in Fig. 3.4A are easy to implement and have become the most employed. Sophisticated structures can be obtained using this method, and also using this method, functional materials can be easily incorporated. Mechanical drilling and heat extrusion have also been employed for preform fabrication. Preforms with high precision and good reproducibility can be made using such mechanized

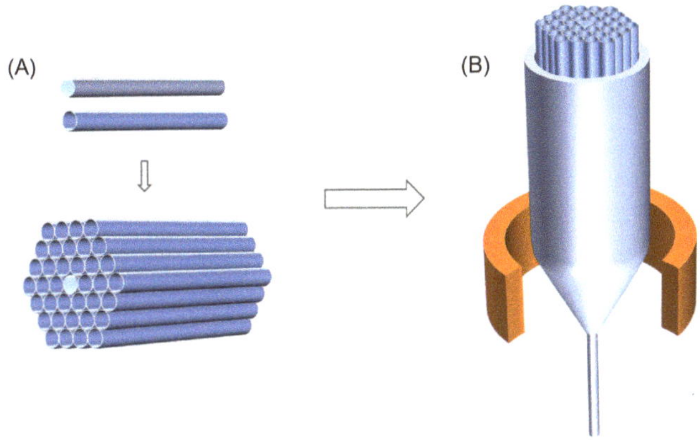

FIGURE 3.4

Schematics of the fiber-drawing process for microstructured fiber fabrication. (A) preform fabrication; and (B) fiber drawing.

production techniques. Secondly, the fiber preform is mounted on a fiber drawing tower and fed into a furnace to heat the preform to above its glass transition temperature. Then, the heated section of the preform shrinks in diameter under the pulling force from the gravity of the lower section. A fiber-drawing equipment then pulls down the thin fiber. During the shrink process, the cross-section geometry of the preform remains unchanged. Thus, optical fibers with desired cross-sections can be obtained. The outer diameter of the fiber can be controlled by adjusting the feeding speed and the pulling speed. Although fiber drawing is a useful technique for microstructured fiber fabrication, the fiber-drawing tower and accessorial equipment required are very expensive and are available only in some laboratories worldwide.

Alternatively, some other easy-to-access techniques, including fiber tapering technology and femtosecond laser micromachining technology, have been explored for microfiber device fabrications. Fiber tapering technology is a traditional method extensively employed for fusion fiber coupler fabrications [24]. Generally, the arranged fibers are fixed by two clamps and heated by a heating element (including oxyhydrogen flame, ceramic heater, and CO_2 laser beam) to above the glass transition temperature. Then, the fiber is pulled in opposite directions by two motorized translation stages. Using this method, lab-in-a-fiber devices can be fabricated from commercially available optical fibers and capillaries. Fig. 3.5A displays a typical lab-in-a-fiber device fabricated through the tapering method.

On the other hand, the femtosecond laser micromachining technology has also been explored for microstructured fiber fabrication in commercially available silica fibers and capillaries [37–39]. The femtosecond laser can be used to remove materials or change a material's properties in transparent substances. On the one hand, the femtosecond laser can be used to fabricate microfluidc channels inside silica optical fibers or silica capipllaries, which severs as the miniature containers

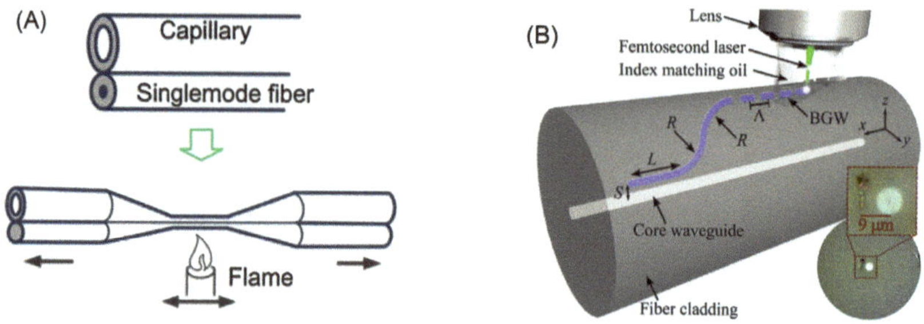

FIGURE 3.5

(A) Schematics of the fiber tapering technique for lab-in-a-fiber device fabrication [24]. (B) Schematics of the femtosecond laser micromachining technique for lab-in-a-fiber device fabrication [37].

for liquid/gasous samples. Microfluidic channels can be fabricaed in two ways: laser ablation and laser inscription with a postetching process. On the other hand, the femtosecond laser can be used to modify the refractive index of the glass material and thus forms waveguides/gratings inside the glass substrates. These waveguide/gratings functions as the sensing element to detect the biochemcial molecules in the samples under test. Fig. 3.5B shows the basic schematics of the microchannel fabrication inside optical fibers using a femtosecond laser. With this technology, researchers managed to fabricate a hybrid waveguide coupler lab-in-a-fiber device from silica capillaries. This highly flexible and programmable fabrication technique opens new opportunities for lab-in-a-fiber biosensors.

3.3 Lab-in-a-fiber sensors for bioapplications

The lab-in-a-fiber technology is a versatile platform for biochemical sensing applications. The light-guiding mechanism, the microfluidic channel, and the fiber designs can be customized based on the demand. More importantly, multitudinous optical effects can be implemented into the microstructured fiber devices, which can significantly diversify the lab-in-a-fiber biosensor family. Based on the sensing mechanism and the fiber design, we have classified them into the following types: in-fiber absorption sensors, in-fiber fluorescent sensors, in-fiber surface-enhanced Raman scattering (SERS) sensors, and in-fiber interferometric biosensors. In this section, we review and summarize the recent developments made for these sensors.

3.3.1 In-fiber absorption sensors

The optical fiber absorption sensor is the most straightforward fiber-optic biochemical sensor inspired by the absorption spectroscopy technique. Absorption spectroscopy works as an analytical chemistry tool to quantify a specific substance in a sample. Depending on the molecules to be measured, the absorption wavelength can range from ultraviolet–visible to infrared. The history of fiber-optic absorption sensors dated back to the 1980s. Paul and Kychakoff proposed the fiber-optic evanescent absorption sensor model [40]. When a specific molecule with a particular absorption cross-section enters the evanescent field of the fiber, it can absorb the evanescent wave and cause attenuation in the output optical power. The fiber-optic evanescent wave absorption sensor was the first fiber-optic biochemical sensor developed. Many optical fiber evanescent wave absorption sensors with various fiber geometries, including the tapered optical fiber, U-bent optical fiber, and the side-polished optical fibers, have been reported since then. These fiber sensors obey the Lambert–Beer law, and the transmitted light at the output end of the fiber can be expressed as [35]

$$P_t = P_0 \times e^{-\gamma L}$$

where P_t and P_0 are the transited optical power collected at the output end and the initial power input into the fiber, respectively. L and γ represent the length of the sensor probe and the absorption coefficient, respectively. From the above equation, one can easily infer that the sensor's sensitivity can be improved by increasing the length of the probe. However, as the proposed sensors are fabricated from commercial telecommunication fibers, the length of the sensing section is limited.

The invention of the microstructured optical fiber has led to the development of fiber-optic absorption sensors. The solid core exposures to a microfluidic channel can be extended to a meter scale or even tens of meters scale so that the light–analyte interaction can be effectively enhanced.

The microstructured fiber absorption sensors were first employed for gas sensing. In 2008, Sun et al. adopted a hollow-core photonic bandgap fiber sensor for absorption-based DNA detection [14]. A 30-cm-long hollow-core photonic crystal fiber with a cladding pitch of 3.8 μm and a core diameter of 10.9 μm was used in their study. Alexa Fluor 700–labeled DNA oligo solution was filled inside the core by capillary force, and the total volume is only 1 μL. The overlap between the guided light and the liquid sample was calculated to be as high as 90%. Using the experimental setup depicted in Fig. 3.6A, they determined the concentration of DNA oligos by tracing the absorbance at 690 nm (Fig. 3.6B). In 2016, Emily et al. employed the suspended core fiber with a core size of 3.3 μm and a length of 0.8 m as the micro-cuvette for human immunoglobulin class G (IgG) antibody detection [15,41]. The structural design of the sensor is shown in Fig. 3.6C. In their study, an LED light source in the 1300–1700 nm range was employed, and the transmitted spectrum shows distinct absorption peaks between 1485 and 1590 nm, which is due to the presence of the antibodies. As shown in Fig. 3.6D, this sensor presents a promising application prospect in biomedicals to quantify the concentrations of IgG antibody with high sensitivity and ultralow sample consumption. The other advantages of the in-fiber absorption sensor are the ultrahigh sensitivity and the ultrasmall sample volume required.

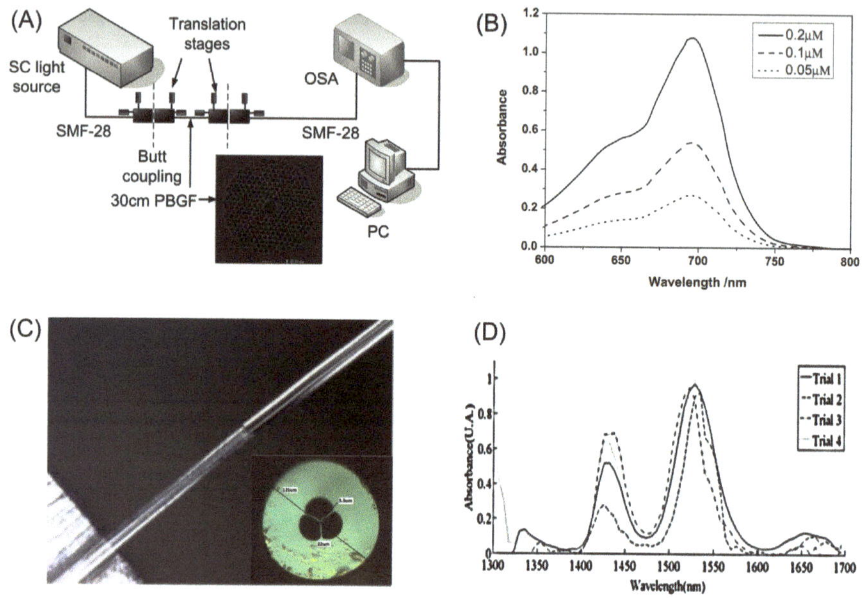

FIGURE 3.6

(A) Schematics showing the experimental setup for in-fiber absorption detection. (B) Spectral response of the hollow-core photonic crystal fiber to DNA oligo solution with different concentrations [14]. (C) Optical microscopy showing the splicing section between the suspended core fiber and single-mode fiber. (The inserted figure shows the cross-section of the suspended core fiber.) (D) Absorption spectra of the sensor to different concentrations of IgG [15].

3.3.2 In-fiber fluorescent sensors

The fluorescent-based assays are widely used in the high-throughput screening of biomolecules due to their high sensitivity, diverse selection of fluorophores, ease of operation, and various readout modes. It has been established as a standard technique in clinical diagnosis and molecular biology. Combining the fluorescent assay and microstructured optical fibers is a clever design, resulting in highly efficient and sensitive biosensing devices. The fiber core can guide the excitation light to the sensing area with low loss and can act as a collector for the emission light.

In 2010, Padmanabhan et al. demonstrated the first photonic crystal fiber-based fluorescent fiber biosensor [16]. The estrogen receptors from an MCF-7 breast carcinoma cell lysate were immobilized inside a hollow-core photonic crystal fiber and detected using anti-ER antibodies with fluorescent dyes. The sample consumption volume was only 50 nL. In 2013, Yang et al. proposed a ring-shaped core fiber and used it for oxygen detection through the fluorescence assay [12].

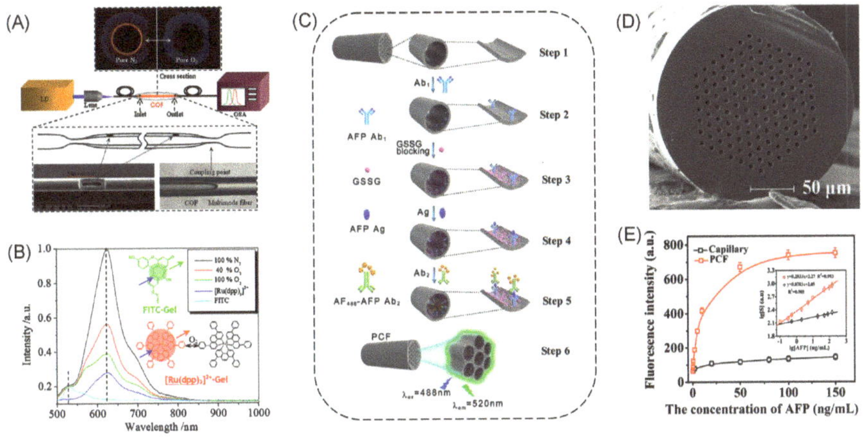

FIGURE 3.7

(A) Cross-section of the ring-shaped core fiber and the experimental setup for in-fiber fluorescent sensing. (B) Spectral response of the sensor to O_2 with different concentrations [12]. (C) Schematics for the surface modification and detection for AFP assay. (D) Cross-section of the solid-core photonic crystal fiber. (E) Sensitivity and linear range of the immunosensor [17]. *AFP*, Alpha-fetoprotein.

O_2 detection is crucial in various fields of clinical diagnosis, environmental monitoring, and biotechnology. The cross-section of the fiber and the experimental setup are shown in Fig. 3.7A. In the study, they coated the inner surface of the ring-shaped core with a thin layer of oxygen ratiometric luminescence-sensitive organosilicon gel. Hence, the excitation laser in the ring core can excite the dyes and collect the emission light. Simultaneously, the concentration of O_2 can be measured. The spectral response to O_2 with different concentrations is shown in Fig. 3.7B. Later in 2017, Liu et al. developed a fluorescence immunosensor to detect alpha-fetoprotein (AFP) [17]. As shown in Fig. 3.7C, they immobilized the first antibody on a solid-core photonic crystal fiber (Fig. 3.7D) to capture the AFP antigen. Then, dye-labeled secondary antibodies are introduced and exited by the evanescent wave of the excitation light. The concentration of the antigen can be inferred by the intensity of the collected emission light. A detection limit as low as 0.1 ng/mL was demonstrated (Fig. 3.7E).

3.3.3 In-fiber SERS sensors

The Raman spectroscopy utilizes the Raman scattering effect when light interacts with a chemical bond in material and has become a widely applicated technique for determining the chemical structure, phase, polymorphy, and crystallinity of a material sample. SERS is an emerging technique that enhances Raman signal intensity by several orders of magnitude using electromagnetic enhancement by nanoroughened metallic surfaces or nanoparticles. Combining the SERS technology with the microstructured optical fiber can make a flexible fiber-optic SERS probe, performing fast and specific detections in hard-to-reach environments. In these sensors, the excitation laser and the Raman signal are transmitted by the fiber core.

To develop lab-in-a-fiber SERS sensors, noble metal nanoparticles that provide strong local field enhancement must be incorporated into the fiber. The nanoparticles can be immobilized on the inner surface of a hollow core or on the outer surface of a solid core or suspended in the liquid sample in the free state. The typical sensing scheme is depicted in Fig. 3.8A. An excitation laser with a wavelength of 532, 633, or 785 nm is launched into the microstructured fiber, and the light excites the Raman signals of the analytes that fall into the local enhanced electrical field of the nanoparticles. U. S. Dinish and Malini Olivo have carried out a series of studies in this field to enable biomedical measurements. They have investigated various kinds of fibers, including hollow-core photonic crystal fibers,

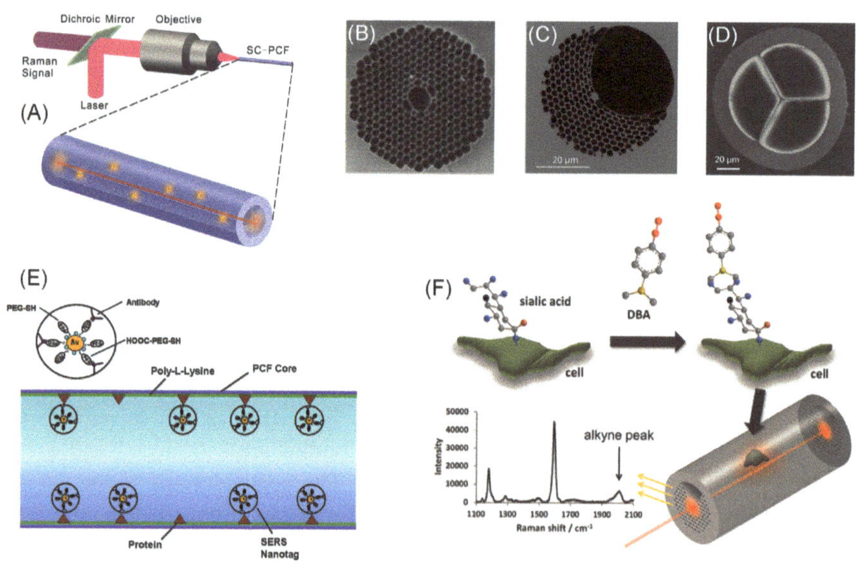

FIGURE 3.8

(A) Illustration of the detection method [20]. (B–D) Cross-section of the hollow-core photonic crystal fiber [43], side-channel solid-core photonic crystal fiber [20], and the suspended core fiber. (E) Schematic of the binding of SERS nanotags to immobilized biomarkers inside the hollow core fiber for multiplex detection [42]. (F) Schematic of single-cell detection using side-hole photonic crystal fiber [21]. *SERS*, Surface-enhanced Raman scattering.

side-channel solid-core photonic crystal fibers, and suspended core fibers (Fig. 3.8B–D), to detect cancer biomarkers through SERS nanotags [18–22,42]. As shown in Fig. 3.8E and F, they have been successful in carrying out multianalyte detection in the same lab-in-a-fiber device [19] and single-cell analysis [21]. The length of the fiber was also optimized both theoretically and experimentally to improve the sensing performance [21] (Fig. 3.9).

In 2019, the Yuan group developed a unique suspended core fiber that has a high refractive index core suspended on the inner hole of a capillary. The cross-section of the suspended core fiber and the sensing system are shown in Fig. 3.9A. The scanning electron microscope image of silver nanoparticle-coated fiber core is depicted in Fig. 3.9B. They demonstrated the feasibility of the development of this fiber by decorating silver nanoparticles on the surface of the

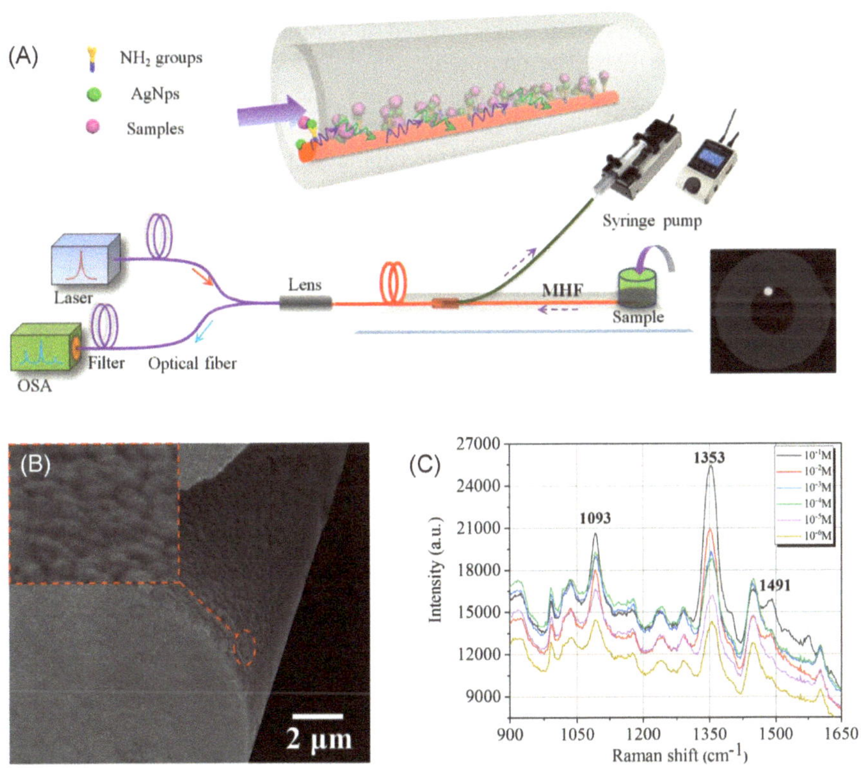

FIGURE 3.9

(A) Schematic diagram of the suspended core fiber optofluidic SERS sensor. (B) SEM images of the fiber core that is coated with silver nanoparticles. (C) SERS spectra for a ceftriaxone solution with different concentrations [44]. *SERS*, Surface-enhanced Raman scattering. *SEM*, Scanning Electron Microscope.

suspended core for in-line SERS sensing and successfully implemented this technique for antibiotic sensing (Fig. 3.9C) [44]. Later, they employed this sensor for the detection of uremic toxin analytes. The results show a detection limit of 10^{-4} M for urea with good linearity, demonstrating good development prospects in the field of clinical biomedicine [45].

3.3.4 In-fiber interferometric biosensors

The traditional optical fiber devices such as gratings, optical interferometers, and fiber couplers can serve as biosensing platforms via the refractive index measuring mechanism. Such devices can facilitate label-free and real-time monitoring of biospecimens. Researchers have endeavored to transplant these classical fiber-optic devices into microstructured fibers. We will introduce these studies with an emphasis on their prospects for biomedical sensing applications.

In 2018, the Guan group demonstrated a capillary optofluidic sensor for microRNA detection [24]. The optofluidic sensor is fabricated using the tapering technology from a close-packed silica capillary and standard optical fiber. As illustrated in Fig. 3.10A, the sensor mainly consists of a microfiber and a thin-wall silica capillary. An SEM image of the capillary optofluidic sensor is depicted

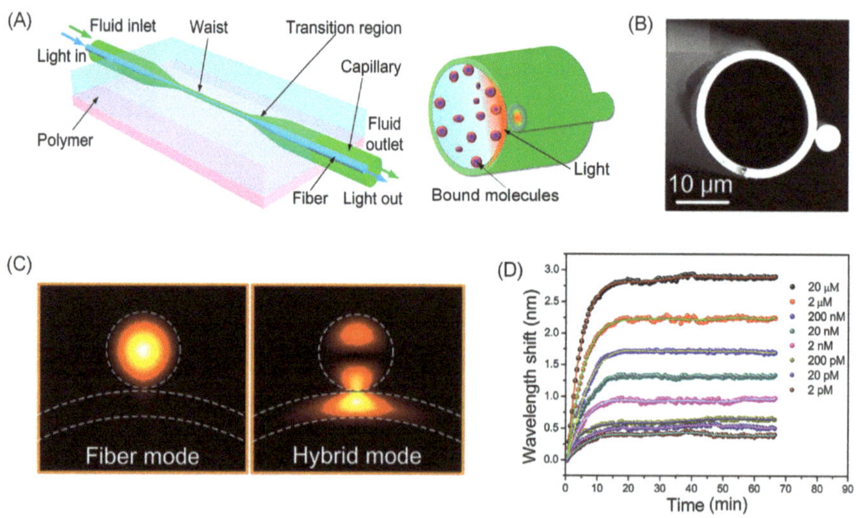

FIGURE 3.10

(A) Schematics of the capillary optofluidic sensor. (B) SEM image showing the cross-section of the capillary optofluidic sensor [23]. (C) Calculated modal profiles of the two fundamental optical modes. (D) The real-time response of the sensor to miRNA-let7a with different concentrations [24]. *SEM*, Scanning Electron Microscope.

in Fig. 3.10B. The light guided by the optical fiber can couple to the wall of the silica capillary at the taper section, and excite new optical modes. As shown in Fig. 3.10C, the interference between two fundamental guide modes in the newly formed hybrid structure waveguide results in an oscillating transmission spectrum, which can be used for biodetection. Using this sensor, the authors managed to detect microRNA-let7a, with a low detection limit of 212 pM (Fig. 3.10D). The same technology was applied to the assay of a well-recognized cancer biomarker p53. An ultralow detection limit of 10 fg/mL was achieved, demonstrating its potential for use in point-of-care testing instrumentations in medical diagnostics [23].

The Yuan group demonstrated an in-fiber Mach–Zehnder interferometric biosensor by splicing a section of twin-core hollow fiber between single-mode fibers with two sections of multimode fiber as the buffer [25]. The fiber contains a suspended core on the inner surface of the capillary and a berried core in the capillary wall. The design of the fiber and the sensing system is shown in Fig. 3.11A. The two cores serve as the two arms of the interferometer with one exposed to the analyte. By measuring the shift of the peaks in the transmitted spectrum, the concentration of biomolecules can be determined (Fig. 3.11B).

In the same year, Nan et al. reported an in-fiber-Sagnac interferometer-based biosensor with a specially designed side-hole photonic crystal fiber. The cross-section and the sensing setup are shown in Fig. 3.12. The large air hole, along with the solid core, running along the whole length of the fiber provides a compact platform for biomolecule analysis. By modifying the surface of the fiber core with specific antibodies for cTnT antigen, the sensor can enable label-free detection of cTnT antigen with a detection limit of 1 pg/mL (Fig. 3.12C and D). This in-fiber biosensor is a good prospect for use as a portable device for heart disease diagnosis [26].

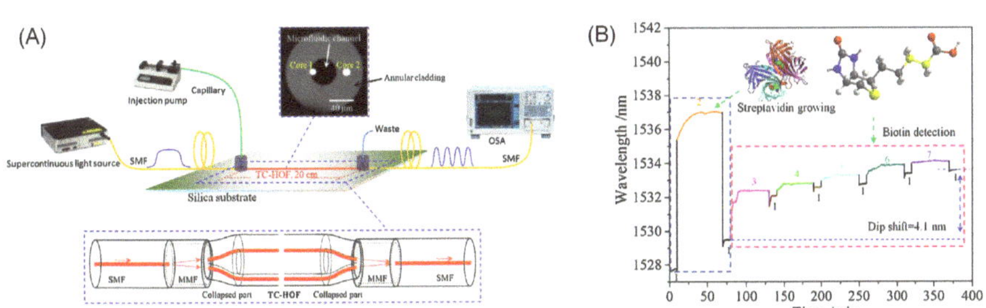

FIGURE 3.11

(A) Schematic diagram for the twin-core hollow optical fiber Mach–Zehnder interferometric biosensor. (B) Real-time spectral response of the sensor during the surface functionalization followed by sequential biotin detection processes [25].

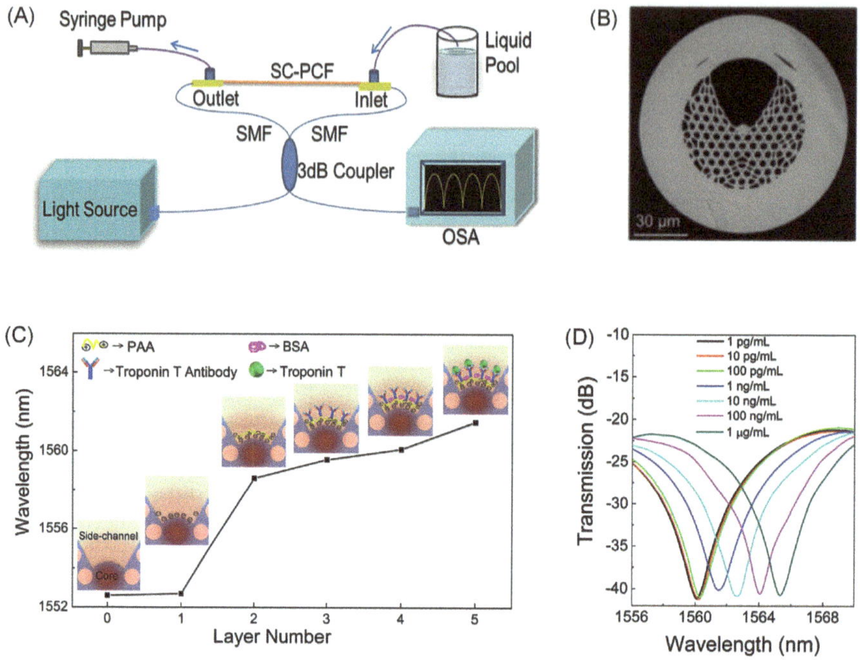

FIGURE 3.12

In-fiber-Sagnac interferometric biosensor with a side-hole photonic crystal fiber. (A) Schematics of the experimental setup. (B) SEM image showing the cross-section of the fiber. (C) The shift of the resonance wavelength is monitored at various surface treatment steps. (D) Responses of the sensor to cTnT antigen with different concentrations [26]. *SEM*, Scanning Electron Microscope.

3.4 Conclusions and outlooks

This chapter first reviewed the basics of microstructured optical fibers, including the guiding mechanisms and fiber designs, and the fabrication techniques. Then, we reviewed and summarized the recent developments in the lab-in-a-fiber biosensor area. Different types of lab-in-a-fiber sensors, including absorption-based sensors, fluorescent sensors, SERS sensors, and in-fiber interferometric biosensors, were reviewed in this chapter. The lab-in-a-fiber biosensor technology, with the unique merits of compact size, strong light–analyte interaction, small sample consumption, and versatile structure design, demonstrated a promising technique for portable point-of-care testing applications.

Meanwhile, the lab-in-a-fiber biosensors are also facing some challenges. On the one hand, the infiltration of the liquid sample and the coupling of the light are sophisticated and more often need sophisticated setups, with advanced interface technology that allows the infiltration of liquid samples and light coupling in an easy way is needed. On the other hand, the functionalities that have been incorporated in the fiber device are still limited. More functionalities like electric modulation, light detection, and sample pretreatment may be incorporated in a single lab-in-a-fiber device with advances in fiber fabrication technology.

Funding information

National Natural Science Foundation of China (No. 62005101 and No. 51905155); the Basic and Applied Basic Research Foundation of Guangdong Province (No. 2022A1515010235).

References

[1] X.D. Wang, O.S. Wolfbeis, Fiber-optic chemical sensors and biosensors (2015–2019), Anal. Chem. 92 (1) (2020) 397–430. Available from: https://doi.org/10.1021/acs.analchem.9b04708.

[2] Y. Wu, X. Deng, F. Li, X. Zhuang, Less-mode optic fiber evanescent wave absorbing sensor: Parameter design for high sensitivity liquid detection, Sens. Actuators B Chem. 122 (1) (2007) 127–133. Available from: https://doi.org/10.1016/j.snb.2006.05.026.

[3] W. Zhou, et al., Ultrasensitive label-free optical microfiber coupler biosensor for detection of cardiac troponin I based on interference turning point effect, Biosens. Bioelectron. 106 (2018) 99–104. Available from: https://doi.org/10.1016/j.bios.2018.01.061.

[4] S.R. Garden, G.J. Doellgast, K.S. Killham, N.J.C. Strachan, A fluorescent coagulation assay for thrombin using a fibre optic evanescent wave sensor, Biosens. Bioelectron. 19 (7) (2004) 737–740. Available from: https://doi.org/10.1016/j.bios.2003.08.002.

[5] P.R. Stoddart, D.J. White, Optical fibre SERS sensors, Anal. Bioanal. Chem. 394 (7) (2009) 1761–1774. Available from: https://doi.org/10.1007/s00216-009-2797-6.

[6] K. Li, T. Zhang, G. Liu, N. Zhang, M. Zhang, L. Wei, Ultrasensitive optical microfiber coupler based sensors operating near the turning point of effective group index difference, Appl. Phys. Lett. 109 (10) (2016). Available from: https://doi.org/10.1063/1.4961981.

[7] K. Li, G. Liu, Y. Wu, P. Hao, W. Zhou, Z. Zhang, Gold nanoparticle amplified optical microfiber evanescent wave absorption biosensor for cancer biomarker detection in serum, Talanta 120 (2014) 419–424. Available from: https://doi.org/10.1016/j.talanta.2013.11.085.

[8] X. Shu, L. Zhang, I. Bennion, Sensitivity characteristics of long-period fiber gratings, J. Light Technol. 20 (2) (2002) 255–266. Available from: https://doi.org/10.1109/50.983240.

[9] N.M.Y. Zhang, et al., Hybrid graphene/gold plasmonic fiber-optic biosensor, Adv. Mater. Technol. 2 (2) (2017). Available from: https://doi.org/10.1002/admt.201600185.

[10] S. Pissadakis, Lab-in-a-fiber sensors: a review, Microelectron. Eng. 217 (November 2018) (2019). Available from: https://doi.org/10.1016/j.mee.2019.111105.

[11] B. Eggleton, C. Kerbage, P. Westbrook, R. Windeler, A. Hale, Microstructured optical fiber devices, Opt. Express 9 (13) (2001) 698. Available from: https://doi.org/10.1364/oe.9.000698.

[12] X. Yang, Y. Zheng, S. Luo, Y. Liu, L. Yuan, Microfluidic in-fiber oxygen sensor derivates from a capillary optical fiber with a ring-shaped waveguide, Sens. Actuators B Chem. 182 (2013) 571–575. Available from: https://doi.org/10.1016/j.snb.2013.03.076.

[13] T.M. Monro, et al., Sensing with suspended-core optical fibers, Opt. Fiber Technol. 16 (6) (2010) 343–356. Available from: https://doi.org/10.1016/j.yofte.2010.09.010.

[14] J. Sun, C.-C. Chan, Y.-F. Zhang, P. Shum, Analysis of hollow-core photonic bandgap fibers for evanescent wave biosensing, J. Biomed. Opt. 13 (5) (2008) 054048. Available from: https://doi.org/10.1117/1.2983676.

[15] J. Rabah, A. Mansaray, R. Wynne, M. Duran, Steering wheel photonic crystal fiber for human IgG detection, Proc. IEEE Sens. 2014 (2014) 1403–1406. Available from: https://doi.org/10.1109/ICSENS.2014.6985275.

[16] S. Padmanabhan, V.K. Shinoj, V.M. Murukeshan, P. Padmanabhan, Highly sensitive optical detection of specific protein in breast cancer cells using microstructured fiber in extremely low sample volume, J. Biomed. Opt. 15 (1) (2010) 017005. Available from: https://doi.org/10.1117/1.3302810.

[17] X. Liu, X. Song, Z. Dong, X. Meng, Y. Chen, L. Yang, Photonic crystal fiber-based immunosensor for high-performance detection of alpha fetoprotein, Biosens. Bioelectron. 91 (November 2016) (2017) 431–435. Available from: https://doi.org/10.1016/j.bios.2016.12.058.

[18] U.S. Dinish, C.Y. Fu, K.S. Soh, B. Ramaswamy, A. Kumar, M. Olivo, Highly sensitive SERS detection of cancer proteins in low sample volume using hollow core photonic crystal fiber, Biosens. Bioelectron. 33 (1) (2012) 293–298. Available from: https://doi.org/10.1016/j.bios.2011.12.056.

[19] U.S. Dinish, G. Balasundaram, Y.T. Chang, M. Olivo, Sensitive multiplex detection of serological liver cancer biomarkers using SERS-active photonic crystal fiber probe, J. Biophotonics 7 (11–12) (2014) 956–965. Available from: https://doi.org/10.1002/jbio.201300084.

[20] N. Zhang, et al., Side-channel photonic crystal fiber for surface enhanced Raman scattering sensing, Sens. Actuators, B: Chem. (2016). Available from: https://doi.org/10.1016/j.snb.2015.09.087.

[21] T. Gong, et al., Highly sensitive SERS detection and quantification of sialic acid on single cell using photonic-crystal fiber with gold nanoparticles, Biosens. Bioelectron. 64 (2015) 227–233. Available from: https://doi.org/10.1016/j.bios.2014.08.077.

[22] U.S. Dinish, F. Beffara, G. Humbert, J.L. Auguste, M. Olivo, Surface-enhanced Raman scattering-active photonic crystal fiber probe: towards next generation liquid biopsy sensor with ultra high sensitivity, J. Biophotonics 12 (11) (2019) 415–534. Available from: https://doi.org/10.1002/jbio.201900027.

[23] L. Liang, L. Jin, Y. Ran, L.P. Sun, B.O. Guan, Fiber light-coupled optofluidic waveguide (FLOW) immunosensor for highly sensitive detection of p53 protein, Anal. Chem. 90 (18) (2018) 10851–10857. Available from: https://doi.org/10.1021/acs.analchem.8b02123.

[24] L. Liang, L. Jin, Y. Ran, L.P. Sun, B.O. Guan, Interferometric detection of microRNAs using a capillary optofluidic sensor, Sens. Actuators B: Chem. 242 (2017) 999–1006. Available from: https://doi.org/10.1016/j.snb.2016.09.153.

[25] X. Yang, et al., Optofluidic twin-core hollow fiber interferometer for label-free sensing of the streptavidin-biotin binding, Sens. Actuators B: Chem. 277 (2018) 353–359. Available from: https://doi.org/10.1016/j.snb.2018.09.032.

[26] N. Zhang, et al., Ultra-sensitive chemical and biological analysis: via specialty fibers with built-in microstructured optofluidic channels, Lab. Chip 18 (4) (2018) 655–661. Available from: https://doi.org/10.1039/c7lc01247k.

[27] P. Kaiser, H.W. Astle, Low-loss single-material fibers made from pure fused silica, Bell Syst. Tech. J. 53 (6) (1974) 1021–1039. Available from: https://doi.org/10.1002/j.1538-7305.1974.tb02780.x.

[28] J.C. Knight, T.A. Birks, P.S.J. Russell, D.M. Atkin, All-silica single-mode optical fiber with photonic crystal cladding: errata, Opt. Lett. 22 (7) (1997) 484. Available from: https://doi.org/10.1364/ol.22.000484.

[29] H. Han, H. Park, M. Cho, J. Kim, Terahertz pulse propagation in a plastic photonic crystal fiber, Appl. Phys. Lett. 80 (15) (2002) 2634–2636. Available from: https://doi.org/10.1063/1.1468897.

[30] J. Fatome, et al., Linear and nonlinear characterizations of chalcogenide photonic crystal fibers, J. Light. Technol. 27 (11) (2009) 1707–1715. Available from: https://doi.org/10.1109/JLT.2009.2021672.

[31] M. Alexander Schmidt, A. Argyros, F. Sorin, Hybrid optical fibers – an innovative platform for in-fiber photonic devices, Adv. Opt. Mater. 4 (1) (2016) 13–36. Available from: https://doi.org/10.1002/adom.201500319.

[32] K. Kuriki, et al., Hollow multilayer photonic bandgap fibers for NIR applications, Opt. Express 12 (8) (2004) 1510. Available from: https://doi.org/10.1364/opex.12.001510.

[33] J.R. Hayes, et al., Antiresonant hollow core fiber with an octave spanning bandwidth for short haul data communications, J. Light. Technol. 35 (3) (2017) 437–442. Available from: https://doi.org/10.1109/JLT.2016.2638205.

[34] L. Htein, Z. Liu, D. Gunawardena, H.-Y. Tam, Single-ring suspended fiber for Bragg grating based hydrostatic pressure sensing, Opt. Express 27 (7) (2019) 9655. Available from: https://doi.org/10.1364/oe.27.009655.

[35] A.K. Sharma, J. Gupta, I. Sharma, Fiber optic evanescent wave absorption-based sensors: a detailed review of advancements in the last decade (2007–18), Optik (Stuttg.) 183 (December 2018) (2019) 1008–1025. Available from: https://doi.org/10.1016/j.ijleo.2019.02.104.

[36] K. Li, W. Zhou, S. Zeng, Optical micro/nanofiber-based localized surface plasmon resonance biosensors: fiber diameter dependence, Sensors (Switz.) 18 (10) (2018). Available from: https://doi.org/10.3390/s18103295.

[37] J.R. Grenier, L.A. Fernandes, P.R. Herman, Femtosecond laser inscription of asymmetric directional couplers for in-fiber optical taps and fiber cladding photonics, Opt. Express 23 (13) (2015) 16760. Available from: https://doi.org/10.1364/oe.23.016760.

[38] R.R. Gattass, E. Mazur, Femtosecond laser micromachining in transparent materials, Nat. Photonics 2 (4) (2008) 219–225. Available from: https://doi.org/10.1038/nphoton.2008.47.

[39] C. Lin, et al., Optofluidic gutter oil discrimination based on a hybrid-waveguide coupler in fibre, Lab. Chip 18 (4) (2018) 595–600. Available from: https://doi.org/10.1039/c8lc00008e.

[40] P.H. Paul, G. Kychakoff, Fiber-optic evanescent field absorption sensor, Appl. Phys. Lett. 51 (1) (1987) 12–14. Available from: https://doi.org/10.1063/1.98888.
[41] J. Rabah, A. Mansaray, R. Wynne, M. Duran, Human immunoglobulin class G (IgG) antibody detection with photonic crystal fiber, J. Light. Technol. 34 (4) (2016) 1398–1404. Available from: https://doi.org/10.1109/JLT.2016.2518921.
[42] U.S. Dinish, et al., Highly sensitive SERS detection of cancer proteins in low sample volume using hollow core photonic crystal fiber, Biosens. Bioelectron. 33 (2012) 293–298. Available from: https://doi.org/10.1016/j.bios.2011.12.056.
[43] X. Yang, A.S.P Chang, B. Chen, C. Gu, T.C Bond, et al., High sensitivity gas sensing by Raman spectroscopy in photonic crystal fiber, Sens. Actuators B : Chem. 176 (2013) 64–68. Available from: https://doi.org/10.1016/j.snb.2012.09.004.
[44] D. Gao, et al., Optofluidic in-fiber integrated surface-enhanced Raman spectroscopy detection based on a hollow optical fiber with a suspended core, Opt. Lett. 44 (21) (2019) 5173. Available from: https://doi.org/10.1364/ol.44.005173.
[45] D. Gao, et al., In-fiber optofluidic online SERS detection of trace uremia toxin, Opt. Lett. 46 (5) (2021) 1101. Available from: https://doi.org/10.1364/ol.418602.

CHAPTER

Flexible and mountable microfluidics for wearable biosensors

4

Lingyin Meng[1], Itthipon Jeerapan[2,3,4] and Wing Cheung Mak[5,6]

[1]*Biosensors and Bioelectronics Centre, Division of Sensor and Actuator Systems, Department of Physics, Chemistry and Biology (IFM), Linköping University, Linköping, Sweden*

[2]*Center of Excellence for Trace Analysis and Biosensor, Prince of Songkla University, Hat Yai, Songkhla, Thailand*

[3]*Division of Physical Science, Faculty of Science, Prince of Songkla University, Hat Yai, Songkhla, Thailand*

[4]*Center of Excellence for Innovation in Chemistry, Faculty of Science, Prince of Songkla University, Hat Yai, Songkhla, Thailand*

[5]*Department of Biomedical Engineering, The Chinese University of Hong Kong, Shatin, Hong Kong*

[6]*Division of Sensor and Actuator Systems, Department of Physics, Chemistry and Biology (IFM), Biosensors and Bioelectronics Centre, Linköping University, Linköping, Sweden*

4.1 Introduction

Microfluidics is a well-consolidated multidisciplinary technology with the capability of precise control and manipulation of fluids at the micron-scale [1]. This unique feature endows microfluidic-based devices with several merits compared to conventional laboratory-based macroscale techniques. These merits include much smaller sample sizes and reagent volume requirements, high-throughput, low-cost, high-resolution separations, and sensitivity, enabling numerous opportunities in various biomedical applications such as organ-on-a-chip, cell biology, material fabrication, drug delivery, and micro-total analysis systems [1–4]. In recent years, microfluidics has played a vital role as one of the key components integrated with electronics and optical transducers in the fabrication of biomedical analytical devices such as biosensors for sample collection and handling, as well as for signal transduction and amplification [5]. On the other hand, flexibility and portability have gained significant interest in the field of device fabrication, particularly in wearable sensors and biosensors to track an individual's physiological information in decentralized locations at a physical fitness level or for medical treatment [6–8]. The advancement in flexible materials and wearable electronics in recent years has driven the development of the next generation of flexible and wearable devices that can be intimately attached to human skin and tissue externally or implanted internally to deliver accurate, continuous, and remote

Microfluidic Biosensors. DOI: https://doi.org/10.1016/B978-0-12-823846-2.00005-5

measurement of physiological parameters or biomarkers [6]. This also brings new opportunists in the field of flexible and mountable microfluidics, migrating from conventional rigid to emerging flexible materials for high body compliance, further serving as an excellent sample–device interface in wearable sensors and biosensors.

In this chapter, we aim to cover the research and development in flexible and mountable microfluidics for wearable biosensors. Hence, the chapter commences with the properties and functions of flexible materials in conformable microfluidics as well as the fabrication methods for various flexible microfluidics. We then move to the state-of-the-art wearable and skin mountable microfluidic biosensors available in the form of epidermal patches, contact lenses, smart bandage/dressing, and wearable mechanical sensors, and discuss their designs, working mechanisms, and analytical performances comprehensively.

4.2 Flexible materials for microfluidics

4.2.1 From rigid to flexible materials for high body compliance

Conventional materials used for the fabrication of microfluidics are typically inorganic materials, such as silicon, glass, and ceramic, due to their abundant availability, high chemical resistance and mechanical strength, as well as high aspect ratio and high accuracy on the virtue of well-developed microelectromechanical systems (MEMS) or advanced micromachining process with etching techniques [9,10]. Silicon was among the first materials chosen and has been extensively used for the fabrication of microfluidic devices [10]. A precise definition of nano-/micro-channels, pumps, valves, and layered structures can be realized by subtractive methods, such as wet etching by HF/HCl and dry etching by reactive ion etching or laser drilling, etc. [10]. Silicon can be further modified with other functional materials (e.g., metals, dielectrics, or insulating coatings) using additive methods. In addition, the well-established silicon surface chemistry via silanol group (−Si−OH) endows the silicon with tunable surface properties including hydrophobicity, hydrophilicity, surface charge, and immobilization of biomolecules. However, silicon also faces several limitations as a substrate material for microfluidics, including a narrow window of mechanical properties with brittle characteristic-induced overall fragility, high cost and opacity, and high requirements for chemicals (e.g., harsh acids), instruments (e.g., photolithography and etching), and environments (e.g., clean room). Sharing the similarities with silicon in mechanical properties and chemical resistance, glass has emerged as an alternative material for the fabrication of microfluidic devices, especially with the merits of lower cost, optical transparency, and compatibility. Beyond these, low-temperature co-fired ceramic materials offer a promising alternative to silicon and glass for the monolithic integration of complex structures, ranging from inner microchannels and reservoirs to various electronic components (e.g., insulators, dielectrics, resistors, and conductors) [11]. What's more, hybrid systems, such as silicon–glass, silicon–polymer, glass–polymer, ceramic–polymer, are also commonly used to overcome the limitations of a single material with respect to the optical properties, thermal conductivity, bonding, and sealings.

Processing of these inorganic material-based substrates usually requires complicated techniques for patterning and bonding, such as wet/dry etching, micromachining, micromilling, sandblasting, and anodic and adhesive bonding, which are usually costly, time-consuming, and require trained personnel in special facilities [12]. On the other hand, despite the high pressure tolerance, and thus warranted performance, of these inorganic material-based microfluidic devices for high-pressure demanding applications ranging from capillary electrophoresis [13] to gas chromatography [14], their special mechanical properties such as high rigidity and high Young's modulus (~GPa) limit their applications in the emerging flexible electronics.

The developments of flexible electronics have proposed new requirements for the manufacturing of the next generation of microfluidic devices for precise control and manipulation of fluids with greater flexibility to deform, bend, and stretch to satisfy different application scenarios. This has led to the research for the convergence of flexibility and microfluidics with new flexible materials and new techniques. Key mechanical properties for the materials needed in flexible microfluidics include low bending stiffness and low Young's modulus values. The bending stiffness is a quantitative metric that refers to the resistance against bending deformation, which determines the comfort of user experiences. The Young's modulus is a measure of material's stretch stiffness under strain. Therefore, low Young's modulus is necessary to avoid the mismatch between the flexible microfluidic device and skin/tissue (~kPa) [15]. Various flexible materials have been used as compliant substrates to construct the flexible microfluidic devices, and these can be briefly categorized into elastomers, thermoplastics, fabrics, and papers.

Polydimethylsiloxane (PDMS), as an elastomer material with an inorganic –Si–O– backbone and organic component side chain, is the most commonly used and commercially available material in flexible microfluidics with advantages including optical transparency, gas permeability, chemical inertness, precise molding accuracy, biocompatibility, and stability. PDMS also possesses high elasticity with low Young's modulus value ranging from 0.57 to 3.7 MPa according to cross-linking degree with different base/curing agent ratio [16,17]. Therefore, PDMS has been intensively employed for molding and prototyping of microfluidic systems (e.g., microchannels and microchambers). In addition, PDMS also plays an important role as the compliant substrate for immobilization of molecules via surface treatment and modification as well as the integration with electronic components [6]. Thermoplastics are made up of polymer resins having a high molecular weights that are moldable at temperature higher than their corresponding glass transition temperature (T_g) and then are resolidified while cooling. A wide range of thermoplastics, such as polymethyl methacrylate (PMMA), polyethylene terephthalate (PET), polystyrene (PS), polyvinyl chloride (PVC), polycarbonate (PC), etc., have been used for the fabrication of microfluidics by the virtue of low-cost fabrication methods for high-throughput production, in spite of a relatively high Young's modulus value in the magnitude of ~GPa [18]. Polyurethane (PU), composed of organic units joined by carbamate (urethane) links, is a class of versatile polymers with high tailorability of the resulting physicochemical properties based on the starting monomers. PU can be adjusted from rigid to flexible and proceeded to elastomer with similar characteristics as PDMS [19]. Especially, thermoplastic PU provides the combined merits of thermoplastics and PU via copolymerization of alternating sequences of hard and soft segments, showing processability as thermoplastics while heated, as well as flexibility and elasticity as PDMS while cooled [20]. In addition to these synthetic polymers, natural polymers, such as textiles, fabrics, threads, and papers, are promising materials for the application in microfluidics due to their light weight, flexibility, and stretchability for comfortable fitting of human body. The flow of liquids through these materials relies on the capillary flow through the gaps between directionally interconnected fibers known as the passive fluid transportation force without the need for an external power source [21]. The control of the flow can also be engineered via weaving, wax printing, plasma treatment, and chemical treatment. Paper has shown its wide application in analytical chemistry, especially for the lateral flow assay. It is also a good substrate in microfluidic paper-based analytical devices for liquid processing via patterning, surface treatment, and functionalization using inkjet, wax, or screen printing [22]. Fig. 4.1 illustrates some selective flexible materials used for microfluidics. Different microfluidic components with flexibility can be realized by using flexible materials as both cover layer and bottom layer, including microchannels, micropumps, microvalves, and micromixers. Behind the exploration of new materials or the innovative engineering of existing materials, the unique properties or functionalities also contributed to the development of various techniques [e.g., three-dimensional (3D) printing] for microfabrication, which is discussed in detail in Section 4.2.3.

Nowadays, wearable devices integrated with various electronic, sensing, and biosensing modules have attracted extensive attention in both academic research and industry. The convergence of flexibility and microfluidics also benefits the advances in wearable devices, especially for personal healthcare management in the form of sensors and biosensors. Good mechanical and adhesive properties with light weight originating from the compliant substrates possess the potential to fulfill the requirements of mechanical behavior for intimate interfacing with soft, deformable, and complex biological tissue internally. Flexible microfluidics can mimic body organs and allow for in-depth in-vitro analysis of the cells. The attachment of such devices onto the human body also provides the promising potential for noninvasive or less-invasive continuous monitoring externally. In addition, flexibility, stretchability, durability, and user-friendly simplicity of the assembled devices are crucial for long-term wearing without conspicuous discomfort. Flexible microfluidics, as an important part of such devices, can realize various functions, including sample collection, pretreatment, handling, and storage. In

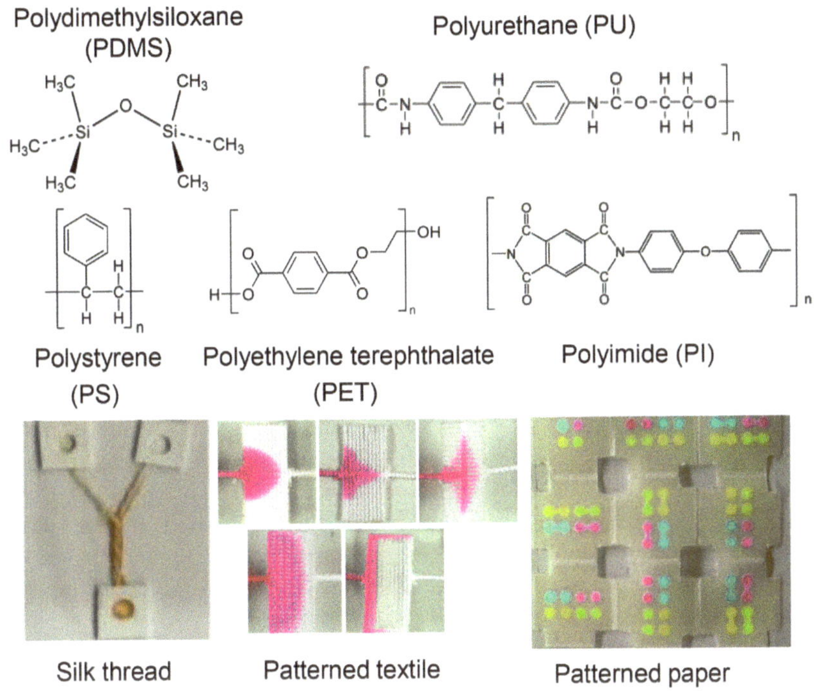

FIGURE 4.1

Some selective flexible materials for microfluidics: silk thread [23], patterned textile [21], and patterned paper [24].

Silk thread—reproduced from S.S. Banerjee, A. Roychowdhury, N. Taneja, R. Janrao, J. Khandare, D. Paul, Sensor. Actuat. B Chem. 186 (2013) 439–445 with permission from copyright 2013 Elsevier B.V.; patterned textile—reproduced from Y. Li, R. Fischer, R. Zboray, P. Boillat, M. Camenzind, C. Toncelli, et al., ACS Appl. Mater. Interfaces 12 (2020) 29908–29916 with permission from copyright 2020 American Chemical Society; patterned paper—reproduced from H. Liu, R.M. Crooks, JACS 133 (2011) 17564–17566 with permission from copyright 2011 American Chemical Society.

addition, the micrometer size of channels and elaborate microstructures ensure the fast measurement with high sensitivity and high accuracy, as well as the multiplexed sensing and biosensing with integrated multichannel electronic and optical transducer modules. The exploration of suitable materials and techniques for microfluidic fabrication also enables the fabrication of wearable devices in the manner of low cost and large-scale production.

4.2.2 Functions of flexible materials in microfluidics

A full microfluidic analytical device consists of many operational components [25]. Beyond the material flexibility and the sensing capability within the microfluidic platform, it is important to engineer several operational components to provide particular functions, including fluid carriage, separation, mixing, and flow regulation (Fig. 4.2). The significant operational elements of the microfluidic device are introduced in this section.

The primary unit that must be considered is a microchannel that links to the inlet track to collect the sample fluid (such as biofluids containing analytes) and transport it to other parts of the device toward the detection system (such as electrochemical transducers that measure electrical signals or spectroscopic readers that record spectrum or colored signals). The microchannel can be visualized as a miniaturized tube, generally created by soft PDMS materials [26]. The small dimension of the fluidic channel does not only make the whole device tiny but also provides a unique behavior of fluid mechanics. In practice, the microfluidic device for on-body applications may rely on the wick and the microfluidic channel. The design can employ the microscopic and hydrophilic nature of porous materials (e.g., paper or fabric) [27–29]. When scaling down to the microsize, the narrow channel allows the sucking pressure. The rate of the fluid flow is affected by the permeability of the porous material used in the device. Young–Laplace law is applied to design the system of the capillary-driven flow in the microfluidic devices. The nature of surface tension to the curvature in the surface of the fluid determines its flow [30]. The larger the size of the capillary-like channel, the smaller the capillary pressure [27]. Tuning the cross-sectional area of the microchannel can regulate the capillary pressure and hence the flow of the fluid. The use of such microporous designs or extremely small channels allows the self-driving force. This is beneficial for wearable applications as there is no need to carry extra complex pumps or energy sources. Note that the

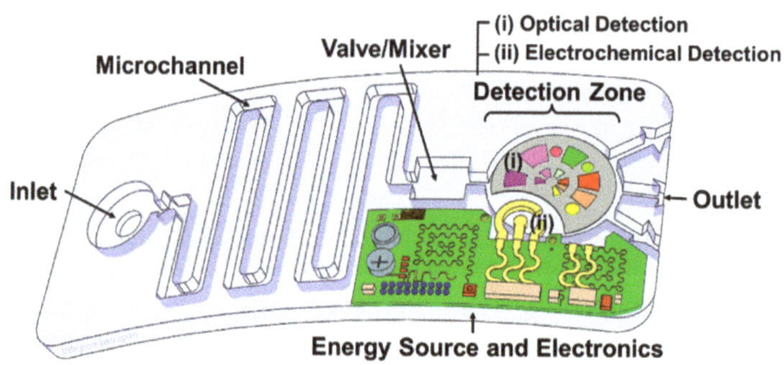

FIGURE 4.2

Operational components of flexible microfluidic-based sensing device for wearable sensors. Main components contain an inlet (to collect the biofluid); a microchannel, a valve, and a mixer (to deliver and manipulate the sample or reagents); detection systems, such as electrochemical or optical sensors (to obtain the analytical output); and electronics that are powered by external energy sources or integrated energy-harvesting devices (to control the fluid flow or sensing units).

deformation of the fluidic channel can be expected when using the device for in-vivo settings. The change of the structure will inevitably alter fluidic physics. Therefore, the mechanical deformation and dimension of the channel should be kept in mind. Key approaches to fabricate the microfluidic channels are explained in Section 4.2.3.

Another crucial unit is a miniaturized valve. This unit is integrated into the device to control the flow of the biofluids or reagents. We can classify the valving systems into two classes: passive and active valves. The capillary pressure and bursting pressure are important to govern the working of passive valve. The passive valve without the assistance of extra instrumentation can be a stand-alone system, making it attractive to wearable settings. In contrast, the active valve relies on additional instrumentation. However, some microfluidics for sensing applications require the control of the flow to attenuate/stop the liquid flow or switch the direction of the flow. For instance, different flow rates of the sample can cause variable signals when using the amperometric technique, because the sensor current response is proportional to the flux of analyte molecules onto the sensor surface [31]. Without the manipulation of the fluid flow rate, the deviation of the signals can occur due to the inconsistent advective and diffusive analyte transport to the sensor surface and that may cause the distortion of biomarker analysis. Therefore, researchers have tried to engineer several types of valves for the control of flow, including the valves relying on thermoresponsive materials [31], pH-responsive materials [32], and electrowetting [33].

A micromixer is another significant unit in soft microfluidics. When we design a microfluidic device, the nano/microworld of the system must be considered by following the Reynolds number ($Re = \rho \nu L/\mu$) [34]. The Reynolds number is directly proportional to: (1) the fluid density, ρ; (2) the fluid speed, ν; (3) the diameter of the tube, L; and is inversely proportional to fluid viscosity, μ. Therefore, a smaller diameter of the channel gives a smaller Reynolds number. The small channel will introduce the domination of the viscous force, compared to the inertial force. In the microfluidic world, the Reynolds number goes down to a small value (<2300), representing the domination of the laminar flow. The mixing of different fluids in microfluidics depends on turbulence, which can be achieved at a high Reynolds number (>4000). Moreover, biofluids usually have a higher viscosity than water. Therefore, in the same setting and at the same speed, while the water flow tends to be turbulent, the flow of the viscous fluid can be laminar. This laminar flow causes a challenge because the chaotic mixing is trivial. Therefore, it is required to design a micromixing system to achieve the efficient mixing. The passive configuration based on a long serpentine microtrack can be adapted to enhance the mixing [35]. In addition, reliable active systems with external energy input for efficient mixing effect can be used, such as acoustic/ultrasonic, dielectrophoretic, electrokinetic time-pulse, pressure perturbation, electro-hydrodynamic, magnetic, or thermal techniques [36]. Note that the simple and stand-alone device is preferred for on-body applications.

The detector is another important unit to gain the analytical signal. After converting one form of chemicals (related to the analyte) into another, the measurable signal is observed at the detection unit to reveal the level of the target. Different physicochemical transduction modules can be integrated with the microfluidic devices. Once the sample delivered through the channel reaches the detector unit, the detector should provide a sensitive analytical signal. The measurable signal can be briefly classified into optical (e.g., absorption, transmission, emission, scattering, and reflection) and electrical signals (e.g., current, voltage, charge, capacitance, and impedance) based on the usage of optical and electrochemical sensing modules, respectively. The optical sensing modules allow the detection of (bio)chemical reactions via a range of techniques, such as fluorescence, luminescence, ultraviolet (UV)/visible absorption, surface-enhanced Raman scattering, and surface plasmon resonance, etc. [37,38]. However, the abovementioned optical sensing modules usually need complex optical components including light source, light guide, and photodetector, which limit the applications in microfluidic-based devices due to (1) expensive optics, (2) difficulty in miniaturization and integration of these optics, and (3) systematic or environmental influence on the background. Beyond these, colorimetry has been widely studied and integrated into wearable microfluidic devices as an optical transducer for signal transduction by the merits of simple signal readout by naked eyes or smartphones [37,39]. In addition to optics-based approaches, electrochemical methods are among useful classes of detection designs by using various electrochemical techniques such as potentiometry [40], voltammetry [41], amperometry [42], and impedimetry [43]. In general, the detection unit consists of three electrodes: working, reference, and counter electrode, while the potentiometric sensor can be simple as to rely only on the indicator and reference electrodes. Moreover, to miniaturize the system, the simplified two-electrode system attached into the microfluidic chamber can also be used [31]. The electrode should provide good matching mechanical properties of the skin and the microfluidic chambers. The soft electrochemical sensing electrode has been prepared through several strategies (such as coating, screen printing, inkjet printing, laser-assisted patterning, and photolithographic patterning processes), allowing mechanical compliance, small size, and seamless integration with the microfluidics [44]. Except the soft electrodes, the advances in miniaturization of microelectronics also provide superior advantages for the integration of electrochemical sensing module into microfluidics. However, considerations should also be taken into account, such as the drifts, biofouling, and interferences caused by ions, proteins, and other electroactive molecules present in biofluids.

In addition to key microfluidic components discussed earlier, the energy unit and electronics are crucial items when developing microfluidic devices. Preferably, wearable microfluidic patches should be functional independently. However, old-style electronics rely on external energy sources, leading to unsatisfactorily large or heavy devices and the need for recharging of devices. An effective solution is developing self-powered electronics coupled with skin-worn sensors [45,46]. For instance, Bandodkar et al. [47] reported a sweat-activated battery technology that can be embedded within a soft microfluidic platform, which is based on the conversion of the chemical energy present in chemicals (such as glucose and lactic acid) into electricity. The harvested electricity is then used to supply integrated electronics or act as self-powered signals for analytical chemistry purposes.

4.2.3 Fabrication of flexible microfluidics on compliant substrates

Unlike the rigid inorganic materials requiring the planar photolithographic technique and chemical etching adapted from MEMS, the employment of flexible materials with specific functionality as compliant substrates has promoted various other fabrication methods and techniques for the manufacturing of flexible microfluidics. Soft lithography, widely known as non-photolithographic strategy initiated from 1998 [48], usually uses soft polymeric materials to fabricate an elastomeric mold to create micro-sized or sub-micro-sized features onto a surface. Soft lithography has attracted much interest because of its low cost, high throughput, and accessibility compared to other lithography techniques (e.g., photolithography). Fig. 4.3 illustrates the comparison of the general procedures needed for

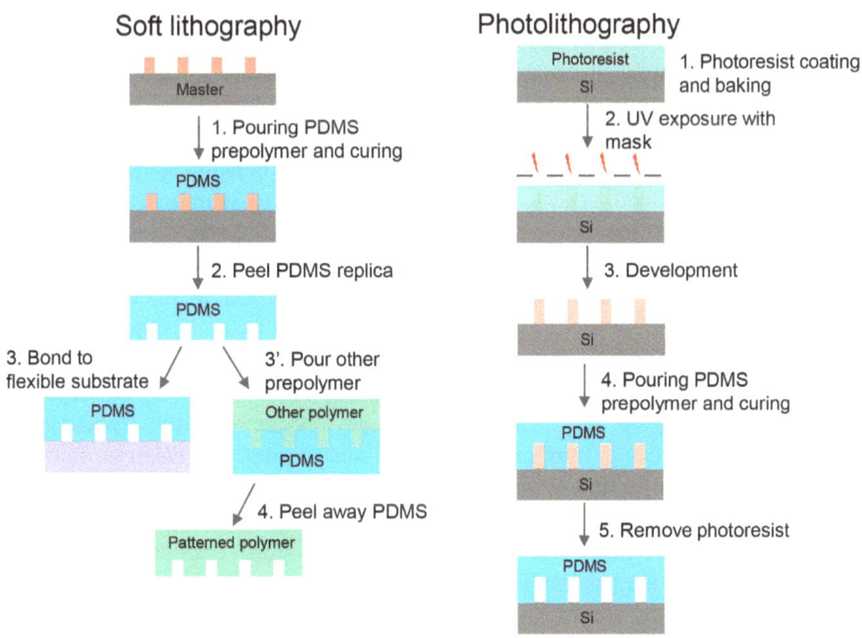

FIGURE 4.3

Comparison of general procedures needed for soft lithography and photolithography (negative PR). *PR*, Photoresist.

the photolithography and soft lithography. PDMS is one of the most commonly used materials for soft lithography due to its low glass transition temperatures and high compatibility. In soft lithography, the PDMS stamp or mold is prepared by poring over a predesigned master with corresponding relief structure on it, followed by the curing and peeling. The resulting PDMS stamp can be directly boned to another flexible substrate by oxygen plasma treatment, or the PDMS mold is used as the replica model to transfer the pattern with other soft materials by various techniques, including replica molding (REM), micro-molding in capillaries, soft embossing, micro-transfer molding, micro-contact printing, and solvent-assisted micro-molding [48,49]. As a comparison, the photolithography relies on a light-sensitive polymer (i.e., photoresist, PR) coated on the substrate wafer with the subsequent exposure to UV light through a mask in the manner of that either PR is dissolved (positive PR) or remains by cross-linking (negative PR) when exposed to light. The patterned PR is then used as the mold for the fabrication of PDMS-based microfluidics.

The outstanding advantages, such as low-cost, simplicity, and applicability to a wide range of soft materials, enable the fabrication of flexible microfluidics via soft lithography combined with other techniques for wearable electronics. Huang et al. [50] designed a serpentine microfluidic antenna based on soft lithography. The Ecoflex substrate ($\sim$2 mm thickness) patterned by a mold was bonded onto another half-cured Ecoflex substrate, forming a microstructured channel. The further injection and sealing of liquid metal inside microfluidic channels with epoxy resin showed reversible stretchability and designed mechanical frequency modulation. Similarly, Chen et al. [51] reported a low-cost, pump-free, and capillary flow-driven microfluidic chip via UV-nanoimprint lithography (UV-NIL) for the control of flow in microchannels, as well as an electrochemical transducer via inkjet printing for electrochemical biosensing. The PDMS stamping template was prepared by photolithography and then used to prepare the microfluidic channels on flexible PET films by a UV curable thiol-ene polymer based on NIL. The inkjet-printed three-electrode system and electrowetting valves photonically sintered on second flexible PET film were bonded directly to the channel-containing layer to form sealed fluidic device as cost-effective biosensors.

On the other hand, laser micromachining has been intensively used in the rapid manufacturing of rigid glass microfluidic devices with the advantage of mask-free, direct writing, minimal mechanical and thermal deformation, low-cost and ease of implementation in industry [52]. The laser micromachining via a CO_2 or UV laser can also be applied for the manufacturing of flexible microfluidics with compliant substrates, including thermoplastics [53] and papers [54]. As shown in Fig. 4.4, the process is based on the direct laser irradiated ablation of the material with high versatility, which can be extended to cutting, drilling, marking, turning and threading, etc. Several parameters can be tuned for lasers to achieve the varying structural depths, aspect ratios and structures, including laser wavelength, laser pulses, power, scan speed, distance-to-focus, and resolutions. For instance, higher power intensity can penetrate into deeper locations of the

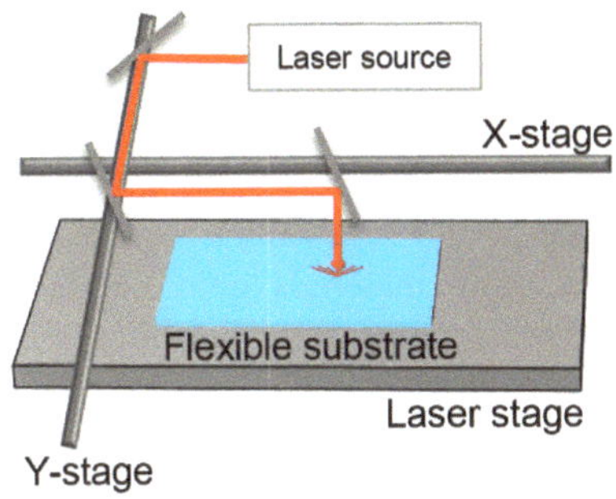

FIGURE 4.4

Illustration of a laser for laser micromachining.

materials with an increase in depth, while defocus can increase the spread of laser energy over a wider region with a decreased power density, resulting in shallower and wider channels [55]. The laser ablation also brings the potential of contamination by the ejected particles that are deposited or spread on the surrounding surface [56]. Nevertheless, the laser micromachining or laser cutting technique can directly pattern microfluidic features from a computer-aided design software on a sequence layer of substrates in a short time, and the subsequent lamination can realize the assemble and stacking of different layers into a 3D microfluidic architecture. For instance, Lin et al. [57] reported the laser cutting of two-dimensional (2D) embodiments with microchannels and vertical interconnected access on tape- and PET-based substrates. With proper alignment of the individual microchannels and vertical interconnected access, 3D microfluidic structures with fluidic connections in the vertical direction were achieved. Eriksson et al. [58] demonstrated the application of laser engrave for geometric flow control on nitrocellulose membrane-based lateral flow immunoassay device (LFID) with an significant improvement in sensitivity and limit of detection as well as reduced reagent consumption, providing not only a new dimension to LFID optimization but also a cost-effective and practical innovation to the field.

Plasma treatment is another simple and cost-effective technique for the patterning of microfluidic structures, especially on paper-based substrates. Usually, the paper is first hydrophobized by a cellulose hydrophilization agent, alkyl ketene dimer, and then treated with plasma through a metal mask. Unlike the laser micromachining with the removal of materials from the substrate, the exposure of the unmasked paper area to plasma introduces strong water wettability while remaining its original flexibility [59]. The hydrophilicity allows the transport of aqueous solutions along and within the patterned structures via capillary penetration. This technique enables mass production of paper-based flexible microfluidics with low-cost and environmental-friendly reagents, while it has the barrier for the submillimeter size [60].

The past few decades have witnessed the advancements in 3D printing and its potential for fabrication of microfluidics since the development of stereolithography (SL) by Charles Hull in 1986 [61,62]. 3D printing is an additive manufacturing technique that constructs a 3D object by progressively adding materials, usually layer-by-layer, via a printer following a designed model under precise digital control. The direct printing materials onto the build platform possess the intrinsic advantage of less waste, versatility in design and construction compared to that of subtractive manufacturing techniques. The representative techniques for microfluidic device fabrication in 3D printing are SL, fused deposition modeling (FDM), and inkjet printing [61,62]. In SL, the fluid photosensitive resin material in a vat is cured and hardened layer-by-layer from the bottom via a UV laser beam with the designed structure until the 3D object is eventually created. This process endows SL with the high surface resolution and precision on the order of 10 μm [62,63]. FDM uses thermoplastic materials as the raw materials that are heated to melt and extruded from the nozzle head to the platform, which solidifies by spontaneous cooling immediately after the extrusion. A wide range of thermoplastics with low-cost and high biocompatibility can be chosen, including polycaprolactone, polylactic acid, polybutylene terephthalate, and polyglycolic acid. Several limitations also appear with the surface roughness, mechanical weakness and weak sealing owing to interlayer distortion. In inkjet printing, each layer is deposited onto a tray via an injection nozzle, which solidifies via rapid polymerization under UV exposure for liquid photopolymer-based materials or via adhesives for powder-based materials, respectively. The comparison of these three representative 3D printing techniques is summarized in Table 4.1 [62,64]. 3D-

Table 4.1 Comparison of these three representative three-dimensional (3D) printing techniques [62,64].

3D printing techniques	Materials	Advantages	Limitations
SL	Photocurable resin/ polymer	High resolution, fine detail, smooth surface	Expensive machine and resin, limited lifespan, limited choice of the materials
FDM	Thermoplastics	Cheap and accessible materials and instruments, easy-to-use	Weak mechanical properties, problems with warping and shrinking, rough surface need post-polishing
Inkjet printing	Photocurable resin/ polymer, soft elastomers, wax-based inks, liquid suspensions	High resolution, smooth surface, fast build speed, multi material printing	Removal of support materials, layer-by-layer finish

FDM, Fused deposition modeling; *SL,* stereolithography.

printed microfluidics have been intensively researched for the application in different biomedical applications, such as scaffolds for tissue engineering, cell culture chamber, surgical models of organs, personalized prosthetics and implants, and biodegradable drug-delivery devices [64]. More and more versatile 3D printing techniques are under development for the direct fabrication of microfluidic devices. However, it is still challenging for 3D printing and integration of hybrid materials, especially for the compliant substrates and functional electronic elements, because the consumer 3D printer usually prints one material at a time. Alternatively, 3D printing is used for the construction of sacrificial structures to support the fabrication of microfluidics. Guo et al. [65] reported a simple method to fabricate microchannels based on an inkjet printing liquid template. A Y-shape pattern is printed onto the PDMS prepolymer and then is wrapped by PDMS spontaneously, while the subsequent heating cures the prepolymer and evaporates the pattern. Saggiomo et al. [66] demonstrated the intricate PDMS microfluidic device using a 3D printed scaffold as the sacrificial structure for PDMS curing and removal of the scaffold by acetone.

Except for the one-step fabrication of microfluidic structure, layer transfer and lamination via sealing and bonding are commonly needed to fabricate flexible microfluidic devices. Usually, the open geometries are fabricated by laser ablation or soft lithography, which are subsequently coved with a laminated cover layer. Therefore, the sealing and bonding strength at the interface between the bottom and the cover layer is of importance for the performance of microfluidic devices. A few techniques have been developed for the sealing and bonding of microfluidic devices that can be categorized into indirect and direct routes. The substrate material properties, surface chemistry, dimensions and intended operating conditions (temperature and fluid pressure), etc., must be taken into consideration for the selection of suitable techniques [67]. Indirect bonding techniques using additional materials, such as epoxy and adhesives, provide a high bond strength [67]. Direct bonding techniques can be chosen from different surface activation (e.g., plasma treatment), thermal fusion, solvent bonding and ultrasonic welding, etc., which provide the strong bonding of the substrates without additional materials added to the interface. For instance, PDMS with open channels can be easily bonded to other substrates by plasma surface treatment with introduction of polar silanol groups for the surface hydrophilicity of PDMS.

4.3 Epidermal microfluidic patches

4.3.1 Considerations when developing epidermal patches

An epidermal microfluidic patch is a piece of a soft biomedical device that can adhere to the human skin. Attaching the epidermal patch aims to directly receive the biofluid (e.g., sweat) from the skin. The biofluid containing analytes enters the inlet, then is delivered toward the detection zone to show the analytical output. To achieve an epidermal microfluidic sensing patch, several key requirements should be considered. Traditional microfluidic systems based on stiff materials (such as PMMA, PC, and glass) are used to perform automated chemical analysis [68]. The system allows the injection of the sample containing the analyte into a flowing stream, driven by an external controller, toward the analytical detector. To translate the rigid and bulky system to a wearable device (such as an epidermal patch), adaptability to miniaturization and compatibility to human skin are primary considerations [69,70].

First, the epidermal patch should match the skin characteristics. The skin of the human body, consisting of epidermis, dermis (contains sweat glands secreting biofluids to the skin surface), and subcutaneous tissue [71], is deformable to endure changes in length when applying lengthwise tension or compression. The Young's modulus of our skin is around 10–500 kPa [72,73]. Therefore, the epidermal patch should match irregular and curved characteristics of the biological layer. Stiff materials possessing too high modulus (over GPa) are inappropriate for building the epidermal patch. In addition, normal activities in daily life include multidirectional movements, which will affect the stability of the wearable devices. Stiff materials used for building fluidic channels and sensing compartments can crack due to the wearer's movements, causing severe impairment of the entire system. Recently, soft, flexible, or stretchable epifluidic devices have attracted considerable attention. Researchers have engineered materials and designs of the skin-attachable patch to match the skin's properties and the wearer's activities, such as the usage of PDMS (Young's modulus of 0.4–3.7 MPa) and Ecoflex (125 kPa) [16,17].

Another point is that the patches covering the epidermis should be able to transfer heat and fluids from the wearer's skin to sustain natural homeostasis. Therefore, the epidermal patches should meet the capability to be breathable. The device should also offer comfort during wearing the device. Importantly, special attention should be paid to biosafety by avoiding skin irritation. The biofluids and reagents in the microfluidic should not leak back to the human skin. The mandatory requirement is to keep desirable fluid manipulation flow in a precise direction. Unlike the stiff fluidic device used on a well-control-laboratory bench, the epidermal microfluidic patch is expected to experience an additional external strain, which will cause the deformation of the device. Therefore, the key components including microchannels should not collapse to ensure the stable fluidic transport. To meet the mentioned requirements, it is essential to consider the intrinsic properties of materials along with strategies to apply the selected materials for building the epidermal patch. Key materials that have been used include polymeric elastomer [74], cellulose-based materials [75], and textiles [76]. These become attractive candidates because of their inherent compatibility to the skin and decent wearability after the integration.

PDMS has been extensively applied for developing skin-attachable microfluidic patches due to favorable properties, stretchability, gas permeability, transparency, biocompatibility, and high stability. It can also act as flexible electric insulation when integrating electronics and electrochemical sensors. The inherent hydrophobic nature of PDMS can protect the units, such as electronics, from solutions that can destroy the device. However, the hydrophobicity can be tuned, which is crucial for realizing desirable fluidic channels. Considering a small volume of sweat that can be collected using microfluidics, it is required to design accurate microfluidic modalities. The evaporation and transportation should be well-controlled [25]. The intrinsic structure of PDMS can store the biofluid only in a brief time due to its high vapor permeability. It is, thus, crucial to avoid the permeation of water vapor; engineered materials are needed. However, we still need to consider biocompatibility in terms of mechanical properties and biosafety. For instance, Heo et al. [77] reported a collagen-PDMS composite material for extended fluid retention in the skin-interfaced microfluidic devices (Fig. 4.5A). The collagen-PDMS film containing 30 wt.% collagen reduced the water evaporation significantly by 80.2%, while maintaining deformable and stretchable properties required for skin-interfaced microfluidics with modulus of 1.80 MPa. Sweat collection device using collagen-PDMS as top and channel layers increase the sweat retention by 45% through 9 hours compared with pure PDMS.

In addition to elastomeric polymers, cellulose or cellulose fiber-based paper is another class of suitable substrates for microfluidics due to not only the low-cost, biocompatibility, light weight, and flexibility of the materials, but also high throughput rapid prototyping, ease of fabrication, and ease of use for the paper-based microfluidic devices. An important advantage of cellulose or cellulose fiber-based paper is the possibility for passive wicking of fluids by capillary forces without external pumps or complicated power sources. The flow can be further manipulated via the regulation of channel dimensions as well as the tunning of the fluid-substrate contact angle and the surface hydrophilicity. Moreover, reagents such as electrolytes, color-developing agents, enzymes, and biorecognition elements can be patterned on paper by printing. As shown in Fig. 4.5B, Cao et al. [78] developed a paper-based microfluidic patch with a 3D folded structure for perspiration analysis. A commercial Whatman#4 cellulose filter paper was used as the flexible substrate material with further wax printing to create a patterned hydrophobic barrier onto filter paper to guide the flow of fluid samples. The patterned cellulose filter paper was folded into a 3D paper-based epidermal microfluidic, resulting the fluid flows vertically alongside a sweat collector, that consisted of vertical channel, transverse channel, electrode layer, and encapsulating layer. Such design forms a sweat evaporator asscessable to air allowing sweat evaporation, and maintaining the fluid flow. Among these, the electrode layer is a thin screen-printed electrode on PET film containing ion-selective electrodes (ISEs), which are connected to a customized smartwatch with an open circuit board for sweat electrolyte sensing. Silva et al. [79] used a microbial nanocellulose, an exceptional skin-substitute natural material, produced from Gluconacetobacter xylinum as a flexible substrate to print electrode designs. The film is biocompatible, breathable, and skin conformable, which are key requirements when building epidermal sensing patches. This system offers the

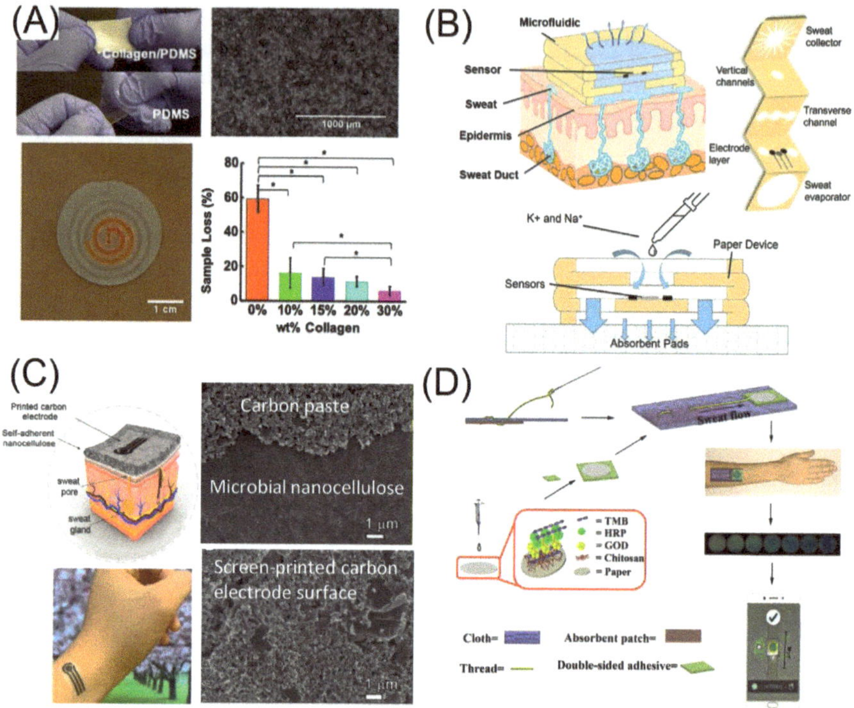

FIGURE 4.5

Selected examples of materials for developing epidermal microfluidic patches. (A) Collagen-PDMS composite for developing microfluidic device, (B) cellulose filter paper for developing paper-based microfluidic electrolyte sensor, (C) microbial nanocellulose for developing a skin-worn sensing patch for the detection of heavy metals, uric acid, and 17β-estradiol, (D) cellulose-based yarn and paper for developing a skin-worn glucose sensor. *PDMS*, Polydimethylsiloxane.

(A) Reproduced from Ref. [77] with permission from copyright 2019 Elsevier B.V. (B) Reproduced from Ref. [78] with permission from copyright 2020 Wiley-VCH GmbH. (C) Reproduced from Ref. [79] with permission from copyright 2020 Elsevier B.V. (D) Reproduced from Ref. [80] with permission from copyright 2019 Springer Nature B.V.

opportunity to fabricate a wearable sensor platform adherent to human skin for monitoring various analytes in sweat, such as uric acid, 17β-estradiol hormone, and heavy metals (Fig. 4.5C).

Fabric or textile-based microfluidics is another unique platform due to their potential to be woven into clothing and common accessories for daily wearing. However, special transport channels must be engineered to regulate the biofluidic transport via the manipulation of the wettability. This can be realized by various surface modification techniques to increase the hydrophobicity/hydrophilicity, such as wet chemistry coating with low-surface-energy materials [81,82] and plasma treatment for etched rough surface micro-/nano-textures or introduction functional groups [83]. Sweat can self-propel in an open-channel form of the hydrophilic track following Laplace pressure gradients due to differences in the degrees of hydrophobicity/hydrophilicity. Xiao et al. [80] developed a microfluidic analytical device for sweat analysis, which contains an oxygen plasma-treated cotton thread and functionalized filter paper (Fig. 4.5D). The oxygen plasma treatment effectively improve the hydrophilicity and wicking ability for sweat harvest, while the filter paper was sequentially functionalized with chitosan, glucose oxidase (GOx), horseradish peroxidase (POD), and 3,3,5,5-tetramethylbenzidine (TMB) for colorimetric sensing of glucose with a blue color readout by the naked eyes or the color-reader application in a smartphone.

4.3.2 Epidermal microfluidics for detecting electrolytes and nutrients

Sweat contains a variety of markers, linking to insightful information of the individual health and status during physical exercise and emotional stress [84]. Tracking the concentration of electrolyte species during activity can give useful information for improving a person's sport and fitness. Examples of significant markers include sodium ion (Na^+), potassium ion (K^+), chloride ion (Cl^-), and sweat pH. For example, changes in skin pH can indicate the pathogenesis of skin diseases (e.g., dermatitis and acne [85]). Monitoring Na^+ concentrations in sweat is a way to estimate Na^+ loss in the body. Sodium deficiency is the major cause of heat cramps, and can link to diseases, for example, hyponatremia in severe situations [86]. Microfluidics can improve the quality of biofluid detection by providing sequential and continuous detection while controlling sample evaporation or reabsorption back to the skin. The key to realizing microfluidic sensors is to attach the transducer onto the skin-worn microfluidic.

A main practical transducer for electrolyte detection relies on ISEs, which is a membrane-based potentiometric device that can measure the concentration of ions in biofluids. This sensor measures an electrochemical cell's voltage between two electrodes (i.e., the working electrode and the reference electrode) under static circumstances. On the working electrode, an important compartment is an ion-sensitive membrane. This working electrode is crucial to be sensitive and responsive to the target ions, whereas the reference electrode is essential to define the reference potential. The ideal potential response to monovalent ions or group ions (e.g., Na^+, K^+, NH_4^+, Cl^-, and H^+ for pH) gives the Nernstian slope of 59.2 mV [87,88]. The response for the determination of divalent ions (e.g., Ca^{2+}) gives the slope of 29.6 mV [89].

From an application point of view, simplicity and wearability must be taken into account. Therefore, the development of a transducer for the detection of ions in practical settings is desired to be in a solid-state device without an internal filling solution and other components used in traditional potentiometric sensors (e.g., glass-based pH sensors used in a common laboratory). A solid-state transducer is usually composed of conductive electrodes (such as carbon, platinum, or gold) [87]. The working and references electrodes in a solid-state platform can be used to integrate with microfluidic devices. Ma et al. [90] demonstrated the integration of a solid-state transducer with a flexible microfluidic manipulation unit into a wearable microfluidic device for continuous sensing of Na^+ in perspiration. The transducer is an all-solid-state potentiometric electrochemical sensor based on a screen-printed graphite electrode on PET film consisting of (1) an indicator electrode with subsequent electrodeposition of a conducting polymer poly(3,4-ethylenedioxythiophene) polystyrene sulfonate (PEDOT:PSS) as the ion-to-electron transducer and deposition of a layer of Na^+-selective membrane (Na ionophore X), (2) a reference electrode with a Ag/AgCl layer and a polyvinyl butyral-salt layer to provide a stable reference potential without potential drift caused by electrolyte. As shown in Fig. 4.6A, the resulting solid-state sodium sensor on the PET

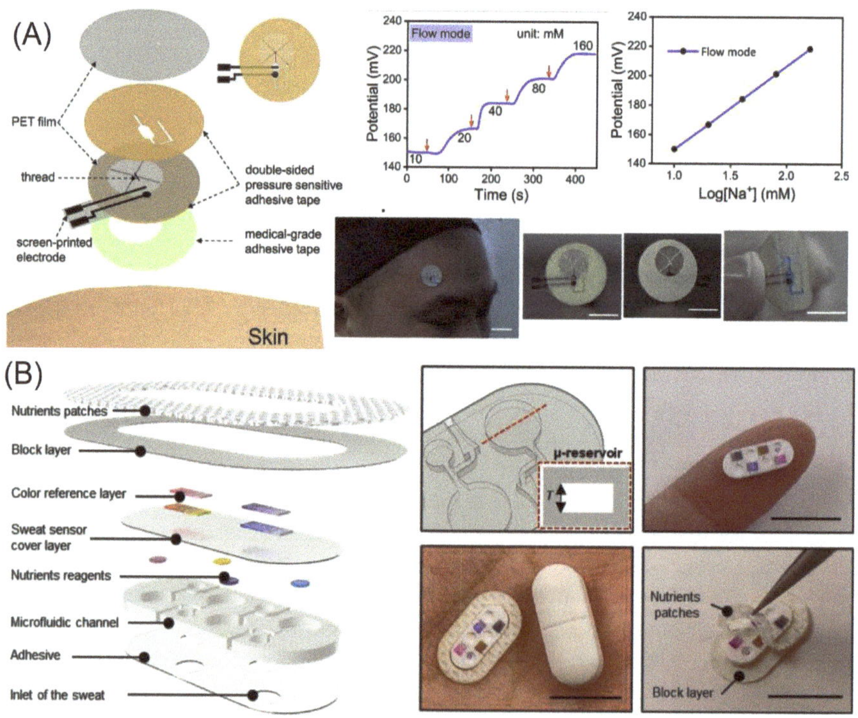

FIGURE 4.6

Epidermal microfluidics for detecting electrolytes and nutrients. Microfluidic devices for monitoring (A) sodium ions [90] and (B) vitamin C [91].

(A) Reproduced from B. Ma, J. Chi, C. Xu, Y. Ni, C. Zhao, H. Liu, Talanta 212 (2020) 120786 with permission from copyright 2020 Elsevier B.V. (B) Reproduced from J. Kim, Y. Wu, H. Luan, D.S. Yang, D. Cho, S.S. Kwak, et al., Adv. Sci. (2021) 2103331.

film was then sandwiched with (1) a patterned medical-grade adhesive tape with a circular opening (10 mm in diameter) for device attachment on skin, (2) microfluidic channel patterned double-sided pressure sensitive adhesive tape, and (3) another layer of PET film to seal to channel and provides a punched hole as outlet. Moreover, a biocompatible and hydrophilic thread was also added to facilitate the sweat collection from the skin through capillary absorption. The epifluidic design enables the flow rate reaching 5 μL/min. To test the potentiometric sodium sensor, the thread was introduced into the channel to deliver a NaCl testing solution through the continuous flow. The obtained sensitivity was 56.7 mV, correlating to the Nernstian response over a concentration range of 10–160 mM NaCl. This allowed the on-body real-time monitoring of the sweat sodium covering the real dynamic sweat sodium of 11–40 mM Na^+ and with an operation time of around 30 minutes.

For multiplexing monitoring, epidermal microfluidics can be integrated with various sensors for monitoring different ions. Paul et al. [92] reported a multisensing sensor array integrated wearable microfluidic sweat sensing patch for the detection of Na^+ and K^+ ions and pH. Such epidermal microfluidics ions and pH sensors include an iontophoretic system for sweat stimulation, PET-based microfluidic platform for sampling and fluid transport, and a potentiometric sensors array embedded into the microfluidic detection zone. The multisensing potentiometric sensor array was fabricated via screen printing on a flexible PET film with subsequent modification with ion-sensitive membrane of ionophore for Na^+ and K^+ ions and pH-sensitive membrance of polyaniline for pH. The potentiometric sensors exhibit sensitivity close to Nernstian behavior and wide linear range for Na^+ ions (10–160 mM), K^+ ions (2–32 mM), and pH (4–8).

In addition to electrolytes, nutrients are essential for sustaining key physiological functioning and reducing illness. An example of a compact device enables convenient and rapid colorimetric analyses of numerous vital elements in sweat, including ascorbic acid (vitamin C), calcium, zinc, and iron ions by Kim et al. [91] (Fig. 4.6B). This soft and skin-worn microfluidic has a miniaturized dimension of 1.0×0.5 cm^2, which can collect and transport the fluid through a system of microchannels, passive microvalves, and reservoirs. An oval-shaped medical grade adhesive layer with openings adhered on the skin for collecting of secreted sample in a defined area (~ 10 sweat glands). Simultaneous screening of the concentration for the four analytes was realized on the specific layout including four repeated sets of microchannels and μ-reservoirs, in which colorimetric sensing reagents were deposited. The readout in the color of pink for ascorbic acid [93], violet for calcium ion [94], yellow color for zinc ion [95], and blue for iron ion [96] can be collected as digital images for further quantitative extraction of the color information for concentration values for each nutrient using an appropriate calibration factor. Beyond the skin-interfaced, miniaturized microfluidic analysis of these vital elements, nutrient delivery system was also integrated for passive, sustained delivery of these species to the body throughout a period of wear for maintaining proper nutritional balance and homeostasis.

4.3.3 Biochemical wearable microfluidic biosensors

In addition to electrolytes, metabolites in sweat are also important health indicators. For example, sweat glucose levels can fluctuate in response to diet and exercise, providing data on prediabetic conditions [97]. Sweat lactate levels suggest metabolic problems under anaerobic respirations [98]. High energy needs during intense exercise that cannot be fulfilled by aerobic respiration cause anaerobic metabolism, which raises the content of lactic acid in the tissue. Hence, lactic acid is a significant biomarker for sports medicine and clinical diagnosis, for example, pressure ischemia [99]. Therefore, it is critical to include multimodal sweat biosensors in a soft, skin-worn microfluidic platform for monitoring intermediates or end products of metabolism.

Amperometry is one of the most common approaches used for detection of metabolites. It is a technique for measuring current at a fixed voltage. Clark created one of the earliest amperometric biosensors to measure dissolved O_2 in blood in 1950s [100]. This initial concept of O_2 sensor translates into a glucose sensor, for example a commercial *YSI* enzyme electrode with a PC outermost layer for constant diffusion of glucose and oxygen, an immobilized GOx layer for specific enzymatic catalysis of glucose to gluconolactone and hydrogen peroxide, and a cellulose acetate innermost layer for semi-permeability of hydrogen peroxide onto a Pt anode. In amperometry, the current is measured at a fixed potential, under which the electroactive analyte molecules are oxidized/reduced at the working electrode. The measured current is proportional to the concentration of the metabolites [101]. It should be emphasized that for amperometric biosensing, we need to couple biocatalysis such as enzymes that convert the target metabolite into a detectable electrochemical active species (such as hydrogen peroxide) that can be oxidized or reduced to generate a measurable current. Hydrogen peroxide is used as a mid-product for readout in many biosensors, such as for lactic acid (relying on lactate oxidase reaction), ethanol (relying on alcohol oxidase reaction), and glutamine (relying on glutaminase or glutamine oxidase reactions) [102,103].

Ultimately, a single patch of microfluidics should be designed to combine multiple microchannels for a parallel monitor of various analytes. An example of electrochemical microfluidic multiplexed sensors was demonstrated by Vinoth et al. [42] as presented in Fig. 4.7A. The 2.5-mm inlet was connected to a 0.3-mm channel, then splitting up into branches guiding sample fluid into four detection chambers. Multiplex sensing relied on four sensors (one amperometric lactate sensor and three potentiometric sensors for Na^+, K^+, and pH). Each potentiometric sensor has two electrodes, relying on the basic principles discussed in the previous section. The lactate sensor is based on amperometric biosensor principle. The working transducer contains lactate oxidase and a Prussian blue nanocatalyst for a relay conversion of lactic acid into hydrogen peroxide, and then into electrical current when applying a constant potential of -0.17 V. The epidermal microfluidics biosensors allows rapid loading of sample into the multiple detection zones within 40 seconds with a flow rate of 3.75 μL/s. The surface treatment is

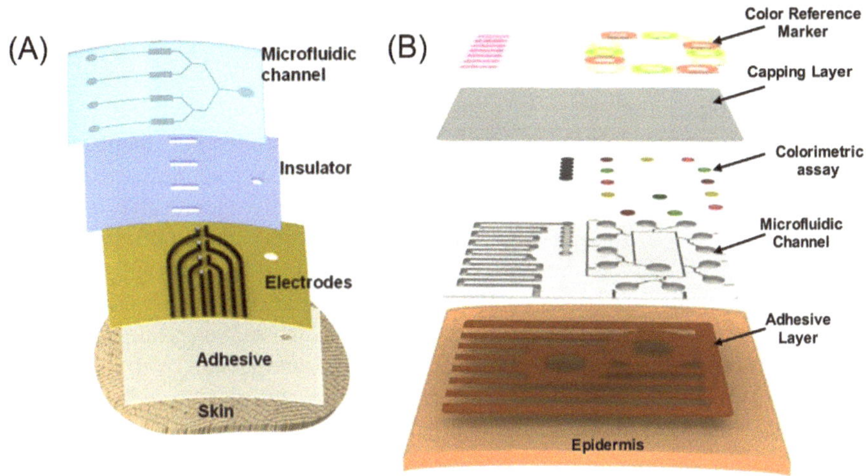

FIGURE 4.7

Epidermal microfluidics for detecting metabolites and relevant markers. (A) Wearable microfluidic electrochemical sensing devices for monitoring lactic acid and electrolytes. (B) Wearable microfluidic colorimetric sensing devices for monitoring lactic acid, glucose, electrolytes, and temperature.

(A) Reproduced from R. Vinoth, T. Nakagawa, J. Mathiyarasu, A.V. Mohan, ACS Sens., 6 (2021) 1174–1186 with permission from copyright 2021 American Chemical Society. (B) Reproduced from J. Choi, A.J. Bandodkar, J.T. Reeder, T.R. Ray, A. Turnquist, S.B. Kim, et al., ACS Sens. 4 (2019) 379–388 with permission from copyright 2021 American Chemical Society.

also essential to ensure the hydrophilicity of the microfluidic surface, enabling the effective capillary-driven sample loading and fluid flow. The surface treatment of the silicon rubber layer on the flexible PI substrate with O_2 plasma and silane solutions could reduce the surface contact angle from ~117 to ~82 degrees even 4 months after the treatment. Therefore, stable hydrophilicity could facilitate capillary force for the autonomous fluidic flow. This epidermal microfluidics biosensors could be attached under arm and on the upper back to test the on-body performances throughout stationary biking. It proved that the on-body device provided the real-time measurement of Na^+ (covering ~80 mM), K^+ (~7–18 mM), pH (~6), and lactate (fluctuating around 15 mM).

Another example is a skin-worn microfluidic device for multiplex colorimetric analysis of lactic acid, glucose, electrolytes, and temperature through a color-shading readout, as shown in Fig. 4.7B [104]. The skin adhesion layer (which is located between the device's base and the skin's surface) contains an aperture that is aligned with bottom inlet ports in the microfluidic. The sweat moves from the skin into the microtracks (serpentine channel and a collection of reservoirs disconnected by valves). This movement is pushed by the pressures driven by the eccrine gland, without the use of active pumps. Sweat secreted from the gland enters the device's inlet regions before entering into the serpentine track or microchannels that join with reservoirs via a set of passive valves. The valves help to collect the sweat in a way that tracks the order in which sweat-secretion events arise. In principle, the sweat fills the chamber chronologically, following the order of the pressure [26]. For example, when the first collection chamber which has a low pressure is full, the sweat pressure bursts the connecting valve, permitting the sweat to move to the next chamber. Thanks to these valve-integrated microfluidic features, multiple biomarkers can be tested in real-time using a single platform. For colorimetric detection, assay reagents are immobilized in the detection chambers. For glucose, pH, and lactate analysis, chemical and enzymatic reagents are loaded into the reservoirs. For glucose and lactate assays, the colorimetry relies on enzymatic reactions (e.g., GOx and lactate oxidase) coupled with color-developing assay kits. Universal pH dye and nitrophenyl octyl ether are mixed in the PVC dissolved by an organic solvent to homogenize the composite before loading at the detection zone. These entrapping reagents in the polymer matrix enable color changes over a wide range of pH. Moreover, colorimetric detection modules for monitoring sweat rate and total sweat loss were designed by monitoring the Cl^- within sweat. For example, silver chloranilate is entrapped by using poly(hydroxyethyl methacrylate) (PHEMA) for sweat Cl^- assay. In addition to colorimetric assays, the detection reservoirs can be loaded with thermochromic materials for skin temperature sensing. This example shows a way to apply soft multimodal microfluidic devices for capturing, storing, and colorimetrically analyzing several parameters (including the concentrations of sweat biomarkers, sweat rate, and temperature).

4.4 Flexible microfluidic contact lenses

4.4.1 Basics of soft contact lenses

According to the "World report on vision" by the World Healthcare Organization in 2019, at least 2.2 billion people have vision impairment or blindness [105]. Vision impairment occurs when an eye condition affects the visual system and one or more of its vision functions, including refractive errors and age-related conditions such as glaucoma and cataract. Among these, refractive errors, due to an abnormal shape or length of the eyeball, is the most common cause of vision impairment for the poor visual performance as near-sightedness or far-sightedness [105,106]. To address this, spectacles or contact lenses are extensively used as an effective compensation for their vision impairment. Contact lenses are thin ocular prosthetic devices with high transparency and hemispherical shape. They can be worn directly onto the outer layer of the eye, cornea, for vision corrections, cosmetics, or therapeutics. Contact lenses also possess several advantages over spectacles, such as conformation to the curvature of the eyeballs, no wearing limitations by exerting, and weather caused fog up.

The advances in material science have evolved the field of contact lenses. By the inspiration of the concept of contact lenses introduced by Leonardo da Vinci to submerge patient's head in a bowl filled with water to alter the vision in 1508, it has witnessed the development of contact lenses from brittle glass in the 1800s to rigid thermoplastic, PMMA in 1930s, and then to silicone hydrogels in recent decades [107]. Several criteria are needed to be fulfilled from the aspects of materials for the fabrication of contact lenses. Conventional PMMA paved the way for the manufacture of contact lenses by its low cost and durability; however, it also has obvious limitations regarding its rigidity and non-gas-permeability. Then several emerging gas permeable materials have been used for the rigid gas-permeable lenses such as cellulose acetate butyrate, pure silicone, silicone acrylate and fluorosilicate acrylic [106,108]. Soft contact lenses, usually made of hydrogels that are water-swollen polymeric materials, nowadays are proving to be the most dominant kind of soft contact lenses material today due to their high oxygen permeability (indicated as *Dk* value) and water content. These hydrogel materials are usually based on PHEMA, silicone monomer containing hydrogels, and poly(vinyl alcohol). In addition, the mechanical properties are also important for the contact lenses with respect to their comfortability and durability under compression and tension. Rigid contact lens materials have high Young's modulus, while soft materials have low modulus value. What's more, the optical transparency and the refractive index are also crucial for the visual performance. Table 4.2 lists the brief comparison of these criteria for several commercially available contact lenses.

Primary applications of contact lenses are the correction of refractive disorders, prosthetics, and therapeutics [106]. Besides this, the wearing of contact lenses directly onto eyeballs also provides the continuous interaction between the lenses and eyeball as well as tear fluid, which allows the noninvasive monitoring of physical and physiological state continuously in real time. This feature has propelled the development of "smart contact lenses" for health care monitoring. Various parameters can be collected via the smart contact lenses ranging from physical signs [e.g., intraocular pressure (IOP) and motion] for the eyeball and extended to physiological biomarkers (e.g., pH, different ions, various metabolites including glucose and lactate) of the tear fluids. To achieve this, the integration of microfluidics onto soft contact lenses is essential not only for the precise tear fluid processing including collection, storage and preparation, but also for the accommodation of sensing elements for accurate and real-time analysis.

4.4.2 Microfluidics in contact lenses for pressure monitoring

Among all the vision impairments, the neurodegenerative disease glaucoma is the second most cause of irreversible and progressive loss of vision. Unregulated IOP is a primary contributing factor in glaucoma. Usually the normal IOP values fluctuate over the range of 10–21 mmHg. The continuous monitoring of IOP is essential for the effective management of glaucoma. For example, capacitive

Table 4.2 Brief comparison of different commercially available contact lenses.

Base material	Commercial name	Manufacturer	Oxygen permeability (*Dk*[a])	Water content (wt.% 20°C)	Optical transparency	Refractive index (hydrated)
Silicone acrylate	Boston IV	Bausch & Lomb	26	<1%	>90%	1.47
Fluorosilicate acrylic	Fluorex 300	GT Laboratories	30	<1%	77%–89%	1.47
Silicone hydrogel	Definitive 74	Contamac Ltd	60	74%	99%	1.51
PVA[b]	Dailies AquaComfort Plus	Alcon	26	69%	>92%	1.38
PHEMA	SofLens 38	Bausch & Lomb	8.4	38.6%	86%–98%	1.43

PHEMA, Poly(hydroxyethyl methacrylate); *PVA*, poly(vinyl alcohol).
[a]Dk *unit* = *10^{-11} *(cm^2/s)(mL O_2/mL*mmHg).*
[b]*Polymer of poly(vinyl alcohol) partially actualized with* N-*(formylmethyl)acrylamide.*

sensor-based soft contact lenses have been reported by researchers by using reconfigurable capacitive sensor interface circuitry and wirelessly powered radio-frequency identification system for sensor control and data communication [109]. Beyond this, it is attractive to design the microfluidic channel-integrated soft contact lenses, which can convert the high IOP into the fluidic volume expansion or displacement of the liquid interface. Photolithography, soft lithography and bonding techniques mentioned in Section 4.2.3 are intensively used for the fabrication of microfluidic channels integrated soft contact lenses. For instance, An et al. [110] reported the fabrication of microfluidic contact lens sensors for the detection of IOP based on the volume variance and dyed liquid's interface displacement caused by the deformation of the sensing layer under pressure. As shown in Fig. 4.8, the fabrication process consists of soft lithography, salinization of the substrate, plasma-treated bonding, thermal molding, and liquid injection. For the micropatterned PDMS sensing layer, conventional photolithography was used to prepare the SU-8 master, which was then used as the replica for the casting, curing, and peeling of PDMS via soft lithography. Hydrophilic PET treated via oxygen plasma and (3-aminopropyl)triethoxysilane salinization served as the reference layer, which was bonded to the PDMS micropattern via oxygen plasma activation and lamination.

Typically, the manufacture of contact lenses into the specific curvatures is achieved by injection molding, spin-casting, or lather cutting [106,107]. In injection molding, the molten plastic or liquid monomers are injected into the "female"

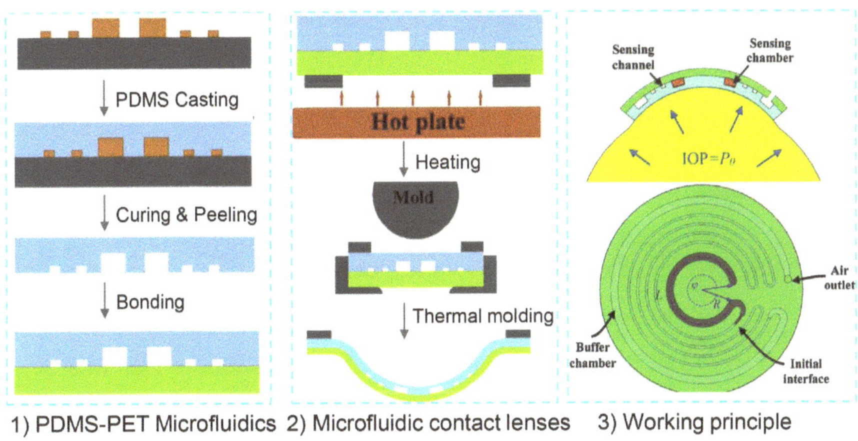

FIGURE 4.8

Schematic illustration of the fabrication microfluidics in soft contact lenses. (1) PDMS-PET microfluidics, (2) thermal molding of microfluidic contact lenses, and (3) sensing principle. *PDMS*, Polydimethylsiloxane; *PET*, polyethylene terephthalate.

Reproduced from H. An, L. Chen, X. Liu, B. Zhao, H. Zhang, Z. Wu, Sens. Actuators A Phys. 295 (2019) 177–187 with permission from copyright 2019 Elsevier B.V.

mold with the excessive polymer squeezed out via a "male" complementary convex mold. Followed with the polymerization under UV light, the contact lenses are peeled off and subsequently polished and hydrated in saline solution. The spin-casting method uses the spinning of mold with a specific curvature to distribute the poured polymer solution and cure it into a lens-shaped piece. Unlike the additive manufacturing methods such as injection molding and spin-casting, lather cutting is a kind of subtractive technique, in which a polymer cylinder is cut via a diamond tool in both the front and back side into the lens shape with a specific thickness. However, these techniques discussed above are not suitable for the fabrication of microfluidic contact lenses as shown in Fig. 4.8. Therefore, a post-molding procedure was used by thermal molding, in which the patterned PDMS-PET microfluidic was fixed on a square clip and heated at 170°C. The transformation onto a terrace die (curvature radius: 8.0 mm) realizes the spherical microfluidic contact lenses. As shown in Fig. 4.8, the microfluidic contact lenses comprise a sensing chamber filled with dyed liquid, a long sensing channel for the flow of the dyed liquid and a large buffer chamber. With the increasing of IOP value, the sensing chamber volume decreased causing the liquid to move from the chamber to the sensing channel. The displacement of the liquid interface inside the sensing channel can be captured optically via a smart-phone camera and it is linear to the IOP value in a specific range. Similarly, Agaoglu et al. [111] reported the passive integrated microfluidic sensor with the dilatometric sensing mechanism by converting small strain changes into a large fluidic volume expansion in the sensing channels. For the guide liquid filled into the liquid reservoir, oil is a good agent because of its low surface energy for hysteresis-free flow. Adsorption test (Fig. 4.9) showed the good compatibility of two commercially available oleophobic materials (Clearflex and NOA65) for the fabrication of soft contact lenses without significant adsorption of colored oil. On the other

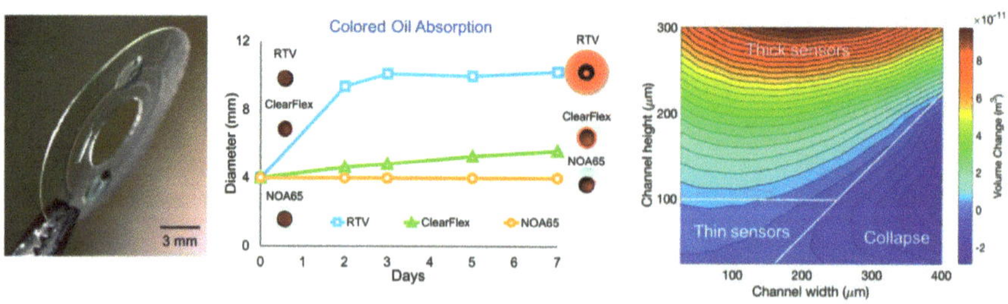

FIGURE 4.9

Selected soft contact lenses designed for the monitoring of IOP. Digital image of the wearable microfluidic contact lenses, dyed oil absorption experiment for different materials and modeling of the architecture by COMSOL Multiphysics software simulations for volume change with respect to channel height and width. *IOP*, Intraocular pressure.

Reproduced from S. Agaoglu, P. Diep, M. Martini, K. Samudhyatha, M. Baday, I.E. Araci, Lab. Chip 18 (2018) 3471–3483 with permission from the Royal Society of Chemistry.

hand, the architecture of the microfluidic channels is important when exposed to the biaxial stress that is typical for cornea. The COMSOL Multiphysics software simulations found that the collapse and thick sensor zones are not suitable for the fabrication of microfluidics due to reduced sensitivity and unstable interface movement, as well as reduced comfort and functionality, respectively. Therefore, thin sensor zone was chosen for the design of the width and height for dictated comfort and high volume change.

4.4.3 Laser ablation microfluidics in contact lenses

As discussed above, the conventional fabrication of microfluidics embedded in soft contact lenses requires time-consuming, laborious, and high-cost techniques, including photolithography, soft lithography, and injection or thermal molding. Compared to these, laser ablation is supposed to be an alternative and simple way to fabricate the microfluidics inside the soft contact lenses. Direct exposure to the laser can cause the ablation of polymer materials via photothermal/photochemical process, while the laser intensity exceeds the binding energy of crosslinked polymer backbone. Hydrogels are widely used in the fabrication of soft contact lenses. It is important to understand the effect of the laser exposure to hydrogel polymer matrix with respect to laser power intensity, laser speed, and exposure time. In the report from Jiang et al. [112], a time-domain finite element method was applied to simulate beam–polymer interaction using a hydrogel matrix (decomposition temperature of 360°C) and CO_2 laser ablation (wavelength of 10.64 μm, power of 30 W). The ablation depth and width increased with a nonlinear process while the exposure time prolonged from 0 to 10 milliseconds (Fig. 4.10A). The function of laser beam power exhibited a positive linear relationship, whereas the laser beam speed has a negative linear relationship for the width and depth. A patterned microfluidic array was successfully prepared under the optimized laser parameters with one main channel divided into four channels connected to the microcavity (Fig. 4.10B). In addition to the simple laser micromachining, the fiber templating method is also convenient to build channels in contact lenses (Fig. 4.10C). A silica fiber template was sandwiched between two contact lenses layers bound by the UV-initiated free-radical polymerization of poly(ethylene glycol) diacrylate. The extract of the template forms a microfluidic channel whose diameter can be controlled by the size of the template. The fluidic dynamic test showed the laminar flow within the microchannel, which can be visualized by fluorescent microbeads in artificial tear fluid. Except for the CO_2 laser, the ultrafast laser writing is found to be a precise procedure to engrave microfluidic arrays or networks into commercial contact lenses. The utilization of a femtosecond laser can result in small features down to 30 μm of the width, and the ultra-short pulses (<400 fs) will not cause thermal fingerprint on the contact lens by avoiding residual stresses and microstructural changes in the contact lenses material and thus preserving their characteristics. Various functional microfluidic elements can be produced via the femtosecond laser system onto an engraved lens with

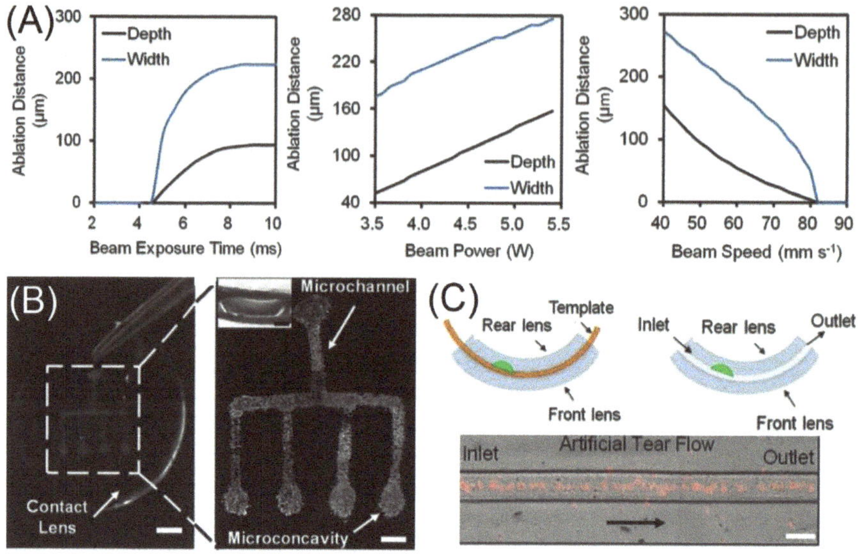

FIGURE 4.10

Microfluidic contact lenses fabrication via laser ablation and template methods [112]. (A) Simulated CO_2 laser ablation of a hydrogel matrix with the parameters of exposure time, laser beam power and laser bean speed; (B) photograph of a microfluidic contact lens array by the laser patterning; (C) fiber templates to form microchannels and the corresponding laminar flow visualized by fluorescent microbeads in artificial tear fluid.

subsequent sealing of the channels via another layer of curved contact lens, such as flow valves, flow resistors, multi-inlet geometries, and splitters [113].

4.4.4 Multiplexed sensing with microfluidic contact lenses

Apart from the monitoring of physical signs of IOP, as a typical biofluid, tears can be used for the evaluation of ocular health by the monitoring of the physiological levels for various biomarkers. Tears comprise a mixture of electrolytes (Na^+, K^+, Cl^-, HCO_3^-, Ca^{2+}, etc.), proteins, lipids, peptides, and different metabolites (e.g., ascorbate, glucose, lactate, etc.). Concentrations of these compositions in tears are different from that of blood (golden standard) [106,112,114]. However, researches have demonstrated that there is some specific correlation for biomarkers in tears compared to that of blood/serum to some extent, such as similar levels for Na^+, Cl^-, HCO_3^-, and urea, relative higher level of K^+ and lactate, as well as relative lower level of glucose [114,115]. Therefore, contact lens-based sensing and biosensing system have been widely researched and fabricated, especially for multiplexed platforms targeting various biomarkers. Multiplexed microfluidic plays an important role in tear fluidic processing. Moreddu et al. [116]

reported the design and fabrication of multiplexed microfluidics by laser ablation. As shown in Fig. 4.11A, the microfluidics consists of a central ring with four branches connected to the terminal microcavities. Inside the four microcavities, biosensing elements are embedded inside for the detection of pH, glucose, protein, and nitrite, respectively. The teat pH sensor was based on the mixture of methyl red, bromothymol blue, and phenolphthalein, which shows a pH-dependent color variation ranging from dark yellow at 6.0 to blue at 8.0. The tear glucose sensor relied on a two-step enzymatic method with GOx/POD for the oxidation of glucose into hydrogen peroxide and in turn oxidation of TMB under POD catalysis, displaying a color shift from yellow to different shades of green according to the level of glucose. The existence of protein in tear can result in the binding of the embedded protein dye with free amino groups, inducing a color change from beige to light blue. The level of nitrite level can be adjusted by the

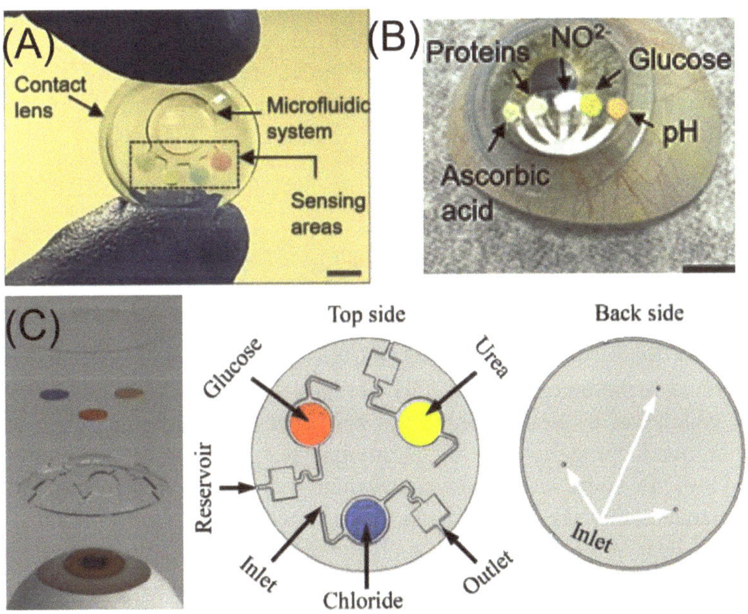

FIGURE 4.11

Multiplexed microfluidic contact lenses. (A) Microfluidic contact lenses for the colorimetric sensing of tear metabolites [116]. (B) Microfluidic system in a contact lens for embedding of paper microfluidic chips [117]. (C) Illustration of the top and back sides of the microfluidic contact lenses [118].

(A) Reproduced from R. Moreddu, J.S. Wolffsohn, D. Vigolo, A.K. Yetisen, Sensor. Actuat. B Chem. 317 (2020) 128183 with permission from Elsevier B.V. (B) Reproduced from R. Moreddu, M. Elsherif, H. Adams, D. Moschou, M.F. Cordeiro, J.S. Wolffsohn, et al., Lab. Chip 20 (2020) 3970–3979 with permission from the Royal Society of Chemistry. (C) Reproduced from X. Yang, H. Yao, G. Zhao, G.A. Ameer, W. Sun, J. Yang, et al., J. Mater. Sci. 55 (2020) 9551–9561 with permission from the Springer Nature.

intensity of pink color based on the reaction of nitrite ions with sulfanilamide to form a diazonium salt that further binds *N*-(1-naphthyl)-ethylenediamine dihydrochloride. In the microfluidic channel, tear fluid could reach all the four sensing sites (microcavities) within 560 milliseconds with small amount volume of 5 μL. The colorimetric readouts were collected with a smartphone camera and then processed with MATLAB algorithm based on the nearest neighbor model to output the concentration of the corresponding analytes. The same group reported another platform (Fig. 4.11B) for multiplexed detection of clinically relevant analytes including H^+, proteins, glucose, nitrites, and L-ascorbic acid using laser-ablated microfluidics in contact lenses for flow processing and laser-cut paper matrices for immobilization of biosensing elements [117]. Layer-laminated multiplexed microfluidic contact lenses in Fig. 4.11C consist of an inner contact lens with microchannels, a colorimetric analysis module embedded in the cavity of the inner lens and an outer lens in contact with the eyelid that formed an open-loop microchannel and reservoirs together with the inner lens. For the inner contact lens, three identical microchannels were distributed including an inlet, the detection zone, and reservoir connected to the outlet. Tear fluid flow alongside the microchannels via capillary force and flowing back of the tears during blinking can be eliminated by the curved microchannel design.

4.5 Flexible microfluidic-based smart bandages

4.5.1 Chronic wound

Wound is an injury of the skin tissue that breaks its continuity due to violence. Acute wound is more common in daily life caused by cuts, scratches, surgery, or trauma, which can be healed within the predicted time-frame and expected rate with four overlapping hemostasis, inflammatory, proliferative, and remodeling phases [119]. In some cases, an acute wound can be developed into a chronic wound that seldomly proceeds beyond the inflammatory phase for months, years, or never to heal. A chronic wound is also commonly seen as a complication for diabetes mellitus, obesity, pressure ulcers, and age-related vascular diseases. About 1%–2% of the population in many countries have been estimated to experience a chronic wound in the course of their lifetime, and older persons aged more than 60 years are more vulnerable to chronic wound and problematic healing, which is painful for individuals and brings heavy burden for both individuals and health care system [120].

Conventionally, the chronic wound status is clinically checked by clinicians via assessment of the local factors for the wound via either visual inspection or lab testing. The visual inspection can fulfill a quick diagnosis for the status; however, it is subjective and qualitative. Although lab testing (e.g., exudate swab) can realize the objective and quantitative diagnosis, it is painful, expensive, and time-consuming. The gross symptoms for a chronic wound are yellow exudate and red inflammation. Except these, there are various other parameters that can be used as potential markers for the diagnosis. Table 4.3 lists the potential markers, including physical, biochemical, and biological markers, with the proposed monitoring mode and the evaluations of chronic wound status based on the markers [121–123].

Many wound bandages/dressings have been developed for the personal and clinical management of chronic wound. Wound dressings can provide a barrier between the wound bed and the external environment with a well-maintained oxygen-permeable, warm, and moist environment required for healing. Excess wound exudate or bioburden released from the wound bed can be absorbed by the wound dressing. In addition, the primary concern that needs to be taken into consideration is the risk of bacterial infection. Therefore, biocompatible and antimicrobial materials are usually used. Table 4.4 summarizes some of the representative materials used for wound dressings and wound bandages [121]. For individual management of the chronic wound using wound dressings, it is still challenging to determine the frequency of the changing of dressings to keep the balance and optimal environment for wound healing and to prevent bacterial infection. The integration of flexible microfluidics, as well as sensing and biosensing elements

Table 4.3 Chronic wound markers and monitoring mode [121–126].

Category	Markers	Monitoring mode	Evaluations
Physical	Temperature	Electrical/ Infrared	Indication of infection by a temperature spike in peri-wound
	Moisture/ exudate	Visual/humidity impedance	Indication of infection/assessment of dressing efficacy
	Pressure	Pressure-induced electrical signal	Adequate or excessive pressure
Biochemical	Oxygen	Electrical/optical	Lack of oxygen prevents tissue growth
	pH	Electrical/ colorimetric	Early alert of infection: pH of skin between 4 and 6, while wound exudate has fluctuate between 7.15 and 8.90, >10 when bacteria proliferation
	Nitric oxide	Indirect nitrate measure	Low level in nonhealing wounds
Biological	Uric acid	Electrochemical/ colorimetric	High level inhibit wound healing
	Protein	Immunoassays	Bacterial infection
	Cytokines	Bioassay	Inflammation, high level in nonhealing wounds
	Enzyme	Bioassay	Bacterial infection
	Bacterial	Swab, culture	Bacterial infection

Table 4.4 Summary of natural and synthetic materials for wound dressings and wound bandages.

Category	Polymers	Advantages	Disadvantages
Natural	Cellulose	Hydrophilic, water uptake capacity, permeability, and tensile strength	No resorption in tissues
	Chitosan	Strong adhesion, inertness, nontoxicity, biodegradability, biocompatibility, antimicrobial properties, and hemostatic effects	Extensive swelling in water
	Collagen/gelatin	Promoting tissue granulation, bacterial inhibition, biocompatible and degradable	Not absorptive
	Hyaluronic acid	Tissue integrity, lubrication, and water absorption	No wound repair
Synthetic	Poly(lactide-*co*-glycolide)	Degradable and biocompatible, mechanical strength, for drug delivery	Need antimicrobial substances
	Polyurethanes	Delivery route, semipermeable membrane, absorbent materials	Need contact layer
	Polyethylene glycol	Hydrophilic, flexible, and compatible	Too adhesive
	Polycaprolactone	Biocompatible and biodegradable	No antimicrobial effects

inside the wound dressings, could potentially provide the quantitative data for individuals and clinicians for a more effective management of chronic wounds in a manner of precise, continuous, and wireless monitoring and therapy.

4.5.2 Smart bandages for chronic wound monitoring

In wound management for individuals, the monitoring of different parameters in chronic wounds can provide information regarding the healing status and conditions of the wound microenvironment for early identification and real-time alters of complications such as infection. Subsequent control of the frequency for the changing of wound dressings and bandages contributes to the reduction of healing time and burden. The markers can be categorized into two aspects with the physical parameters of the wound bed and environment including temperature, strain, humidity and oxygen level, as well as the biochemical/biological markers in wound exudate including pH, uric acid, lactate and bacterial, etc. Flexible microfluidics are used for chronic wound monitoring via accommodation of sensing elements or wound exudate collection and process.

For sensing elements accommodated in flexible microfluidics, Escobedo et al. [127] reported a smart bandage with wireless strain and temperature sensors and a battery-less near-field communication tag. The strain sensor is based on a flexible microfluidic channel filled with a solution processable conducting polymer PEDOT:PSS in PDMS substrate. The strain sensor showed a 3-order (Δrde_0 ~1250) increase in the electrical resistance for 10% strain (~180 kΩ) from the original resistance of ~140 Ω owing to the formation of defects and electrical discontinuities in the polymer. The electrical resistance change ($\Delta R/R_0$) is linearly proportional to the strain up to 30%, which is the typical strain limit for human skin at different body parts, such as fingers, wrists, knees, and elbows for the collagen fibers and tissues to become straitened. Ota et al. [128] reported a robust platform on PDMS as a liquid-to-liquid "heterojunction" device that is responsive to external stimuli for temperature, humidity and oxygen. As shown in Fig. 4.12A, two microchannels filled with Galinstan liquid are connected via a series of "heterojunction" microchannels (30 mm wide, 31 mm high, and 200 mm long). The "heterojunction" microchannels introduce high flow resistance preventing the interpenetrating of two Galinstan liquids. The liquid heterojunctions are electrically active and can be configured as a close circuit with a few resistors in series with the two Galinstan regions serving as the source/drain metal electrodes and the ionic liquid "heterojunction" microchannels as the sensing element. By using different ionic liquids, different sensors have been achieved by measuring

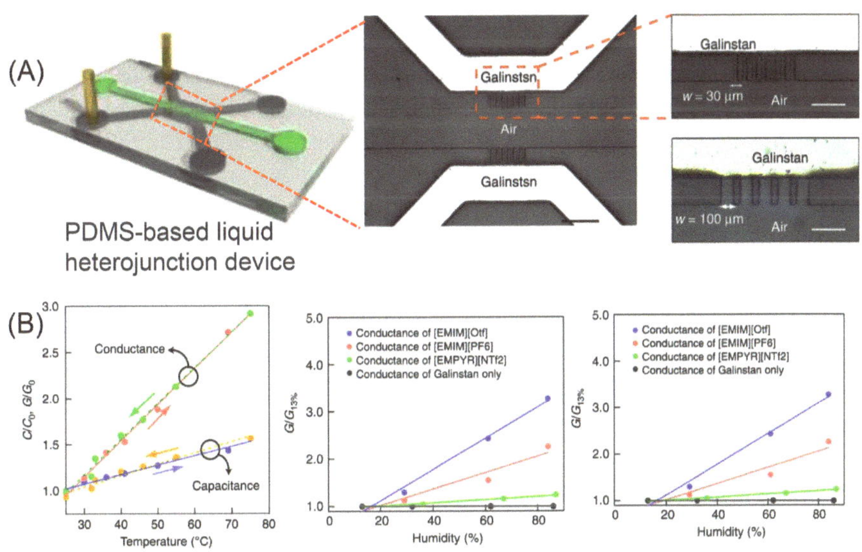

FIGURE 4.12

Sensing element accommodated in flexible microfluidics. (A) Highly deformable liquid-state heterojunction sensors and the mechanism, (B) Liquid-state heterojunction sensors for temperature, humidity, and oxygen level detection.

(A) and (B) Reproduced from H. Ota, K. Chen, Y. Lin, D. Kiriya, H. Shiraki, Z. Yu, et al., Nat. Commun. 5 (2014) 1–9 with permission of copyright 2014, Nature Publishing Group.

the conductance or capacitance, showing excellent performance characteristics (Fig. 4.12B) even under deformation for the detection of temperature, humidity, and oxygen in the linear range of 25°C–69°C, 13%–87% humidity, and 0%–21% oxygen, respectively. The mechanical deformability of the liquid phase sensing element, the use of liquid-state device technology and flexible substrates provide the potential for the integration with wound dressing for chronic wound parameters detection.

Assessment for the chronic wound status can be done by the quantitative profiling of the biochemical/biological parameters in the wound exudate. Flexible microfluidics can function as wound exudate fluid collectors. Gao et al. [129] reported a microfluidic wound exudate collector to enable the directional liquid transport of exudate from the wound bed to the sensing area of the integrated sensor array (Fig. 4.13A). Efficient wound fluid collection and delivery is essential for accurate in-situ biomarker detection. Inspired by the skin of Texas horned lizard (*Phrynosoma cornutum*) with the predetermined directional fluid flow toward it's snout defying gravity by the network of special capillary, an annular pattern was designed and fabricated, which is composed of a polar array of interconnected half-open, sawtooth-shaped capillary channels with a decreasing width of 200–160 μm (outer to inner). Based on this, the exudate can be collected through the perforated wound contact layer and continuous flow in the forward direction from the outer to the inner side. In the central region of the microfluidic wound exudate collector, a multiplexed sensor array was placed and embedded, including a temperature sensor, pH sensor, several immunosensors for detection of tumor necrosis factor-α (TNF-α), TNF-β1, interleukin-6 (IL-6), IL-8, and bacteria *Staphylococcus aureus*. The integration of flexible microfluidics and flexible electronics enabled the wireless detection of the wound physical parameters and biomarkers with smartphone-based data readout. Flexible microfluidics can also function for fluid processing for the sensing sample preparation. Yang et al. [130] constructed a bandage-like wearable flexible microfluidic recombinase polymerase amplification sensor for the detection of nucleic acids (Fig. 4.13B). Various soft materials were used for the fabrication of flexible microfluidics and wearable sensor, including a skin adhesive layer made from Ecoflex with a Young's modulus of 125 kPa for conformation on the human body surface, and an Ecoflex material substrate sealed by PDMS with a Young's modulus of ~3 MPa for efficient laminar flow mode from the inlet to outlet. The isothermal DNA amplification can be triggered by human body heat (30°C–37°C), subsequent intercalation of fluorescent probe with double strand DNA produced an obvious visible fluorescence signal under UV–vis in the reservoir. This provided a suitable wearable sensor integrated with flexible microfluidics for in-situ pathogen detection at wound sites. Potential sample purification inside the microfluidic channel may also be possible for isolation and amplification of ionic compounds by several folds in raw wound exudate by the integration of isotachophoresis chip together with microfluidics and microelectronics [131].

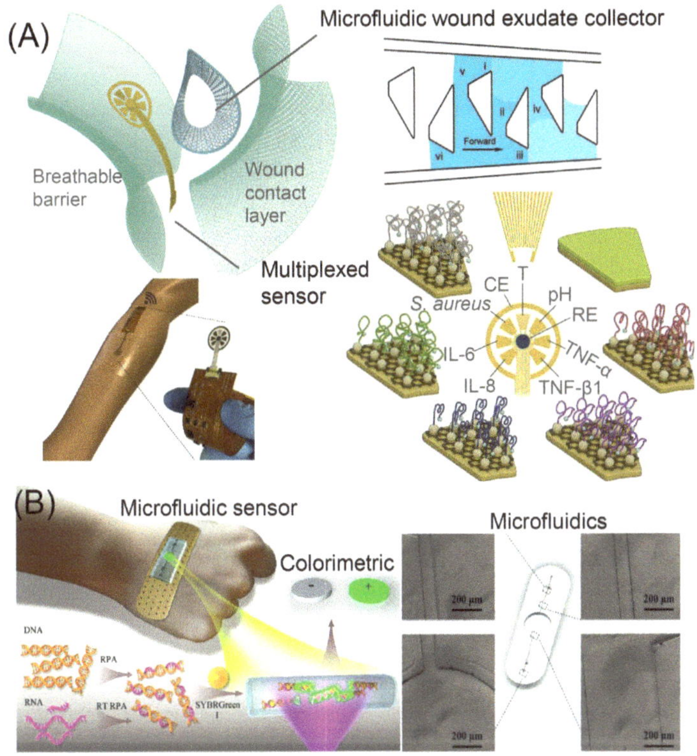

FIGURE 4.13

Flexible microfluidics for biochemical/biological parameters detection in wound exudate. (A) Multiplexed sensor array for chronic wound monitoring with flexible microfluidics for wound exudate collection. (B) Bandage-like wearable microfluidic sensor for rapid visual detection of nucleic acids.

(A) Reproduced from Y. Gao, D.T. Nguyen, T. Yeo, S.B. Lim, W.X. Tan, L.E. Madden, et al., Sci. Adv. 7 (2021) eabg9614. (B) Reproduced from B. Yang, J. Kong, X. Fang, Talanta 204 (2019) 685–692 with permission from Elsevier B.V.

4.5.3 Microfludic delivery systems for chronic wound management and treatment

Beyond the primary prevention and assessments, a therapeutic approach combined with the smart dressing could contribute to the systematic chronic wound management and treatment. However, the wound healing process is dynamic requiring different amounts of therapeutic agents over time, which cannot be well fulfilled by passive release. Therefore, an active drug release system is needed with the capability for the modulation of drug release amount under the internal or external trigger elements such as pH, temperature, chemical reaction, redox reactions, magnetic field, electric field, light, ultrasound, etc. [132]. Active drug release system with integrated flexible electronics inside the wound dressing or wound bandages has been reported with controlled release rates, such as integrated flexible microheater for thermal-triggered drug and growth factor release [133], PDMS-encapsulated flexible electronics with UV-LED triggered release of antibiotics [134], and smart hydrogel-based wound dressing capable of monitoring of and near infrared light-triggered antibiotic release [135], etc. Usually, these drug-loaded matrixes are exposed to the wound bed directly or with a layer of hydrogel, which may be challenging due to the coexisting passive release. Microfluidic platforms can potentially address this by providing precise control at a desired rate. Recent advances of controlled drug delivery and release using microfluidic platforms have been well reviewed in previous reports [136–138].

Flexible microfluidics, integrated with the flexible electronics in wound dressing, are a promising approach used for the design and fabrication of smart dressing/bandage though there are not many reports on this topic until now. Rajput et al. [139] proposed the usage of COMSOL multiphysics for the design and simulation of microfluidic smart bandage with several blocks for different functions, including an active area with microfluidic channels for drug release; four drug reservoirs for drug storage and mixing through mixers; three sensors for the detection of pH, oxygen, and moisture; and a microcontroller for feedback signals. Derakhshandeh et al. [140] developed a wirelessly controlled smart bandage with 3D-printed miniaturized needle arrays to deliver a variety of drugs into deeper layers of the wound bed, in which microchannel arrays, micropumps, and drug reservoirs were built together by PDMS to ensure the flexibility of bandage and flow of drugs. In addition to these drugs, hyperbaric oxygen therapy is used for the healing process of hypoxic chronic wounds. So the precise and periodic oxygen administration is important to avoid hyperoxygenation of tissue surrounding the wound. Ochoa et al. [141] reported a PDMS microfluidic network bonded to a parchment paper substrate with spatially defined catalytic regions as the oxygen-generating platform. The constant peroxide solution flows from the inlet alongside the microchannels and reaches the catalytic spot resulting in the release of oxygen by the catalytic reaction to the wound site at a constant rate.

4.6 Flexible microfluidic-based wearable mechanical sensors

Apart from the research and development for wearable chemical sensors and biosensors as discussed above, the advancements in flexible electronics and microfluidics also propel the prosperity in human–machine applications ranging from soft-skin robotics for human–machine interface to wearable electronics for healthcare systems. Conventionally, solid-state electronics based on semiconductor materials (e.g., silicon) and metals dominate the field, while they face limitations regarding the mechanical mismatch between the machine and human tissues (e.g., skin and cornea) at the target site due to limited deformability. In the physical aspect, flexibility, stretchability, durability, and user-friendly simplicity of the sensors are crucial for long-term wearing without conspicuous discomfort. The most important structure of a flexible and wearable device is composed of a supporting substrate and a sensing element for signal transduction. A supporting substrate is preferably flexible, stretchable, and highly biocompatible compared to the conventional rigid and bulk substrate materials, such as fabrics, organic polymers, and silicone elastomers. A sensing element can be realized by flexible materials including patterned conductive material in thin film or percolation networks [142] and conductive liquid. The solid conductive material-based mechanical sensors may have hysteresis in the electrical responses and mechanical fatigue due to crack formation and disconnection of patterned networks. Therefore, liquid materials, usually embedded in microchannels, have been introduced because of their high physicochemical stability and ultimate threshold in mechanical deformability. In this section, wearable mechanical sensors based on the active liquid embedded in a flexible substrate are discussed, and the role of flexible microfluidics are introduced.

4.6.1 Mechanical sensing mechanism via flexible microfluidics

There are three primary sensing mechanisms employed in mechanical sensors for the detection of pressure, shear force, and strain, including resistive, capacitive, and piezoelectric. In a resistive sensor, the resistance of the active elements changes according to the external force stimuli, which is commonly applied in the traditional strain gauges. The capacitive sensor is based on the differential capacitance change between a pair of plates with a dielectric medium. A specific voltage is generated in a piezoelectric sensor while the force is applied. Similarly, the flexible microfluidic-based wearable mechanical sensors rely on the detection of electrical resistance, capacitance, inductance, and voltage change toward the mechanical deformation. To achieve this, different kinds of conductive liquids have been used for the fabrication of flexible microfluidics with the conductive liquids embedded inside the microchannels, such as liquid metal (e.g., eutectic gallium indium) [143], carbon material suspensions (e.g., graphene and carbon nanotubes) [144], conducting polymers (e.g., PEDOT:PSS) [145], and electrolytes (e.g., KCl and ionic liquid) [146,147].

The flexible microfluidic mechanical sensor can operate on the basis of resistive sensing mechanism as shown in Fig. 4.14A. When external pressure is applied, the microchannel collapses causing the conductive liquid flow to the end regions, due to which the overall resistance increase. The conductive liquid can flow back to the original state under the release of the pressure. Therefore, the mechanical forces or pressures applied on the device can be quantified by the changes in the electrical resistance of the electrical circuit or by normalized electrical resistance to the initial state ($\Delta R/R_0$) [143,144,148]. In a capacitive sensor, the pressure input is usually transduced into the capacitance change between a

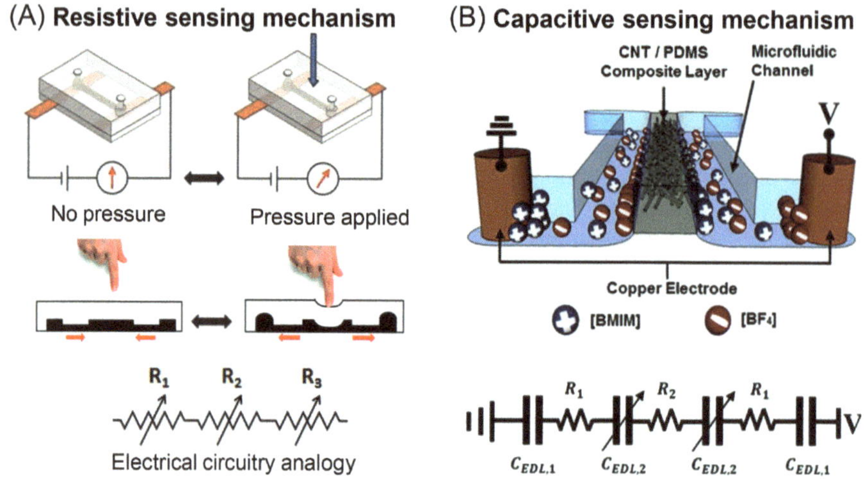

FIGURE 4.14

Different sensing mechanism for flexible microfluidic-based mechanical sensors. (A) Resistive sensing mechanism with the liquid metal inside the microchannel. (B) Capacitive sensing mechanism with the ionic liquid embedded in microchannel.

(A) Reproduced from Refs. [143,144,148] with the permission from American Chemical Society, Royal Society of Chemistry, and WILEY-VCH Verlag GmbH & Co. KGaA. (B) Reproduced from Ref. [147] with the permission from Royal Society of Chemistry.

parallel plate capacitor over a dielectric material between the plates, in which metal and solid dielectric are usually used for electrode plate and dielectric layer, respectively. In a flexible microfluidic-based mechanical sensor sample shown in Fig. 4.14B, two sets of electric double layer (EDL) capacitance (C_{EDL}) are formed at the interfaces between the charged electrode surfaces and ionic liquids [147]. Under external AC field application of 1 V and 1 kHz, vertical forces applied to the middle elastic carbon nanotube (CNT)/PDMS element can decrease the height of the microfluidic channel, thus causing a redistribution of ions at the diffuse layer and a reduction of the charged surface area for C_{EDL2}. In addition to electrical based mechanism, passive readout has attracted extensive interest because the external circuit is not practical for some wearable devices, especially for the smart contact lenses as discussed in Section 4.3.1. The applied pressure or strain is converted into the displacement of gas–liquid interface between the dye-colored liquid and air inside the microchannels. Yan [149] fabricated a wireless, microfluidic pressure sensor composed of PDMS and dyed glycerol for point-of-care glaucoma diagnosis, in which the mechanical displacement of the sensing chamber by internal pressure can be effectively amplified to the outwardly displaced fluid in the sensing channel that is captured via a camera. On the other hand, the pressure-induced deformation of the microchannels is detected by other optical methods, such as confocal microscopy, color intensity, and grating diffraction [150].

4.6.2 Flexible microfluidic-based strain sensors

With the development of human–machine interfaces, robotics, and wearable consumer electronics, there has been an increased interest in the research of strain sensors. Strain is the relative change in length of a subject caused by external mechanical stimuli, such as compression and tension, which is commonly seen in human body motions such as physical bending, twisting of hand joints, fingers, wrist, neck, and chest. Therefore, strain sensors possess the potential to be applied in wearable devices embedded into clothes or even attached directly onto the skin for the monitoring of human motions and personal health as well as for therapeutics. Unlike the conventional strain sensors based on rigid conductive materials with small strain detection of below 5% due to the limited flexibility [151], liquid state-based microfluidics strain sensors with the micropatterned elastomer microstructures have demonstrated their capabilities for wearable or skin-mountable electronic devices with excellent analytical performance with respect to large strain (usually >200%), high sensitivity, quick response, long-term stability, and durability under periodic dynamic load.

Bhattacharjee et al. [145] reported a conducting polymer PEDOT:PSS encapsulated microchannel-based highly sensitive stretchable strain sensor (Fig. 4.15A). The microchannel was fabricated using an REM technique with elastomer PDMS as the substrate and a wire as a sacrificial template, followed by the injection of PEDOT:PSS into the straight-shaped microchannel. The fabricated strain sensor exhibits about three orders of resistance changes ($\Delta R/R_0 \approx 1200$) for 10% applied strain, which can reach a maximum strain of 30%, that is, the standard strain limit associated with human body parts such as fingers and wrists. In addition to the planer and straight-shaped microchannel, helical-shaped or wave-shaped microchannel-based strain sensors can improve the performance by avoiding break and decrease hysteresis. Oren et al. [152] reported a helical-shaped tubular spring formed within microchannel via PDMS casting with a helical metal coil as a template and following injection of graphene suspension as an active element. The helical shape allows the monitoring both tensile and compressive strains in a wide dynamic detection range from 24% compressive strain to 20% tensile strain. Similarly, the wave-shaped microchannel-based strain sensor is proved to withstand a strain up to 320% and its hysteresis performance is much improved with a low hysteresis degree of 4.51% in the lateral strain direction, compared to that of a straight-shaped microchannel-based strain sensor with a strain of 150% and hysteresis degree of 37.49% (Fig. 4.15B). Such wave-shaped microchannel elastomeric matrix (Ecoflex) was prepared via soft lithography with an injection of eutectic gallium indium as the sensing element. In addition to the microchannels fabricated via lithography, 3D printing holds the potential for additive manufacturing of bendable microchannel-based strain sensors. Agarwala et al. [153] showed the application of 3D printing for ease and fabricating objects with embedded serpentine feature, which is filled with silver nanoparticle ink as sensing element for strain sensor based on the change of resistance measured by

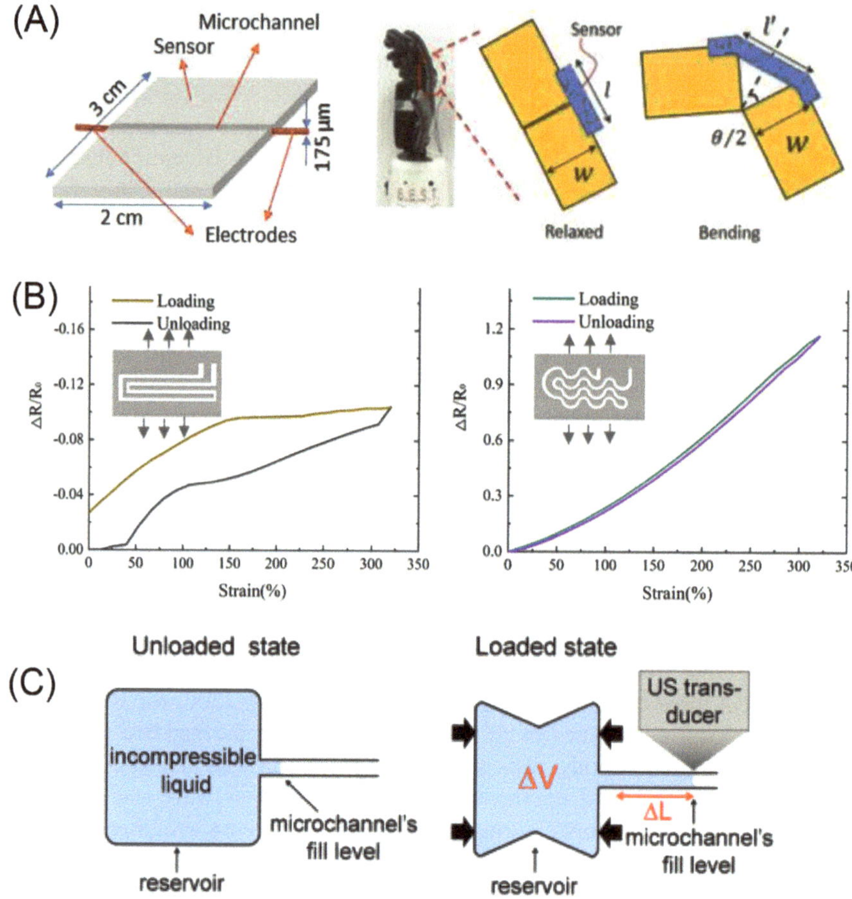

FIGURE 4.15

Flexible microfluidic-based strain sensor. (A) Straight-shaped microchannel for strain sensor. (B) Wave-shaped microchannel-based strain sensors. (C) Wireless passive strain sensor.

(A) Reproduced from Ref. [145]. (B) Reproduced from Ref. [151] with the permission from American Chemical Society. (C) Reproduced from Ref. [154] with the permission from Institute of Physics Publishing.

external circuits. To achieve the wireless strain sensor readout without external circuit assembly, various passive strain sensors have been reported based on the displacement of liquid with subsequent detection via optical methods. Umbrecht et al. [154] reported the use of ultrasound to readout the filling level of a microchannel connected to a reservoir. At the initial state, the microchannel is hardly filled, while for applied strain, the reservoir is compressed with liquid displaced into the attached microchannel with a change of the microchannel's fill level ΔL (Fig. 4.15C).

4.6.3 Flexible microfluidic-based pressure sensors

Pressure sensors can detect the pressure or force distribution over a subject, in which the pressure- or force-induced changes on the sensing elements are transduced into an electrical or an optical signal as discussed in Section 4.6.1. The pressure sensors are promising to serve as an important component for various emerging sensing and biomedical applications, such as electronic skins and wearable consumer electronics, as well as the healthcare monitoring for physiological parameters such as eye pressure. The materials, manufacturing, and progress of pressure sensors have been reviewed by a number of reviews recently [155]. Here, we mainly focus on the discussion of flexible microfluidic-based pressure sensors and their corresponding analytical performance.

Human skin is the most distinct sensory system that can perceive tactile sensation. We can use our fingers to distinguish the size, shape, and texture of the subjects based on the sense-of-touch by multiplex mechanoreceptors inside the skin. Flexible microfluidic-based pressure sensors have been developed to mimic the mechanoreceptors for tactile sensing to perceive the pressure from the contact/interaction between the subject and the environment. The daily activities of the fingers for typing and grasping cover a wide range of pressure over 200 kPa [143]. Yeo et al. [148] developed a "S"-shaped liquid metal alloy-based microfluidics for tactile sensing, which operates according to the resistance change of the liquid metal as the function of mechanical forces or pressures applied. Such a flexible tactile sensor is capable of distinguishing compressive and bending loads with an extremely large range of pressure from 2 to 400 kPa. The special "S"-shaped microfluidics showed higher normalized resistance change ($\Delta R/R_0$) than that of a common straight microchannel under the high pressure range over 70 kPa with a high device sensitivity of 0.02 kPa^{-1}. Except for the efforts devoted to the development of tactile sensors aiming for high pressure range, the resolution of low pressure range for liquid-metal-based sensor is limited. Therefore, Gao et al. [156] reported a diaphragm pressure sensor that is capable of resolving sub-50 Pa changes in pressure. As shown in Fig. 4.16A, the diaphragm sensor is designed as an embedded equivalent Wheatstone bridge circuit, including two sets of tangential sensing grids symmetrically located around the diaphragm center and two sets of radial sensing grids symmetrically located around the diaphragm periphery. On account of the tangential and radial strain fields, high sensitivity of 0.0835 kPa^{-1}, low detection limit below 100 kPa with sub-50 Pa resolution, as well as an ultrafast response time of 90 milliseconds can be realized. This pressure sensor was tested not only for tactile sensing attached to a glove for grasping and releasing action at the thumb, finger, and palm sites, but also for heart-rate monitoring attached to the wrist. Except for the use of liquid metal, graphene oxide nanosuspension was also used for the microfluidic tactile sensor to distinguish various user-applied mechanical forces based on the normalized change of resistance [144].

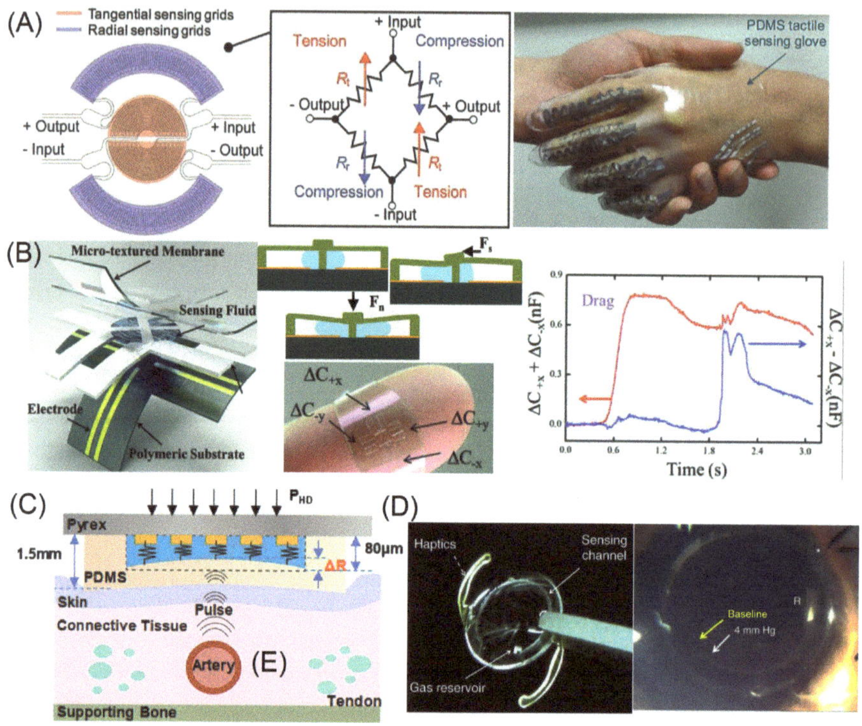

FIGURE 4.16

Flexible microfluidic-based pressure sensors and their applications. (A) Microfluidic tactile diaphragm pressure sensor for tactile touch monitoring. (B) 3D microfluidic tactile sensor for contact force measurement. (C) Microfluidic tactile sensor for post-exercise assessment of arterial parameters. (D) Microfluidic pressure sensor implanted in porcine eyes for IOP readout with a smartphone. *3D*, Three-dimensional; *IOP*, intraocular pressure.

(A) Reproduced from Ref. [156] with permission from WILEY-VCH Verlag GmbH & Co. KGaA. (B) Reproduced from Ref. [157] with permission from the Royal Society of Chemistry. (C) Reproduced from Ref. [158] from the permission form the Biomedical Engineering Society. (D) Reproduced from Ref. [159] with permission from Nature Publishing Group.

In addition to the microfluidic resistive sensing mechanism for tactile sensing mentioned above, microfluidic-based capacitive sensors relying on capacitance vibrations according to geometrical changes of the spaces between electrodes and physical changes of a dielectric layer in an electric field have also been widely studied. Nie et al. [160] reported an iontronic microdroplet array with the ionic droplet sandwiched between two flexible PET substrates patterned with indium–tin-oxide (~100 nm thick) electrodes, creating an EDL capacitance. Upon the external pressure loads, the iontronic microdroplets expand circumferentially according to the mechanical deformation of the flexible substrates, causing the changing of the EDL. The devices achieved an excellent sensing performance with high sensitivity of 0.43 nF/kPa, a minimal detectable pressure of 33 Pa at a $3 \times 3 \times 0.2\ mm^3$ packaging array as well as the high stability of 20,000 cycles of pressure/force loads. This kind of concept was extended for the microfluidic tactile sensor for 3D contact force measurements as shown in Fig. 4.16B [157]. The microfluidic sensing devices accommodate the ionic liquid in the sensing chambers with four pairs of electrodes inside the detection channels. The pressure loads in the z axis (F_n) and the shear force tangent to the surface in the $x-y$ axis (F_s) can cause the symmetric deformation and torque, respectively, leading to a differential change in the contact areas and EDL capacitance in the detection channels for the electrode pairs. As demonstrated, the devices can perceive the drag motion once mounted to a fingertip. The microfluidic capacitive sensor array has been applied to mimic the human skin to realize the artificial electronic skins (E-skins) with various sensory capabilities for the detection of temperature and different body motions including bending, releasing, idling, stretching, and inhalation [147,161].

In addition, flexible microfluidic-based pressure sensors have been applied for health monitoring by providing the physical parameters such as heartbeat rate [156], inhalation [161], arterial pulse [158], and IOP [159]. For instance, Hao et al. [158] reported the application of a microfluidic pressure sensor for arterial pulse signal measurements on real human subjects. The device is based on the PDMS microstructure with an electrolyte-filled microchannel embedded underneath the Pyrex substrate. As shown in Fig. 4.16C, with an external hold-down pressure (P_{HD}) against the artery, the pulsatile pulse signal ($\Delta p(t)$) in an artery deforms the microstructure and converts to a resistance change ($\Delta R(t)$). Such devices achieved the monitoring of pulse signal for simultaneous estimation of three arterial parameters, that is, elasticity (E), viscosity (η), and radius (r_0) of the arterial wall, post exercise for subjects with either obesity or insomnia, offering a promising, affordable, and convenient diagnosis tool for arterial health assessment in the future compared to current techniques (e.g., applanation tonometry and imaging instruments). Passive IOP sensors based on the displacement of air–fluid interface in microfluidics were tested by implanting into the lens capsular bag of enucleated porcine eyes (Fig. 4.16D) [159]. The IOP can be read directly during a routine phacoemulsification cataract surgery to control the accuracy of surgery and reduce patients' discomfort.

4.7 Summary and perspectives

The fabrication of flexible and mountable microfluidics and their integration with flexible electronics have demonstrated various promising applications in wearable sensors and biosensors in which conformable tissue–sample–device interfaces are necessary. In contrast to the conventional rigid materials (e.g., silicon, glass, and ceramic), various flexible materials including thermoplastics, elastomers, hydrogels, fabrics, and papers, have been used as compliant substrates to construct flexible microfluidics for high body compliance via promoted fabrication techniques, such as soft lithography, laser micromachining, and 3D printing. Such flexible microfluidics provide precise control and manipulation of fluids with great flexibility to deform, bend, and stretch to satisfy different application scenarios, ranging from epidermal patches to soft contact lenses and wound bandages. The convergence of flexibility and microfluidics has further benefited the wearable sensors and biosensors via their integration with flexible and wearable electronics and optical transducers. Flexible body-worn sensors based on the flexible microfluidics have been covered in this chapter for sweat analysis with the detection of electrolytes, nutrients, and metabolites; ocular and tear analysis with the detection of IOP and multiple biomarkers; chronic wound monitoring and management; as well as tactile and strain sensing, in which the flexible microfluidics have demonstrated their functions for sample collection, handling, storage, processing, and conversion into readable signals via the integrated electronics.

Until now, we have witnessed numerous proof-of-concept studies on the research and development of flexible microfluidic-based wearable sensors and biosensors. However, there are still certain aspects to meet for the successful translation of the research-based technologies into commercialized devices. First of all, the high cost of fabrication and complicated device integrating and packaging with flexible electronics are main concerns. The future efforts toward low-cost and facile techniques by means of inexpensive devices and fabrication methodologies are needed. Among those techniques discussed in this chapter, 3D printing is a promising technique; however, it is still limited to the availability of materials suitable for 3D printing and the throughput for high volume manufacturing.

From a device material science perspective, a deeper understanding of the essential material properties is important for choosing the ideal base materials for the fabrication of flexible microfluidics. Mechanical properties including flexibility and stretchability define the material's resistance against bending deformation and stretch stiffness, which influences the final device's conformability to soft tissue and long-term wearing without conspicuous discomfort. In addition, other material properties should be taken into consideration, such as light weight for miniaturization of the final device, transparency for optical performance, and esthetics, adhesion and permeability for wearing, biocompatibility and biodegradability for nontoxic, nonirritant, and nonallergenic features to humans and environment, as well as scalability for cost-effective manufacturing [162].

From the sensing and biosensing perspective, the accuracy and stability are the primary concerns. Biomolecules, such as antibodies, enzymes, and nucleic acids, are usually immobilized in the flexible microfluidic-based sensing and biosensing devices for the recognition of biomarkers, while their stability determines the sensitivity of the analytical performance and the long-term practical operation. This needs the research either in the improvement in the stability of biomolecules or the development of robust receptors that are resistant to extreme external conditions with the fluctuations of pH, ionic strength, and temperature. Besides the biorecognition elements, the transducers are also important to provide high sensitivity and stability, especially electrochemical transducers based on electrodes. Recent researches in the field of conducting polymers with soft and flexible material's nature [163,164], laser-induced graphene electrode on flexible polymeric substrates [165], layered 2D materials (e.g., MXene) [166] can serve as effective sensing elements or supporting substrate for diverse sensing and biosensing applications. In addition, conventional point-of-care devices usually provide discrete measurements of analyte in biological samples, while the wearable sensing and biosensing devices need to deliver continuous detection of analytes in real time where conditions may be fluctuating and affecting binding kinetics. This brings a new challenge for flexible microfluidics in their continuous sample collection, processing, and waste storage. What's more, sensors and biosensors designed to measure various key parameters simultaneously for health monitoring are highly desirable. Microfluidics could play a more pivotal role in the fluidic processing for profiling a wide range of biomarkers. Further integration of the flexible microfluidic-based sensing and biosensing devices with power supplies, wireless data transmission, and online back-end cloud servers can provide data analysis, storage, and sharing.

References

[1] P.N. Nge, C.I. Rogers, A.T. Woolley, Chem. Rev. 113 (2013) 2550–2583.

[2] V. Sivagnanam, M.A.M. Gijs, Chem. Rev. 113 (2013) 3214–3247.

[3] K. Ren, J. Zhou, H. Wu, Acc. Chem. Res. 46 (2013) 2396–2406.

[4] S. Marre, K.F. Jensen, Chem. Soc. Rev. 39 (2010) 1183–1202.

[5] D.G. Rackus, M.H. Shamsi, A.R. Wheeler, Chem. Soc. Rev. 44 (2015) 5320–5340.

[6] L. Meng, A.P. Turner, W.C. Mak, Biotechnol. Adv. 39 (2020) 107398.

[7] Y. Chen, Y. Zhang, Z. Liang, Y. Cao, Z. Han, X. Feng, npj Flex. Electron. 4 (2020) 1–20.

[8] T. Someya, Z. Bao, G.G. Malliaras, Nature 540 (2016) 379–385.

[9] Y.-C. Kung, K.-W. Huang, Y.-J. Fan, P.-Y. Chiou, Lab. Chip 15 (2015) 1861–1868.

[10] C. Iliescu, H. Taylor, M. Avram, J. Miao, S. Franssila, Biomicrofluidics 6 (2012) 016505.

[11] W. Zhang, R.E. Eitel, J. Intell. Mater. Syst. Struct. 24 (2013) 1637–1646.

[12] E. Roy, A. Pallandre, B. Zribi, M.C. Horny, F.D. Delapierre, A. Cattoni, et al., Adv. Microfluid. New Appl. Biol. Energy Mater. Sci. (2016) 2464–2471.

[13] E.A. Redman, J.S. Mellors, J.A. Starkey, J.M. Ramsey, Anal. Chem. 88 (2016) 2220–2226.

[14] H. Shakeel, D. Wang, J.R. Heflin, M. Agah, Sensor. Actuat. B Chem. 216 (2015) 349–357.

[15] M. Pawlaczyk, M. Lelonkiewicz, M. Wieczorowski, Adv. Dermatol. Allergol. 30 (2013) 302.

[16] Z. Wang, A.A. Volinsky, N.D. Gallant, J. Appl. Polym. Sci. 131 (2014) 41050.

[17] M. Amjadi, K.U. Kyung, I. Park, M. Sitti, Adv. Funct. Mater. 26 (2016) 1678–1698.

[18] E. Gencturk, S. Mutlu, K.O. Ulgen, Biomicrofluidics 11 (2017) 051502.

[19] K. Domansky, D.C. Leslie, J. McKinney, J.P. Fraser, J.D. Sliz, T. Hamkins-Indik, et al., Lab. Chip 13 (2013) 3956–3964.

[20] Y. Fan, L. Huang, R. Cui, X. Zhou, Y. Zhang, J. Micro/Nanolithogr. MEMS MOEMS 19 (2020) 024501.

[21] Y. Li, R. Fischer, R. Zboray, P. Boillat, M. Camenzind, C. Toncelli, et al., ACS Appl. Mater. Interfaces 12 (2020) 29908–29916.

[22] S. Farajikhah, J.M. Cabot, P.C. Innis, B. Paull, G. Wallace, ACS Comb. Sci. 21 (2019) 229–240.

[23] S.S. Banerjee, A. Roychowdhury, N. Taneja, R. Janrao, J. Khandare, D. Paul, Sensor. Actuat. B Chem. 186 (2013) 439–445.

[24] H. Liu, R.M. Crooks, JACS 133 (2011) 17564–17566.

[25] I. Jeerapan, C. Moonla, P. Thavarungkul, P. Kanatharana, Progress in Molecular Biology and Translational Science, 187, Academic Press, 2022, pp. 249–279.

[26] J. Choi, D. Kang, S. Han, S.B. Kim, J.A. Rogers, Adv. Healthc. Mater. 6 (2017) 1601355.

[27] E. Elizalde, R. Urteaga, C.L. Berli, Lab. Chip 15 (2015) 2173–2180.

[28] B. Liang, Q. Cao, X. Mao, W. Pan, T. Tu, L. Fang, et al., IEEE Sens. J. 21 (2020) 9642–9648.

[29] Q. Cao, B. Liang, T. Tu, J. Wei, L. Fang, X. Ye, RSC Adv. 9 (2019) 5674–5681.

[30] T. Young, Philos. Trans. R. Soc. Lond. (1805) 65–87.

[31] H. Lin, J. Tan, J. Zhu, S. Lin, Y. Zhao, W. Yu, et al., Nat. Commun. 11 (2020) 1–12.

[32] M.R. Romero, G.T. Nieva, J. Vedelago, C.G. Gomez, Soft Matter. 17 (2021) 5941–5949.

[33] A.R. Naik, B. Warren, A. Burns, R. Lenigk, J. Morse, A. Alizadeh, et al., Microfluid. Nanofluidics 25 (2021) 1–9.

[34] E.M. Purcell, Am. J. Phys. 45 (1977) 3–11.

[35] J. Clark, M. Kaufman, P.S. Fodor, Micromachines 9 (2018) 107.

[36] C.-Y. Lee, C.-L. Chang, Y.-N. Wang, L.-M. Fu, Int. J. Mol. Sci. 12 (2011) 3263–3287.

[37] R. Ghaffari, J. Choi, M.S. Raj, S. Chen, S.P. Lee, J.T. Reeder, et al., Adv. Funct. Mater. 30 (2020) 1907269.

[38] Z. Zhao, Q. Li, L. Chen, Y. Zhao, J. Gong, Z. Li, et al., Lab. Chip 21 (2021) 916–932.

[39] J. Kim, Y. Wu, H. Luan, D.S. Yang, D. Cho, S.S. Kwak, et al., Adv. Sci. (2021) 2103331.

[40] R. Ding, G. Lisak, Anal. Chim. Acta 1091 (2019) 103–111.

[41] W. Gao, H.Y. Nyein, Z. Shahpar, H.M. Fahad, K. Chen, S. Emaminejad, et al., ACS Sens. 1 (2016) 866–874.

[42] R. Vinoth, T. Nakagawa, J. Mathiyarasu, A.V. Mohan, ACS Sens. 6 (2021) 1174–1186.

[43] S. Kim, B. Lee, J.T. Reeder, S.H. Seo, S.-U. Lee, A. Hourlier-Fargette, et al., Proc. Natl. Acad. Sci. U.S.A. 117 (2020) 27906–27915.

[44] I. Jeerapan, S. Poorahong, J. Electrochem. Soc. 167 (2020) 037573.

[45] M. Sun, Y. Gu, X. Pei, J. Wang, J. Liu, C. Ma, et al., Nano Energy 86 (2021) 106061.

[46] A.S. Dahiya, J. Thireau, J. Boudaden, S. Lal, U. Gulzar, Y. Zhang, et al., J. Electrochem. Soc. 167 (2019) 037516.
[47] A. Bandodkar, S. Lee, I. Huang, W. Li, S. Wang, C.-J. Su, et al., Nat. Electron. 3 (2020) 554–562.
[48] Y. Xia, G.M. Whitesides, Annu. Rev. Mater. Sci. 28 (1998) 153–184.
[49] D.B. Weibel, W.R. DiLuzio, G.M. Whitesides, Nat. Rev. Microbiol. 5 (2007) 209–218.
[50] Y. Huang, Y. Wang, L. Xiao, H. Liu, W. Dong, Z. Yin, Lab. Chip 14 (2014) 4205–4212.
[51] J. Chen, Y. Zhou, D. Wang, F. He, V.M. Rotello, K.R. Carter, et al., Lab. Chip 15 (2015) 3086–3094.
[52] K.L. Wlodarczyk, D.P. Hand, M.M. Maroto-Valer, Sci. Rep. 9 (2019) 1–13.
[53] S.A.M. Shaegh, A. Pourmand, M. Nabavinia, H. Avci, A. Tamayol, P. Mostafalu, et al., Sensor. Actuat. B Chem. 255 (2018) 100–109.
[54] G. Chitnis, Z. Ding, C.-L. Chang, C.A. Savran, B. Ziaie, Lab. Chip 11 (2011) 1161–1165.
[55] C. Matellan, E. Armando, Sci. Rep. 8 (2018) 1–13.
[56] M. Focke, D. Kosse, C. Müller, H. Reinecke, R. Zengerle, F. von Stetten, Lab. Chip 10 (2010) 1365–1386.
[57] H. Lin, Y. Zhao, S. Lin, B. Wang, C. Yeung, X. Cheng, et al., Lab. Chip 19 (2019) 2844–2853.
[58] E. Eriksson, J. Lysell, H. Larsson, K.Y. Cheung, D. Filippini, W.C. Mak, Research 2019 (2019).
[59] X. Li, J. Tian, T. Nguyen, W. Shen, Anal. Chem. 80 (2008) 9131–9134.
[60] Y. Xu, M. Liu, N. Kong, J. Liu, Microchim. Acta 183 (2016) 1521–1542.
[61] T.D. Ngo, A. Kashani, G. Imbalzano, K.T. Nguyen, D. Hui, Compos. Part B Eng. 143 (2018) 172–196.
[62] M. Padash, C. Enz, S. Carrara, Sensors 20 (2020) 4236.
[63] A.A. Bhanvadia, R.T. Farley, Y. Noh, T. Nishida, Commun. Mater. 2 (2021) 1–7.
[64] N. Bhattacharjee, A. Urrios, S. Kang, A. Folch, Lab. Chip 16 (2016) 1720–1742.
[65] Y. Guo, L. Li, F. Li, H. Zhou, Y. Song, Lab. Chip 15 (2015) 1759–1764.
[66] V. Saggiomo, A.H. Velders, Adv. Sci. 2 (2015) 1500125.
[67] R. Sivakumar, N.Y. Lee, Analyst 145 (2020) 4096–4110.
[68] D.E. Patabadige, S. Jia, J. Sibbitts, J. Sadeghi, K. Sellens, C.T. Culbertson, Anal. Chem. 88 (2016) 320–338.
[69] A.J. Bandodkar, I. Jeerapan, J. Wang, ACS Sens. 1 (2016) 464–482.
[70] S. Coyle, V.F. Curto, F. Benito-Lopez, L. Florea, D. Diamond, Wearable Sensors, Elsevier, 2014, pp. 65–83.
[71] H. Joodaki, M.B. Panzer, Proc. Inst. Mech. Eng. H: J. Eng. Med. 232 (2018) 323–343.
[72] C. Pailler-Mattei, S. Bec, H. Zahouani, Med. Eng. Phys. 30 (2008) 599–606.
[73] Y. Liu, M. Pharr, G.A. Salvatore, ACS Nano 11 (2017) 9614–9635.
[74] K. Raj M, S. Chakraborty, J. Appl. Polym. Sci. 137 (2020) 48958.
[75] T. Akyazi, L. Basabe-Desmonts, F. Benito-Lopez, Anal. Chim. Acta 1001 (2018) 1–17.
[76] D. Agustini, F.R. Caetano, R.F. Quero, J. Fracassi da Silva, M.F. Bergamini, L.H. Marcolino-Junior, et al., Anal. Methods Adv. Methods Appl. 13 (2021) 4830–4857.

[77] B. Heo, M. Fiola, J.H. Yang, A. Koh, Colloid Interface Sci. Commun. 38 (2020) 100301.
[78] Q. Cao, B. Liang, X. Mao, J. Wei, T. Tu, L. Fang, et al., Electroanalysis 33 (2021) 643–651.
[79] R.R. Silva, P.A. Raymundo-Pereira, A.M. Campos, D. Wilson, C.G. Otoni, H.S. Barud, et al., Talanta 218 (2020) 121153.
[80] G. Xiao, J. He, X. Chen, Y. Qiao, F. Wang, Q. Xia, et al., Cellulose 26 (2019) 4553–4562.
[81] Q.-Y. Cheng, M.-C. Liu, Y.-D. Li, J. Zhu, A.-K. Du, J.-B. Zeng, Polym. Test. 66 (2018) 41–47.
[82] S. Xing, J. Jiang, T. Pan, Lab. Chip 13 (2013) 1937–1947.
[83] E. Eslami, R. Jafari, G. Momen, J. Coat. Technol. Res. (2021) 1–24.
[84] A.J. Bandodkar, W.J. Jeang, R. Ghaffari, J.A. Rogers, Annu. Rev. Anal. Chem. 12 (2019) 1–22.
[85] S.M. Ali, G. Yosipovitch, Acta Derm. Venereol. 93 (2013) 261–267.
[86] M.F. Bergeron, J. Sci. Med. Sport 6 (2003) 19–27.
[87] M. Parrilla, R. Cánovas, I. Jeerapan, F.J. Andrade, J. Wang, Adv. Healthc. Mater. 5 (2016) 996–1001.
[88] A.J. Bandodkar, I. Jeerapan, J.-M. You, R. Nuñez-Flores, J. Wang, Nano Lett. 16 (2016) 721–727.
[89] H.Y.Y. Nyein, W. Gao, Z. Shahpar, S. Emaminejad, S. Challa, K. Chen, et al., ACS Nano 10 (2016) 7216–7224.
[90] B. Ma, J. Chi, C. Xu, Y. Ni, C. Zhao, H. Liu, Talanta 212 (2020) 120786.
[91] J. Kim, Y. Wu, H. Luan, D.S. Yang, D. Cho, S.S. Kwak, et al., Adv. Sci. 9 (2021) 2103331.
[92] B. Paul, S. Demuru, C. Lafaye, M. Saubade, D. Briand, Adv. Mater. Technol. 6 (2021) 2000910.
[93] C.K.M. Choy, I.F.F. Benzie, P. Cho, Investig. Ophthalmol. Vis. Sci. 41 (2000) 3293–3298.
[94] J. Stern, W.H.P. Lewis, Clin. Chim. Acta 2 (1957) 576–580.
[95] J. Wang, Y. Niu, C. Zhang, Y. Chen, Food Chem. 245 (2018) 337–345.
[96] M.T. Stauffer, W.E. Weller, K.R. Kubas, K.A. Casoni, Stoichiometry and Research: The Importance of Quantity in Biomedicine, Web of Science, 2012, pp. 311–334.
[97] J. Kim, A.S. Campbell, J. Wang, Talanta 177 (2018) 163–170.
[98] J.M. Green, P.A. Bishop, I.H. Muir, J.R. McLester Jr, H.E. Heath, Int. J. Sports Med. 21 (2000) 556–560.
[99] A. Polliack, R. Taylor, D. Bader, J. Rehabil. Res. Dev. 30 (1993) 250.
[100] L.C. Clark JR, R. Wolf, D. Granger, Z. Taylor, J. Appl. Physiol. 6 (1953) 189–193.
[101] M.Y. Vagin, I. Jeerapan, R. Wannapob, P. Thavarungkul, P. Kanatharana, N. Anwar, et al., Electrochim. Acta 190 (2016) 495–503.
[102] I. Shitanda, M. Mitsumoto, N. Loew, Y. Yoshihara, H. Watanabe, T. Mikawa, et al., Electrochim. Acta 368 (2021) 137620.
[103] J. Kim, I. Jeerapan, S. Imani, T.N. Cho, A. Bandodkar, S. Cinti, et al., ACS Sens. 1 (2016) 1011–1019.
[104] J. Choi, A.J. Bandodkar, J.T. Reeder, T.R. Ray, A. Turnquist, S.B. Kim, et al., ACS Sens. 4 (2019) 379–388.

[105] World Health Organization. (2019). World report on vision. Retrieved from https://www.who.int/publications/i/item/9789241516570.
[106] R. Moreddu, D. Vigolo, A.K. Yetisen, Adv. Healthc. Mater. 8 (2019) 1900368.
[107] Z. Mutlu, S. Shams Es-haghi, M. Cakmak, Adv. Healthc. Mater. 8 (2019) 1801390.
[108] C.S.A. Musgrave, F. Fang, Materials 12 (2019) 261.
[109] J.-C. Chiou, S.-H. Hsu, Y.-C. Huang, G.-T. Yeh, W.-T. Liou, C.-K. Kuei, Sensors 17 (2017) 108.
[110] H. An, L. Chen, X. Liu, B. Zhao, H. Zhang, Z. Wu, Sens. Actuators A Phys. 295 (2019) 177−187.
[111] S. Agaoglu, P. Diep, M. Martini, K. Samudhyatha, M. Baday, I.E. Araci, Lab. Chip 18 (2018) 3471−3483.
[112] N. Jiang, Y. Montelongo, H. Butt, A.K. Yetisen, Small 14 (2018) 1704363.
[113] R. Moreddu, V. Nasrollahi, P. Kassanos, S. Dimov, D. Vigolo, A.K. Yetisen, Small 17 (2021) 2102008.
[114] J. Kim, E. Cha, J.U. Park, Adv. Mater. Technol. 5 (2020) 1900728.
[115] K. Mitsubayashi, T. Arakawa, Electroanalysis 28 (2016) 1170−1187.
[116] R. Moreddu, J.S. Wolffsohn, D. Vigolo, A.K. Yetisen, Sensor. Actuat. B Chem. 317 (2020) 128183.
[117] R. Moreddu, M. Elsherif, H. Adams, D. Moschou, M.F. Cordeiro, J.S. Wolffsohn, et al., Lab. Chip 20 (2020) 3970−3979.
[118] X. Yang, H. Yao, G. Zhao, G.A. Ameer, W. Sun, J. Yang, et al., J. Mater. Sci. 55 (2020) 9551−9561.
[119] B. Mirani, E. Pagan, B. Currie, M.A. Siddiqui, R. Hosseinzadeh, P. Mostafalu, et al., Adv. Healthc. Mater. 6 (2017) 1700718.
[120] C.K. Sen, G.M. Gordillo, S. Roy, R. Kirsner, L. Lambert, T.K. Hunt, et al., Wound Repair Regener. 17 (2009) 763−771.
[121] E. Gianino, C. Miller, J. Gilmore, Bioengineering 5 (2018) 51.
[122] S. O'Callaghan, P. Galvin, C. O'Mahony, Z. Moore, R. Derwin, J. Wound Care 29 (2020) 394−406.
[123] H. Derakhshandeh, S.S. Kashaf, F. Aghabaglou, I.O. Ghanavati, A. Tamayol, Trends Biotechnol. 36 (2018) 1259−1274.
[124] T.R. Dargaville, B.L. Farrugia, J.A. Broadbent, S. Pace, Z. Upton, N.H. Voelcker, Biosens. Bioelectron. 41 (2013) 30−42.
[125] A. McLister, A. Mathur, J. Davis, Curr. Opin. Electrochem. 3 (2017) 40−45.
[126] A. Scalamandré, K.M. Bogie, Innovations and Emerging Technologies in Wound Care, Elsevier, 2020, pp. 225−244.
[127] P. Escobedo, M. Bhattacharjee, F. Nikbakhtnasrabadi, R. Dahiya, IEEE Internet Things J. 8 (2020) 5093−5100.
[128] H. Ota, K. Chen, Y. Lin, D. Kiriya, H. Shiraki, Z. Yu, et al., Nat. Commun. 5 (2014) 1−9.
[129] Y. Gao, D.T. Nguyen, T. Yeo, S.B. Lim, W.X. Tan, L.E. Madden, et al., Sci. Adv. 7 (2021) eabg9614.
[130] B. Yang, J. Kong, X. Fang, Talanta 204 (2019) 685−692.
[131] S. Babikian, G. Li, M. Bachman (2015, May). In *2015 IEEE 65th Electronic Components and Technology Conference (ECTC)* (pp. 685-690). San Diego, CA, USA.

[132] N. Kamaly, B. Yameen, J. Wu, O.C. Farokhzad, Chem. Rev. 116 (2016) 2602–2663.
[133] S. Bagherifard, A. Tamayol, P. Mostafalu, M. Akbari, M. Comotto, N. Annabi, et al., Adv. Healthc. Mater. 5 (2016) 175–184.
[134] Q. Pang, D. Lou, S. Li, G. Wang, B. Qiao, S. Dong, et al., Adv. Sci. 7 (2020) 1902673.
[135] B. Qiao, Q. Pang, P. Yuan, Y. Luo, L. Ma, Biomater. Sci. 8 (2020) 1649–1657.
[136] S.T. Sanjay, W. Zhou, M. Dou, H. Tavakoli, L. Ma, F. Xu, et al., Adv. Drug Deliv. Rev. 128 (2018) 3–28.
[137] R. Riahi, A. Tamayol, S.A.M. Shaegh, A.M. Ghaemmaghami, M.R. Dokmeci, A. Khademhosseini, Curr. Opin. Chem. Eng. 7 (2015) 101–112.
[138] B. Ziaie, A. Baldi, M. Lei, Y. Gu, R.A. Siegel, Adv. Drug Deliv. Rev. 56 (2004) 145–172.
[139] P. Rajput, Turk. J. Comput. Math. Educ. 12 (2021) 1650–1662.
[140] H. Derakhshandeh, F. Aghabaglou, A. McCarthy, A. Mostafavi, C. Wiseman, Z. Bonick, et al., Adv. Funct. Mater. 30 (2020) 1905544.
[141] M. Ochoa, R. Rahimi, T.L. Huang, N. Alemdar, A. Khademhosseini, M.R. Dokmeci, et al., Sensor. Actuat. B Chem. 198 (2014) 472–478.
[142] S.G. Yoon, B.J. Park, S.T. Chang, Sens. Actuators A Phys. 254 (2017) 1–8.
[143] J.C. Yeo, J. Yu, Z.M. Koh, Z. Wang, C.T. Lim, Lab. Chip 16 (2016) 3244–3250.
[144] J.C. Yeo, J. Yu, M. Shang, K.P. Loh, C.T. Lim, Small 12 (2016) 1593–1604.
[145] M. Bhattacharjee, M. Soni, P. Escobedo, R. Dahiya, Adv. Electron. Mater. 6 (2020) 2000445.
[146] F. Wu, S. Chen, B. Chen, M. Wang, L. Min, J. Alvarenga, et al., Small 14 (2018) 1702170.
[147] S.G. Yoon, S.T. Chang, J. Mater. Chem. C. 5 (2017) 1910–1919.
[148] J.C. Yeo, J. Yu, K.P. Loh, Z. Wang, C.T. Lim, ACS Sens. 1 (2016) 543–551.
[149] J. Yan. (2011, January). *In 2011 Annual International Conference of the IEEE Engineering in Medicine and Biology Society* (pp. 2522-2525). Boston, MA, USA.
[150] F. Shen, M. Ai, Z. Li, X. Lu, Y. Pang, Z. Liu, Microfluid. Nanofluidics 25 (2021) 1–31.
[151] J. Chen, J. Zhang, Z. Luo, J. Zhang, L. Li, Y. Su, et al., ACS Appl. Mater. Interfaces 12 (2020) 22200–22211.
[152] S. Oren, H. Ceylan, L. Dong, IEEE Sens. Lett. 1 (2017) 1–4.
[153] S. Agarwala, G.L. Goh, Y.L. Yap, G.D. Goh, H. Yu, W.Y. Yeong, et al., Sens. Actuators A Phys. 263 (2017) 593–599.
[154] F. Umbrecht, P. Wägli, S. Dechand, F. Gattiker, J. Neuenschwander, U. Sennhauser, et al., J. Micromech. Microeng. 20 (2010) 085005.
[155] Y. Huang, X. Fan, S.C. Chen, N. Zhao, Adv. Funct. Mater. 29 (2019) 1808509.
[156] Y. Gao, H. Ota, E.W. Schaler, K. Chen, A. Zhao, W. Gao, et al., Adv. Mater. 29 (2017) 1701985.
[157] B. Nie, R. Li, J.D. Brandt, T. Pan, Lab. Chip 14 (2014) 4344–4353.
[158] Z. Hao, D. Wang, L. Reynolds, Cardiovasc. Eng. Technol. 11 (2020) 295–307.
[159] I.E. Araci, B. Su, S.R. Quake, Y. Mandel, Nat. Med. 20 (2014) 1074–1078.
[160] B. Nie, R. Li, J.D. Brandt, T. Pan, Lab. Chip 14 (2014) 1107–1116.
[161] D.H. Ho, R. Song, Q. Sun, W.-H. Park, S.Y. Kim, C. Pang, et al., ACS Appl. Mater. Interfaces 9 (2017) 44678–44686.

[162] H.R. Lim, H.S. Kim, R. Qazi, Y.T. Kwon, J.W. Jeong, W.H. Yeo, Adv. Mater. 32 (2020) 1901924.
[163] L. Meng, A.P.F. Turner, W.C. Mak, Biosens. Bioelectron. 159 (2020) 112181.
[164] K. Promsuwan, L. Meng, P. Suklim, W. Limbut, P. Thavarungkul, P. Kanatharana, et al., ACS Appl. Mater. Interfaces 12 (2020) 39841–39849.
[165] L. Meng, A.P.F. Turner, W.C. Mak, ACS Appl. Mater. Interfaces 13 (2021) 54456–54465.
[166] A. Bolotsky, D. Butler, C. Dong, K. Gerace, N.R. Glavin, C. Muratore, et al., ACS Nano 13 (2019) 9781–9810.

CHAPTER

Advanced techniques for manufacturing paper-based microfluidic analytical devices

5

Aroosha Faheem[1] and Stefano Cinti[2,3]

[1]*State Key Laboratory of Agricultural Microbiology, College of Life Sciences, Huazhong Agricultural University, Wuhan, P.R. China*

[2]*Department of Pharmacy, University of Naples "Federico II", Naples, Italy*

[3]*BAT Center—Interuniversity Center for Studies on Bioinspired Agro-Environmental Technology, University of Napoli "Federico II", Naples, Italy*

5.1 Introduction

Diagnostic testing has become increasingly popular for diagnosing and monitoring analytes and is utilized in various ways for prognoses and predicting the treatment of multiple diseases. These tests must be conducted under stringent conditions to ensure stable and reproducible results. Compared to traditional wet bench methods, microfluidic analytical device (μFD) based detection has significant advantages of performing analysis in less time, with minimal use of reagents and sample volume. Thus the development of μFDs is an important area of research in the field of diagnosis in clinical, food, water, and environment monitoring. In 2003, the criteria for an ideal test to be used at all healthcare systems in underdeveloped countries were published by the World Health Organization's Programme for Research and Training in Tropical Diseases (WHO-TDR). The aim of the initiative is to aid in policy-making regarding clinical management and treatment of tropical diseases and sexually transmitted infections [1]. The important aspects to consider for an efficient diagnostic test were coined as ASSURED, which means affordable, sensitive, specific, user-friendly, rapid and robust, equipment-free, and deliverable to end-users. Since then, ASSURED criteria had become a standard benchmark for identifying and developing an appropriate test in a resource-deficient environment for point-of-care testing (POCT).

Generally, μFDs were developed with glasses or polymers where microfluidic channels were imprinted with various approaches. Recently, novel strategies were successfully employed to develop μFDs using low-temperature co-fired ceramics [2], silicon [3], and paper [4]. Among these, paper-based analytical devices (μPADs) have captured significant attention owing to ease of use. In 2007,

Microfluidic Biosensors. DOI: https://doi.org/10.1016/B978-0-12-823846-2.00009-2

μPADs were introduced for the first time by Whitesides and colleagues [5]. These devices provide an alternative platform for excellent POCT and disease screening using portable, affordable, disposable, environment-friendly, and economical diagnostic tools that require less sample volume. With the modifications in materials and fabrication methods, these devices can not only provide the capacity to detect multiple analytes but also produce quantitative results to some extent. Hence, the applications of μPADs have a high significance in developing countries, particularly in areas with underdeveloped or low infrastructure facilities for healthcare. The μPADs are typically made of paper with hydrophilic areas enclosed by hydrophobic barriers, which are usually made of polystyrene, photoresist, wax, ultraviolet (UV) polymers, or other materials with similar characteristics. They facilitate in regulating capillary fluid flow in the hydrophilic channels. Several techniques are reported in different studies for the fabrication of microfluidic channels in μPADs [6]. Out of these, wax printing, photolithography, plasma treatment, screen printing, and inkjet printing have been the most prominent ones.

Many μPADs have been introduced to monitor target analytes in biological samples, food [7], water [8] and environment, such as pollutants [9], explosives [10], pesticide residues [11], and metal ions [12]. In addition to the above-mentioned applications, these devices have high potential for use in the healthcare sector as μPADs have been widely utilized for qualitative and semiquantitative diagnostic assays targeting various analytes, including metabolites [13], biomarkers [14], proteins [15], nucleic acids [16] and even pathogens [17]. However, the commercial applications have certain limitations, which need to be overcome considering the large-scale production, specificity, sensitivity, and stability of assay. μPADs are integrated with a number of detection approaches like colorimetric, fluorescence, chemiluminescence, electrochemical, and others [18,19]. Yet, colorimetric detection is a popular method capable of on-site analysis of multiplexed samples in a simple and cost-effective way without requiring a specific instrument. Electrochemical analysis, on the other hand, provides robust detection with high sensitivity compared to the former approach. Nonetheless, the applications and prospects of both approaches are discussed later in this chapter.

5.2 Methods for patterning paper

The μPADs can be fabricated by using two-dimensional (2D) and three-dimensional (3D) fabrication methods enabling transfer of fluids in both vertical and horizontal directions based on the design and application of the diagnostic test. A brief introduction to a few, most commonly used 2D patterning methods is given below. Fig. 5.1 displays the most common methods used.

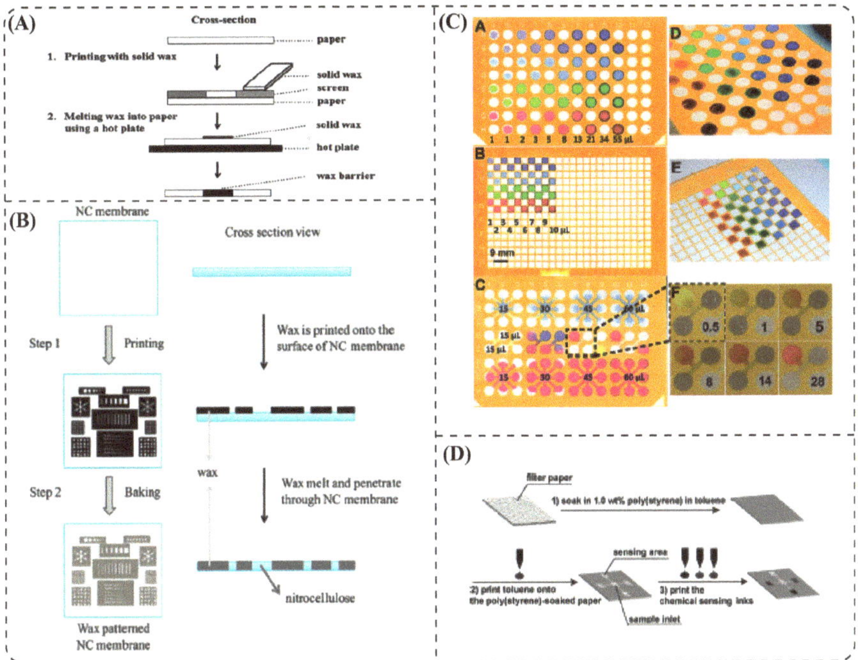

FIGURE 5.1

Schematic illustration of paper patterning methods: (A) screen-printing; (B) wax printing; (C) photolithography; (D) inkjet printing.

(A) Reprinted from W. Dungchai, O. Chailapakul, C.S. Henry, A low-cost, simple, and rapid fabrication method for paper-based microfluidics using wax screen-printing, Analyst 136(1) (2011) 77–82. https://doi.org/10.1039/c0an00406e. ©2020, Analytical Chemistry; (B) Reprinted from Y. Lu, W. Shi, J. Qin, B. Lin, Fabrication and characterization of paper-based microfluidics prepared in nitrocellulose membrane by Wax printing, Anal. Chem. 82(1) (2010) 329–335. https://doi.org/10.1021/ac9020193. ©2010, Analytical Chemistry; (C) Reprinted from E. Carrilho, S.T. Phillips, S.J. Vella, A.W. Martinez, G.M. Whitesides, Paper microzone plates, Anal. Chem. 81(15) (2009) 5990–5998. https://doi.org/10.1021/ac900847g. ©2009, Analytical Chemistry; and (D) Reprinted from K. Abe, K. Suzuki, D. Citterio, Inkjet-printed microfluidic multianalyte chemical sensing paper, Anal. Chem. 80(18) (2008) 6928–6934. https://doi.org/10.1021/ac800604v. ©2008, Analytical Chemistry.

5.2.1 Screen printing

Screen printing is an old method that involves fabricating stenciled patterns using a mesh screen and ink on a flat surface (Fig. 5.1A). A study reported wax screen printing on paper-based analytical devices [20]. The inks used to manufacture electrodes are usually composed of platinum, gold, silver, carbon, or any other material depending on the assay's selectivity and sensitivity. A typical screen-printed electrode system consists of a working electrode for analyte concentration, a reference electrode to apply known potential, and a counter-electrode to pass current and complete the circuit [21]. Screen-printed electrodes combined with microfluidics provide an excellent platform for the electrochemical determination of analytes.

5.2.2 Plasma treatment

Plasma treatment is an emerging paper fabrication approach. Paper is hydrophobized by octadecyltrichlorosilane (OTS) silanization and then plasma-treated with a channeled mask at various sections depending on the pattern . Due to the degradation of OTS by cellulose fibers of the paper, the area exposed to plasma acts as a hydrophilic channel [22]. The dry nature of plasma treatment offers the advantage of engraving complex functional components. One of the known challenges with this approach is that the pattern formed is larger than the mask because of the substrate's outstretching. Nonetheless, this problem can be overcome by optimizing the intensity and time of treatment to obtain the desired channel size.

5.2.3 Wax printing

One of the most widely used high-throughput fabrication methods is wax printing; this method works based on a commercial printer and heating source to construct complicated 2D or 3D structures on paper and is extremely suitable for efficient bulk production (Fig. 5.1B) [23,24] [25]. The three most common wax patterning ways are printing with a wax printer, painting with a wax pen, or printing with an inkjet printer accompanied by painting with a wax pen. The printing process takes no longer than 10 minutes and does not require special reagents and a UV lamp. Thus wax printing has cheap, simple, handy, and eco-friendly applications as wax and paper can be comfortably disposed of by burning.

5.2.4 Photolithography

Photolithography, also known as optical lithography, is another method used to fabricate patterns on paper. A light beam is used to transfer the microfabricated pattern from a photomask to a light-sensitive chemical photoresist (Fig. 5.1C) [26]. Compared to wax printing, photolithography is expensive as it requires specific equipment, costly photoresists, and organic solvents. There are also affordable photoresists that take up to 15 minutes for rapid prototyping, but photolithographically defined channels demonstrate a high background.

5.2.5 Inkjet printing

Inkjet printing is a kind of computer printing where digital images are created by driving ink droplets on plastic or paper substrates in an inkjet printer (Fig. 5.1D) [27]. The printer can fabricate reagents and biomolecules in no time to form precise patterns, yielding particular biological or chemical sensing areas. Inkjet printing offers promising features enabling the development of efficient, economical, paper-based microfluidic devices for commercial applications.

5.3 Application of microfluidics in diagnostics

5.3.1 Colorimetric-based application

The most commonly used method in the application of μPADs is colorimetric detection. The detection approach depends on identifying colored compounds generated by unknown to known (controls) analytes. The assay is easy to perform and provides visual on-site determination and digital off-site quantitative analysis recorded by scanners and cameras according to the intensity of color pigment. A prominent and widely practiced application of this system was reported as developing a blood typing device. The principle behind the working of a text reporting blood typing device is hemagglutination reactions at hydrophilic and hydrophobic fabricated sites to use these text patterns as sites for interaction of RBC and antibodies. The fabricated text patterns are made hydrophilic, and antibody solutions are introduced to corresponding text patterns. The immunoreaction generates text patterns of agglutinated blood that report all blood types in the RhD and ABO systems (Fig. 5.2A). This reporting method significantly enhances the device's capacity to report the results unambiguously to nonprofessional users [28]. A two-layered, 3D yet simple μPAD was developed to detect six metal ions simultaneously from environmental samples. The top layer was used for pretreatment purposes; the samples were added to a reservoir, which was separated into eight channels to carry out the corresponding pretreatment reactions while passing through the relevant zones. The bottom layer, on the other hand, played a role in colorimetric detection (Fig. 5.2B). Hence, double-layered μPADs quantified Fe (III), Ni(II), Cr(VI), Cu(II), Al(III), and Zn(II) with sensitivity reaching as low as 0.2, 0.3, 0.1, 0.03, 0.08, and 0.04 mg/L, respectively [29]. This approach recorded one order of magnitude lower detection limits than previously reported studies for the determination of metal ions.

Recently, a μPAD was designed for multiplex determination of three analytes. The detection zones were treated with various enzymes, substrates, and chromogenic reagents. On adding samples, the enzyme-catalyzed reactions produced distinct colored products. The assay provides an excellent opportunity for qualitative and quantitative analysis as the chromogenic reagents and corresponding substrates turned brown for sucrose, blue for fructose, and pink for glucose (Fig. 5.2C). The color intensity recorded by the computer reported detection limits of 0.9, 0.8, and 0.4 mM for sucrose, fructose, and glucose, respectively. Hence, the results are in agreement with those obtained by traditional HPLC analysis with additional advantages of being easy, inexpensive, and capable of rapid analysis [30]. Later, centrifugal microfluidics presented a novel strategy depending on the application of

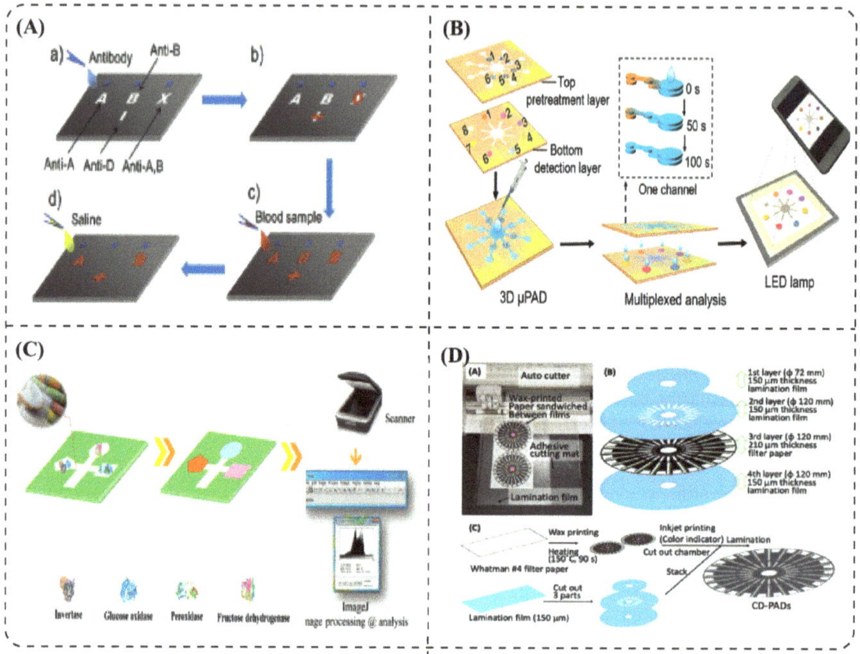

FIGURE 5.2

(A) Fabrication and analysis of the text-reporting blood-typing colorimetric devices. (B) Multiplexed colorimetric detection of Fe(III), Ni(II), Cr(VI), Cu(II), Al(III), and Zn(II) by novel double-layered 3D fabricated μPADs. (C) Simultaneous determination of sucrose, fructose, and glucose using a μPAD. (D) An overview of the fabrication process of centrifugal PADs on CDs. CD, Centrifugally assisted distance-based. *μPAD*, Paper-based analytical device.

(A) Reprinted from M. Li, J. Tian, M. Al-Tamimi, W. Shen, Paper-based blood typing device that reports patient's blood type "in Writing", Angew. Chem. 124(22) (2012) 5593–5597. https://doi.org/10.1002/ange.201201822. ©2012, Angewandte Chemie. (B) Reprinted from F. Li, Y. Hu, Z. Li, J. Liu, L. Guo, J. He, Three-dimensional microfluidic paper-based device for multiplexed colorimetric detection of six metal ions combined with use of a smartphone, Anal. Bioanal. Chem. 411(24) (2019) 6497–6508. https://doi.org/10.1007/s00216-019-02032-5. ©2019, Analytical and Bioanalytical Chemistry. (C) Reprinted from J. Aksorn, S. Teepoo, Development of the simultaneous colorimetric enzymatic detection of sucrose, fructose and glucose using a microfluidic paper-based analytical device, Talanta 207(June 2019) (2020) 120302. https://doi.org/10.1016/j.talanta.2019.120302. ©2020, Talanta. (d) Reprinted from K. Maejima, Y. Hiruta, D. Citterio, Centrifugal paperfluidic platform for accelerated distance-based colorimetric signal readout, Anal. Chem. 92(7) (2020) 4749–4754. https://doi.org/10.1021/acs.analchem.9b05782. ©2020, Analytical Chemistry.

centrifugal forces to regulate the flow of samples into the microfluidic channels irrespective of external pumping. These devices made of filter paper and lamination films provide a centrifugal platform for distance readout analytical devices.

Maejima and co-workers have introduced a colorimetric assay to determine Ni^{2+} with naked-eye interpretation in 1.5 minutes (Fig. 5.2D). The assay was most sensitive to the reporting limit of detection of 44.1 ng of Ni^{2+}. Centrifugal microfluidics offers real-time, sample volume-independent analysis, particularly by a distance readout-based detection motif [31].

5.3.2 Electrochemical-based application

Microfluidic paper-based electrochemical detection has recently attracted a lot of interest due to the collective benefits of high sensitivity, specificity, and broad working range from the luminescence to electrochemical assays. Electrochemical detection using paper-based microfluidic devices being spatially controlled has captured the researcher's attention due to the flexibility offered by these devices in terms of their size, portability, and sensitivity depending on electrode material and detection capability. Paper-based electrochemical devices need a simple instrument consuming less electrical power for field applications. Typical strategies currently used to assemble an electrode in the paper microfluidic channel either use solid electrodes such as metal wire and pencil graphite or employ external electrodes from the dispensable electrochemical cell. The electrodes can be easily integrated on paper by drawing with a pencil or by printing using a screen or inkjet printer, as mentioned earlier. In 2009, Dungchai and co-workers presented electrochemical μPADs for simultaneous detection of glucose, lactate, and uric acid using oxidases (Fig. 5.3A). At first, the electrodes were treated with Prussian Blue (PB) pigment to enhance the selectivity for H_2O_2 since many biosensors depend on the level of detection of peroxide for quantizing analytes. Later, specific enzymes to recognize biomarkers of glucose, lactate, and uric acid from human samples were subjected to the previously modified electrodes [32]. Another paper-based screen-printed electrochemical device was developed for biosensing Paraoxon nerve agent stimulant. The assay was based on the measurement of butyrylcholinesterase enzyme activity to catalyze butyrylthiocholine, and parallel assessments were made irrespective of contaminated samples for the validation of the approach (Fig. 5.3B). Likewise, the sensitivity of the device was enhanced by modifying electrodes with PB or carbon black nanocomposites. The concentration of Paraoxon was successfully identified with a limit of detection, reaching down to 3 μg/L [33]. Overall, the proposed technique easily outperforms and comes handy for use by nonprofessionals, who tend to expand the application of μPADs.

Similar work has also been pursued by [17,34] for simultaneous quantitation of lung cancer tumor biomarkers for early diagnosis and treatment. For this purpose, amino functional graphene–thionin gold nanoparticles and PB-poly (3,4-ethylenedioxythiophene) (PEDOT)–gold nanocomposites were synthesized to modify electrodes. These composites significantly enhanced the electron transfer rate and also provides binding sites for carcinoembryonic antigen (CEA) and neuron-specific enolase (NSE) aptamers (Fig. 5.3C). Hence, this label-free

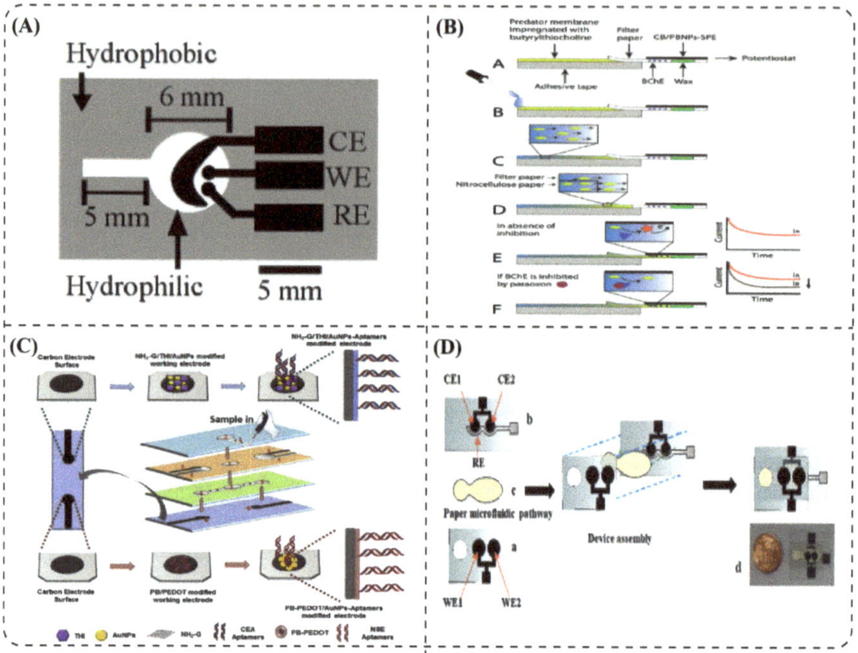

FIGURE 5.3

(A) Basic design of µPADs for electrochemical detection of glucose, lactate, and uric acid. (B) Schematic illustration of a lab-on-chip device for the quantification of neural stimulant. (C) Fabrication of electrochemical µPADs for simultaneous detection of lung cancer biomarkers. (D) Schematic illustration of screen-printed working electrodes designed to detect creatinine and uric acid. *µPADs*, Paper-based analytical devices.

(A) Reprinted from W. Dungchai, O. Chailapakul, C.S. Henry, Electrochemical detection for paper-based microfluidics, Anal. Chem. 81(14) (2009) 5821–5826. https://doi.org/10.1021/ac9007573. ©2009, Analytical Chemistry. (B) Reprinted from S. Cinti, C. Minotti, D. Moscone, G. Palleschi, F. Arduini, Fully integrated ready-to-use paper-based electrochemical biosensor to detect nerve agents, Biosens. Bioelectron. 93(November 2016) (2017) 46–51. https://doi.org/10.1016/j.bios.2016.10.091. ©2017, Biosensors and Bioelectronics. (C) Reprinted from Y. Wang, J. Luo, J. Liu, S. Sun, Y. Xiong, Y. Ma, et al., Label-free microfluidic paper-based electrochemical aptasensor for ultrasensitive and simultaneous multiplexed detection of cancer biomarkers, Biosens. Bioelectron. 136(February) (2019) 84–90. https://doi.org/10.1016/j.bios.2019.04.032. ©2019, Biosensors and Bioelectronics. (D) Reprinted from F.H. Cincotto, E.L. Fava, F.C. Moraes, O. Fatibello-Filho, R.C. Faria, A new disposable microfluidic electrochemical paper-based device for the simultaneous determination of clinical biomarkers, Talanta 195(November 2018) (2019) 62–68. https://doi.org/10.1016/j.talanta.2018.11.022. ©2019, Biosensors and Bioelectronics.

electrochemical approach provides robust POCT with a sensitivity reaching 2 and 10 pg/mL for CEA and NSE, respectively. The assay yielded a notable correlation coefficient with electrochemiluminescence equipment on validation by clinical serum samples. Hence, this system offers a new way for early diagnosis of diseases like cancer, particularly in resource-constrained localities [34]. Afterward, in the same year, another dispensable μPAD was developed for square wave voltammetry based simultaneous electrochemical detection of uric acid creatinine. The device consists of two spot sensors, the first one being modified with graphene quantum dots was used for the oxidation of uric acid. In contrast, the second spot's surface was modified with quantum dots, creatininase, and ruthenium, which act as electrochemical mediators for the oxidation of creatinine (Fig. 5.3D). The novel μPADs demonstrated a remarkable linear working range from 0.010 to 3.0 μmol/L; the analytical device achieved a high sensitivity of 8.4 nmol/L for uric acid and 3.7 nmol/L for creatinine with high precision [35]. Hence, a combination of electrochemical detection with microfluidics offers a low-cost, portable device for analysis using a simple instrument, with supplementary benefits of high specificity and sensitivity.

5.4 Conclusion

With rapid growth in population worldwide, the need to develop affordable, robust, and user-friendly diagnostic sensors for disease screening, and monitoring food, water, and environment quality is the need of the hour. In this context, the significance of μPADs in diagnostics and applications for ASSURED POCT was briefly discussed in this chapter. μPADs have several advantages associated with them: the operation of devices is easy for routine testing and do not require trained professionals, testing requires only a few microliters of sample, multiple analytes can be detected simultaneously, and it does not require specific conditions and infrastructure, and, thus can be integrated for early diagnosis of diseases and infections by the healthcare system of developing countries. The devices are portable, lightweight, energy-efficient, and low-cost testing tools suitable for worldwide acceptance and affordability. However, research to enhance the sensitivity, selectivity, and specificity of μPADs is currently at an early stage, and efforts are needed to make μPADs genuinely useful for commercial applications and to expand their capabilities for POCT. Hence, future work should include research on the structure of the paper, pattern fabrication, printing technologies, and tools for retaining samples on the surface, which may prove highly beneficial.

Acknowledgments

Stefano Cinti acknowledges the MIUR Grant "Dipartimento di Eccellenza 2018–22" to the Department of Pharmacy of University of Naples "Federico II."

References

[1] D. Mabey, R.W. Peeling, A. Ustianowski, M.D. Perkins, Diagnostics for the developing world, Nat. Rev. Microbiol. 2 (3) (2004) 231–240. Available from: https://doi.org/10.1038/nrmicro841.

[2] J. Luo, T. Dziubla, R. Eitel, A low temperature co-fired ceramic based microfluidic Clark-type oxygen sensor for real-time oxygen sensing, Sens. Actuators B: Chem. 240 (2017) 392–397.

[3] Z. Qi, L. Xu, Y. Xu, J. Zhong, A. Abedini, X. Cheng, et al., Disposable silicon-glass microfluidic devices: precise, robust and cheap, Lab Chip 18 (24) (2018) 3872–3880.

[4] T. Tian, Y. Bi, X. Xu, Z. Zhu, C. Yang, Integrated paper-based microfluidic devices for point-of-care testing, Anal. Methods 10 (29) (2018) 3567–3581.

[5] A.W. Martinez, S.T. Phillips, M.J. Butte, G.M. Whitesides, Patterned paper as a platform for inexpensive, low-volume, portable bioassays, Angew. Chem. 119 (8) (2007) 1340–1342.

[6] Y. Xia, J. Si, Z. Li, Fabrication techniques for microfluidic paper-based analytical devices and their applications for biological testing: a review, Biosens. Bioelectron. 77 (2016) 774–789. Available from: https://doi.org/10.1016/j.bios.2015.10.032.

[7] S. He, N. Joseph, S. Feng, M. Jellicoe, C.L. Raston, Application of microfluidic technology in food processing, Food Funct. 11 (7) (2020) 5726–5737.

[8] O. Crahay, J.-P. Raskin, Towards the design of a paper-based microfluidic sensor for water quality monitoring and comparative life cycle assessment, Master Thesis UCLouvain, 2020.

[9] J. Qi, B. Li, X. Wang, L. Fu, L. Luo, L. Chen, Rotational paper-based microfluidic-chip device for multiplexed and simultaneous fluorescence detection of phenolic pollutants based on a molecular-imprinting technique, Anal. Chem. 90 (20) (2018) 11827–11834.

[10] M.J. Kangas, R.M. Burks, J. Atwater, R.M. Lukowicz, P. Williams, A.E. Holmes, Colorimetric sensor arrays for the detection and identification of chemical weapons and explosives, Crit. Rev. Anal. Chem. 47 (2) (2017) 138–153.

[11] L. Jin, Z. Hao, Q. Zheng, H. Chen, L. Zhu, C. Wang, et al., A facile microfluidic paper-based analytical device for acetylcholinesterase inhibition assay utilizing organic solvent extraction in rapid detection of pesticide residues in food, Anal. Chim. Acta 1100 (2020) 215–224.

[12] J.P. Devadhasan, J. Kim, A chemically functionalized paper-based microfluidic platform for multiplex heavy metal detection, Sens. Actuators B: Chem. 273 (2018) 18–24.

[13] Q. Cao, B. Liang, T. Tu, J. Wei, L. Fang, X. Ye, Three-dimensional paper-based microfluidic electrochemical integrated devices (3D-PMED) for wearable electrochemical glucose detection, RSC Adv. 9 (10) (2019) 5674–5681.

[14] E.L. Rossini, M.I. Milani, E. Carrilho, L. Pezza, H.R. Pezza, Simultaneous determination of renal function biomarkers in urine using a validated paper-based microfluidic analytical device, Anal. Chim. Acta 997 (2018) 16–23.

[15] L. Cao, C. Fang, R. Zeng, X. Zhao, Y. Jiang, Z. Chen, based microfluidic devices for electrochemical immunofiltration analysis of human chorionic gonadotropin, Biosens. Bioelectron. 92 (2017) 87–94.

[16] L. Magro, B. Jacquelin, C. Escadafal, P. Garneret, A. Kwasiborski, J.-C. Manuguerra, et al., based RNA detection and multiplexed analysis for Ebola virus diagnostics, Sci. Rep. 7 (1) (2017) 1–9.

[17] C.-H. Wang, J.-J. Wu, G.-B. Lee, Screening of highly-specific aptamers and their applications in paper-based microfluidic chips for rapid diagnosis of multiple bacteria, Sens. Actuators B: Chem. 284 (2019) 395–402.

[18] Paroloi, C. Parolo, A. Sena-Torralba, J.F. Bergua, E. Calucho, C. Fuentes-Chust, et al., Tutorial: design and fabrication of nanoparticle-based lateral-flow immunoassays, Nat. Protoc. 15 (12) (2020) 3788–3816. Available from: https://doi.org/10.1038/s41596-020-0357-x.

[19] S. Cinti, D. Moscone, F. Arduini, Preparation of paper-based devices for reagentless electrochemical (bio) sensor strips, Nat. Protoc. 14 (8) (2019) 2437–2451. Available from: https://doi.org/10.1038/s41596-019-0186-y.

[20] W. Dungchai, O. Chailapakul, C.S. Henry, A low-cost, simple, and rapid fabrication method for paper-based microfluidics using wax screen-printing, Analyst 136 (1) (2011) 77–82. Available from: https://doi.org/10.1039/c0an00406e.

[21] S. Cinti, et al., Preparation of paper-based devices for reagentless electrochemical (bio)sensor strips, Nature Protocols 14 (8) (2019) 2437–2451. Available from: https://doi.org/10.1038/s41596-019-0186-y.

[22] C. Yan, S. Yu, Y. Jiang, Q. He, H. Chen, Fabrication of paper-based microfluidic devices by plasma treatment and its application in glucose determination, Acta Chim. Sin. 72 (10) (2014) 1099–1104. Available from: https://doi.org/10.6023/A14060496.

[23] A. Miglione, et al., Sustainable Copper Electrochemical Stripping onto a Paper-Based Substrate for Clinical Application, ACS Measurement Science Au 2 (2) (2022) 177–184. Available from: https://doi.org/10.1021/acsmeasuresciau.1c00059.

[24] A. Raucci, et al., A Hybrid Screen-Printed Strip for Enhanced Electroanalysis towards Lead and Cadmium in Multi-Matrices, Journal of The Electrochemical Society 169 (3) (2022) 037516. Available from: https://doi.org/10.1149/1945-7111/ac5c98.

[25] Y. Lu, W. Shi, J. Qin, B. Lin, Fabrication and characterization of paper-based microfluidics prepared in nitrocellulose membrane by Wax printing, Anal. Chem. 82 (1) (2010) 329–335. Available from: https://doi.org/10.1021/ac9020193.

[26] E. Carrilho, S.T. Phillips, S.J. Vella, A.W. Martinez, G.M. Whitesides, Paper microzone plates, Anal. Chem. 81 (15) (2009) 5990–5998. Available from: https://doi.org/10.1021/ac900847g.

[27] K. Abe, K. Suzuki, D. Citterio, Inkjet-printed microfluidic multianalyte chemical sensing paper, Anal. Chem. 80 (18) (2008) 6928–6934. Available from: https://doi.org/10.1021/ac800604v.

[28] M. Li, J. Tian, M. Al-Tamimi, W. Shen, Paper-based blood typing device that reports patient's blood type "in Writing, Angew. Chem. 124 (22) (2012) 5593–5597. Available from: https://doi.org/10.1002/ange.201201822.

[29] F. Li, Y. Hu, Z. Li, J. Liu, L. Guo, J. He, Three-dimensional microfluidic paper-based device for multiplexed colorimetric detection of six metal ions combined with use of a smartphone, Anal. Bioanal. Chem. 411 (24) (2019) 6497–6508. Available from: https://doi.org/10.1007/s00216-019-02032-5.

[30] J. Aksorn, S. Teepoo, Development of the simultaneous colorimetric enzymatic detection of sucrose, fructose and glucose using a microfluidic paper-based analytical

device, Talanta 207 (June 2019) (2020) 120302. Available from: https://doi.org/10.1016/j.talanta.2019.120302.

[31] K. Maejima, Y. Hiruta, D. Citterio, Centrifugal paperfluidic platform for accelerated distance-based colorimetric signal readout, Anal. Chem. 92 (7) (2020) 4749–4754. Available from: https://doi.org/10.1021/acs.analchem.9b05782.

[32] W. Dungchai, O. Chailapakul, C.S. Henry, Electrochemical detection for paper-based microfluidics, Anal. Chem. 81 (14) (2009) 5821–5826. Available from: https://doi.org/10.1021/ac9007573.

[33] S. Cinti, C. Minotti, D. Moscone, G. Palleschi, F. Arduini, Fully integrated ready-to-use paper-based electrochemical biosensor to detect nerve agents, Biosens. Bioelectron. 93 (November 2016) (2017) 46–51. Available from: https://doi.org/10.1016/j.bios.2016.10.091.

[34] Y. Wang, J. Luo, J. Liu, S. Sun, Y. Xiong, Y. Ma, et al., Label-free microfluidic paper-based electrochemical aptasensor for ultrasensitive and simultaneous multiplexed detection of cancer biomarkers, Biosens. Bioelectron. 136 (February) (2019) 84–90. Available from: https://doi.org/10.1016/j.bios.2019.04.032.

[35] F.H. Cincotto, E.L. Fava, F.C. Moraes, O. Fatibello-Filho, R.C. Faria, A new disposable microfluidic electrochemical paper-based device for the simultaneous determination of clinical biomarkers, Talanta 195 (November 2018) (2019) 62–68. Available from: https://doi.org/10.1016/j.talanta.2018.11.022.

CHAPTER

Digital microfluidic biosensors

6

Yuye Wang[1,2,3], Shuwen Zeng[2,4,5], Georges Humbert[2] and Aaron Ho Pui Ho[1]

[1]Department of Biomedical Engineering, The Chinese University of Hong Kong, Shatin, Hong Kong

[2]CNRS, XLIM Research Institute, UMR 7252, University of Limoges, Limoges, France

[3]Laboratory of Biomedical Microsystems and Nano Devices, Bionic Sensing and Intelligence Center, Institute of Biomedical and Health Engineering, Shenzhen Institute of Advanced Technology, Chinese Academy of Sciences, Shenzhen, P.R. China

[4]Light, Nanomaterials and Nanotechnologies (L2n), CNRS-ERL 7004, Université de Technologie de Troyes, Troyes, France

[5]Department of Applied Physics and Applied Mathematics, Columbia University, New York, NY, United States

6.1 Introduction

Digital microfluidics (DMF) is a novel technology with the capability of individual droplet manipulation [1–4]. Since the precise definition of DMF may differ in literature [5–7], we only focus on studying and reviewing the recent progress made in integrated DMF systems in which droplet manipulation is carried out on an array of electrodes. In these systems, the transporting, mixing, splitting, and dispensing of individual droplets can be carried out through applying a series of potentials to the electrodes. DMF systems can provide numerous advantages such as reduced reagent consumption and reaction time, fast heat transfer as well as ease of integration. These systems can also effectively reduce cross-contamination due to the use of individual droplets as microreactors. Moreover, the DMF systems do not require external networks, including pumps, tubing, valves, or mechanical mixers. In addition, unlike conventional microfluidic chips, which have specialized channel designs for each specific reaction, droplet operations are reconfigurable and highly adjustable to fit the requirements of different applications in DMF platform. These unique characteristics of a DMF platform have made it a promising tool in a variety of chemical and biological applications.

The basic mechanism of droplet manipulation using DMF devices has been introduced in the following chapters. In this chapter, two typical configurations, parallel-plate and single-plate, have been explained in detail. We have also summarized several fabrication methods for DMF devices and reported

Microfluidic Biosensors. DOI: https://doi.org/10.1016/B978-0-12-823846-2.00003-1

the recent developments made in exploring new substrates. Further, we have focused on the integration of various types of sensing techniques into the DMF systems, namely electrochemical and optical sensors. Based on these integrated systems, a variety of biological applications, including cell-based, DNA-based as well as protein-based applications, have been demonstrated. Finally, the current challenges and possible future research direction have been presented.

6.2 Basics of digital microfluidic devices

6.2.1 Theory

Droplets can be individually controlled to move across an array of actuation electrodes in DMF devices when applying electrical potentials [8–10]. This technique has been named "electrowetting-on-dielectric" (EWOD). In early attempts to explain this phenomenon, thermodynamic theory was proposed. In this theory, the change in contact angle is considered as the cause of interface moving of the droplets [11–13]. In another words, the droplet movement can be regarded as the consequence of the capillary pressure resulted from asymmetric contact angles. In this paradigm, surface tension is considered as the driving force. As a result, electrowetting is described as the electrical control of surface tension. Theoretical analysis of relating the applied voltage to the contact angle is shown in Eq. (6.1) using the Young–Lippmann equation [14,15]:

$$cos\theta(V) = cos\theta_0 + \frac{\varepsilon_r \varepsilon_0}{2\gamma d} V^2 \tag{6.1}$$

where θ_0 is the contact angle without applied voltage, $\theta(V)$ is the contact angle with applied voltage V, ε_r denotes the relative permittivity of the dielectric, ε_0 denotes the vacuum permittivity, γ is the surface tension of the surrounding liquid media, and d is the thickness of dielectric layer. The EWOD force can then be calculated as [16]

$$F = \gamma w(cos\theta(V) - cos\theta_0) = \frac{w\varepsilon_r \varepsilon_0}{2d} V^2 \tag{6.2}$$

where w is the contact length between the droplet and the actuated electrode. However, there are some limitations with the thermodynamic theory. For example, this theory has difficulty in explaining the movement of dielectric droplets [11,12,17] or droplet motion where no apparent changes in contact angle can be observed [18]. Moreover, phenomena such as contact angle saturation (CAS) cannot be adequately explained using this theory [19].

As a result, a more generalized and complete explanation based on electromechanical theory has been proposed [11,20,21]. Unlike thermodynamic theory, the electromechanical perspective considers the contact angle change as the consequence of the electromechanical force rather than the cause of droplet motion. Also, electrostatic force instead of surface tension is regarded as the driving force. It can be calculated through integrating the Maxwell–Stress tensor T_{ij} over any arbitrary surface surrounding the droplet using the following equations [22,23]:

$$\mathbf{F} = \int T_{ij}\mathbf{n}dS \tag{6.3}$$

$$T_{ij} = \varepsilon\left(E_iE_j - \frac{1}{2}\delta_{ij}\mathbf{E}^2\right) \tag{6.4}$$

where $\mathbf{n}$ denotes the unit normal vector, i and j represent the pairs of x, y, and z axes, ε is the dielectric constant of the surrounding fluid media, $\mathbf{E}$ stands for the electric field around the droplet, and δ_{ij} refers to the Kronecker delta. The electromechanical theory can provide a convincing explanation for the movement of dielectric droplets as well as droplets that do not experience a contact angle reduction. It may also provide some useful insights to a full explanation of CAS [19,20,24].

6.2.2 Device configuration and fabrication

Generally, there are two typical configurations of DMF systems [3,10,15]: parallel-plate format (also known as closed systems) and single-plate format (open systems). As shown in Fig. 6.1, in the parallel-plate configuration, droplets are placed between two plates, which are both patterned with electrodes. The top plate is composed of a ground electrode, which is usually made of indium tin oxide (ITO), while the lower one includes an array of actuation electrodes. In the single-plate configuration, droplets are situated on top of a single substrate, which

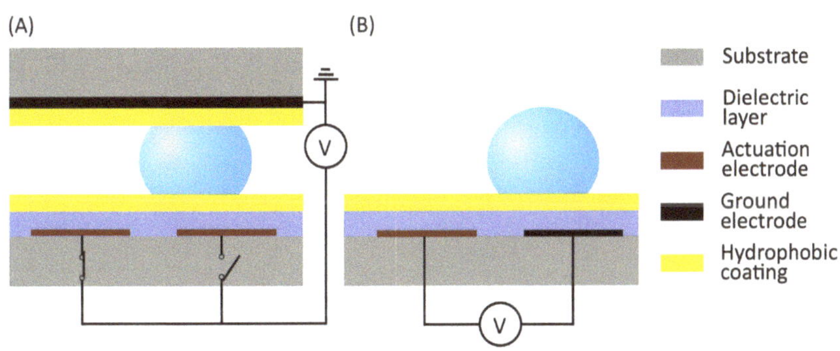

FIGURE 6.1

Schematic diagram of (A) parallel-plate and (B) single-plate DMF configurations (side view). *DMF*, digital microfluidics.

is patterned with an array of ground and actuation electrodes. Both systems have the configuration of depositing a dielectric layer on top of the electrodes. Moreover, all the surfaces in contact with the droplets are covered with a layer of hydrophobic coating in order to increase the contact angle and prevent sample adhesion as well as contact angle hysteresis. The parallel-plate system has the ability of performing multiple functional units, namely transporting, splitting, mixing, merging, and dispensing, which provides more possibilities in developing multiplex bioassays in need of various droplet manipulation skills compared to single-plate systems [2,14]. On the other hand, despite the limitation that droplet splitting and dispensing can hardly be achieved in single-plate systems, open systems can offer the advantages of rapid droplet mixing [25], as well as the flexibility for accommodating pipette-based liquid extraction [26] and analytical instruments such as fluorescence detectors [27,28]. It should also be noted that droplet evaporation happens at a faster rate in a single-plate configuration than parallel-plate configuration, which should be prevented in most applications such as polymerase chain reaction (PCR) [29,30], but may also appear advantageous in specific applications where drying analytes are needed [31].

DMF devices are typically fabricated under clean-room procedures, including photolithography and etching. As introduced in the device configuration part, DMF devices are generally composed of substrates, dielectric layers, electrodes, and hydrophobic coatings. The fabrication process is mainly determined by the choice of substrate materials. Among various materials, silicon and glass have been selected as the substrates of DMF devices due to their chemical inertness. However, the complicated fabrication procedures, low throughput, and the high cost of these materials have greatly hindered the widespread use of DMF systems and thus weakening their superiority over traditional channel-based microfluidic systems. To facilitate the widespread application of DMF devices, many researchers have exploited a printed circuit board (PCB) as the substrate [32–34]. PCB substrates have the advantages of low cost and ease of fabrication. Moreover, the multilayer arrangement enabling electrical access of individual electrodes can be easily enabled through using the mature PCB technology [35,36]. Besides PCB substrates, paper-based substrates have also been developed. Paper-based electronics possess the advantages of low cost, portable size and weight, biocompatibility, well-developed fabrication methods as well as ease of disposal [37–39]. Electrodes are commonly made of metals such as gold, aluminum, copper, or chromium. Other conductive materials, including ITO and doped polysilicon, have also been frequently used. The dielectric layer can be fabricated through various techniques depending on the type of materials used [2,7], for example, the vapor deposition method for parylene, amorphous fluoropolymers, and silicon nitride materials, thermal growth method for depositing silicon oxide as well as spin-coating technique for polydimethylsiloxane or SU-8. As for the hydrophobic coating, it is typically formed by Teflon-AF material using the spin-coating method.

6.3 Integrated digital microfluidic devices with sensing technologies

6.3.1 Integration of sensing techniques into digital microfluidic systems

In recent years, many studies have been carried out on integrating various sensing techniques into DMF systems for biosensing applications. In this section, the detailed information of coupling detection techniques, including optical detection and electrochemical detection with DMF systems, will be described.

6.3.1.1 Optical sensing in digital microfluidic systems

Among various sensing techniques and systems, optical sensing possesses the merits of extremely high sensitivity, being contact free, and having the capability of multiplexed detection, which has drawn a lot of attention in biological or chemical applications. Over the past few years, many studies have been carried out in developing more advanced optical biosensors as well as exploiting them for point-of-care applications. More particularly, their integration with microfluidics [26,40] has been further investigated. To ensure the integration of optical sensors into a DMF device, first, the substrate material should be carefully selected. Normally, the use of transparent substrates as well as electrodes is needed, for example, ITO, to allow for the smooth observation and detection of droplets within an enclosed microfluidic chamber. Moreover, the optical sensing module needs to be compatible with the hydrophobic coating in DMF systems, which ensures effective sensing without interrupting or affecting the electrowetting transport of droplets.

One of the earliest studies in integrating optical sensors into DMF systems focused on the heterogeneous integration of photodetectors [41,42]. In their work, a semiconductor photodetector in the format of thin-film InGaAs has been bonded onto the glass substrate and integrated with the EWOD platform. Based on this integrated device, the test of chemiluminescent compounds in droplets has been successfully demonstrated. The EWOD system was responsible for mixing of two droplets, while the detection of chemiluminescent optical signal resulted from the mixing was made possible by a thin-film photodetector, which clearly showed the advantages of this integrated platform. Later on, some researchers have coupled surface plasmon resonance (SPR) sensors into the EWOD devices to enable label-free biodetection [43,44]. Malic et al. have demonstrated the parallel SPR detection of different samples through combining SPR sensing techniques with an array-based DMF device. In this system, droplets containing different solutions were manipulated using EWOD and four detection spots were enabled for SPR signal detection [43].

More recently, optical fiber based sensing has also been integrated with DMF systems. Choi et al. have developed an integrated DMF platform for absorbance and fluorescence detection based on with fused silica optical fiber waveguides [45].

The optical fibers were inserted and aligned by a custom manifold in the plane of a DMF device, which enables contact-free detection. This integrating technique also allows for convenient removal and insertion of optical fibers. The system can provide the advantages of multistep sample processing and reaction as well as optical detection with high sensitivity, which can be useful in various analytical applications. Moreover, a fiber-optic sensing architecture that allows for $m \times n$ microdroplet sensing spots with only $(m + n)$ control points has been proposed for DMF devices [46]. As shown in Fig. 6.2, m parallel column fiber-optic cables are orthogonally overlapped with n parallel row fiber-optic cables. The detection of microdroplet was based on the change in the total internal reflection angle in contact with fiber-optic cables through lowered optical transmission. This enables sensing the location of microdroplets with correction abilities as well as distinguishing multiple microdroplets.

To enable multiplexed sensing, optical microresonators have been chosen as an effective sensing tool in DMF systems [47–49]. For the first time, Wondimu et al. have successfully integrated an array of whispering gallery mode (WGM) optical sensors into a single cell of a DMF system to enable label-free and multiplexed sensing [49]. As shown in Fig. 6.3, microgoblet sensors are integrated with the top substrate of the DMF system. The sensor chip contained 5000 integrated WGM sensors in total. The optical sensing and readout were made possible through free-space optics, which allows for large-scale integration of sensor

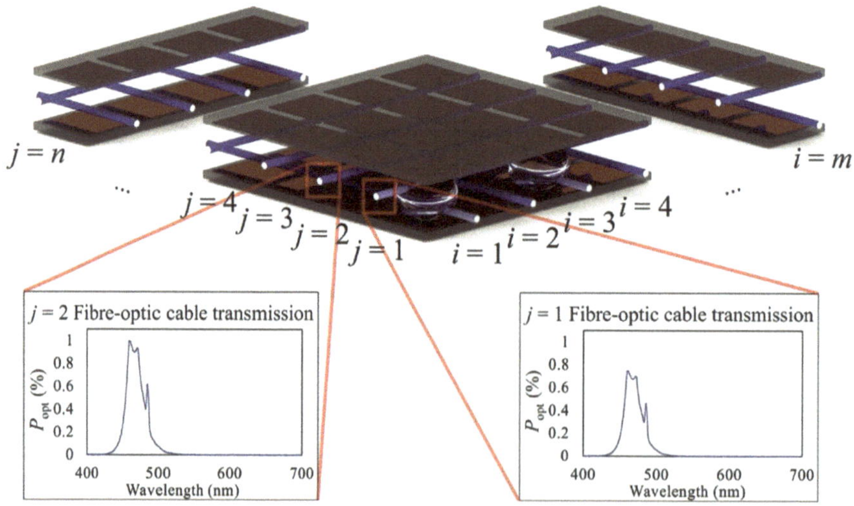

FIGURE 6.2

Fiber-optic sensing architecture integrated with DMF devices. *DMF*, digital microfluidics.

Reproduced with permission I. Spotts, D. Ismail, N. Jaffar, C.M. Collier, Fibre-optic sensing in digital microfluidic devices, Sens. Actuator A Phys 280 (2018) 164–169, https://doi.org/10.1016/j.sna.2018.07.039. Copyright 2018, Elsevier.

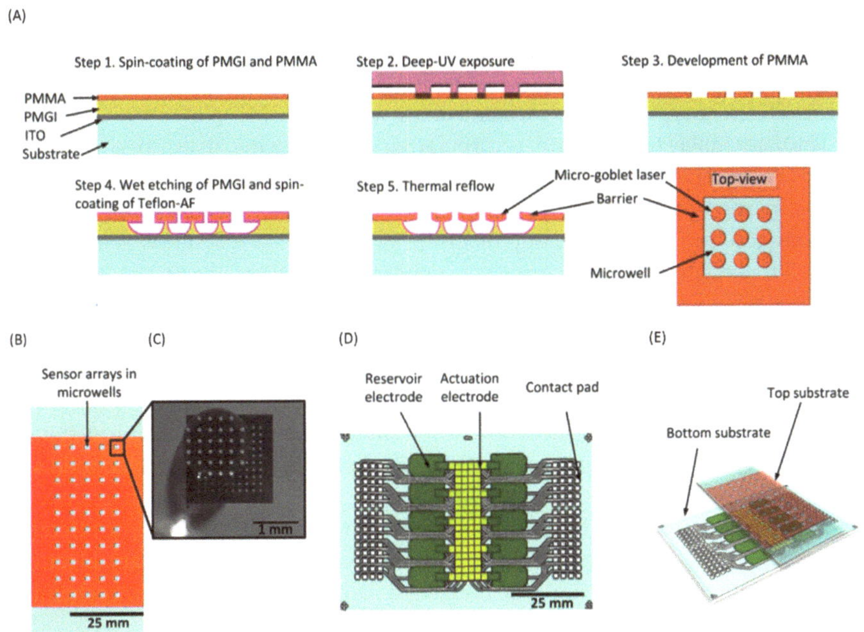

FIGURE 6.3

Fabrication and device integration of the DMF system integrated with WGM sensors. *DMF*, digital microfluidics; *WGM*, whispering gallery mode.

Reproduced with permission S.F. Wondimu, S. von der Ecken, R. Ahrens, W. Freude, A.E. Guber, C. Koos, Integration of digital microfluidics with whispering-gallery mode sensors for label-free detection of biomolecules, Lab. Chip 17 (2017) 1740–1748, https://doi.org/10.1039/c6lc01556e. Copyright 2017, Royal Society Chemistry.

arrays without the need of including extra coupling elements such as prisms or optical fibers. Actuation electrodes were placed on the bottom substrate of the system to manipulate droplets. For demonstration purposes, refractive index sensing with sensitivity up to 28.11 nm per RIU and label-free detection of streptavidin with a sensitivity of 3.83 μg/mL were shown using this device. Exploiting the merits of both DMF and WGM resonators, this system has the capability of in situ label-free sensing and analysis of multiple analytes in a single DMF cell.

6.3.1.2 Electrochemical sensing in digital microfluidic systems

Electrochemical sensors have the ability to detect chemical reactions that happen at the surface of the electrode through measuring certain electrical signals, including current, potential, or capacitance. Compared to optical sensors that require external devices such as light source or detectors, the electrochemical sensors normally require only several electrodes [26,50], which makes them much easier to be integrated with DMF systems. Moreover, due to the similar manufacturing methods of electrochemical sensors and DMF devices, these two are inherently compatible with each other and thus can be easily combined together to form an integrated system [51,52]. Here, we reviewed the recent studies concerning integrating different types of electrochemical sensors with DMF platforms.

The early investigations on incorporating electrochemical sensors into DMF devices were usually based on the external electrodes. For example, Dubois et al. have demonstrated a device containing two catenaries to enable both electrowetting and electrochemical measurements [53]. The electrochemical sensing was based on voltammetric measurements conducted between these two catenaries through connecting the chip to a potentiostat. Using this device, the ionic liquid droplets can be effectively mixed and then transported to the sensing area for online analysis of droplets. The validity of results has been successfully confirmed through comparing the electrochemical measurements with those obtained in microvolume, which shows its potential for enabling convenient chemical synthesis and analysis. Another group has also proposed an external three-electrode system coupled with an EWOD microfluidic device [54]. This system includes a working electrode (Au) where the biochemical reaction takes place, a reference electrode (Ag) to eliminate any external disturbance, and an auxiliary electrode (Pt). Detailed information on this integrated DMF system with electrochemical sensors is given in Fig. 6.4A. In this system, two droplets containing Tris buffer and potassium iodide solution, respectively, were mixed through a T-junction EWOD device and then transported to the three-electrode system suspended at the end of the T junction. The concentration of iodide in the analyte droplets can then be quantitatively analyzed through cyclic voltammetry. The successful analysis of different analyte concentrations in droplets has proven the great potential of the combination of DMF and electrochemical sensing in rapid chemical analysis.

Rather than using external electrodes, in recent years, many researchers have focused on developing on-chip electrochemical measurements integrated with DMF systems [55,56]. Yu et al. have proposed a fully integrated DMF system with electrochemical sensing by transferring the sensor to the top plate [55]. The top plate consists of patterned ground electrodes with an optimized configuration of the electrochemical cell, as shown in Fig. 6.4B, which has the capability of achieving rapid and sensitive electrochemical response. The fabrication processes, including reactive ion etching as well as lift-off patterning, were integrated circuit compatible, which facilitates mass manufacturing. This integrated platform has been successfully exploited to conduct automatic analysis of ferrocenemethanol and dopamine at different concentrations. More recently, Farzbod et al. have proposed an integrated DMF platform with potentiometric electrochemical sensors [51]. Instead of building the platform with isolated prefabricated electrochemical sensor elements, this group has proposed a novel on-demand fabrication method in which Ag/AgCl electrodes can be fabricated during the operation of a DMF platform. Fig. 6.4C shows the schematic diagram of the integrated DMF electrochemical platform. In this configuration, the top plate not only acts as the ground electrode for the DMF device but also plays a role in electroplating of Ag.

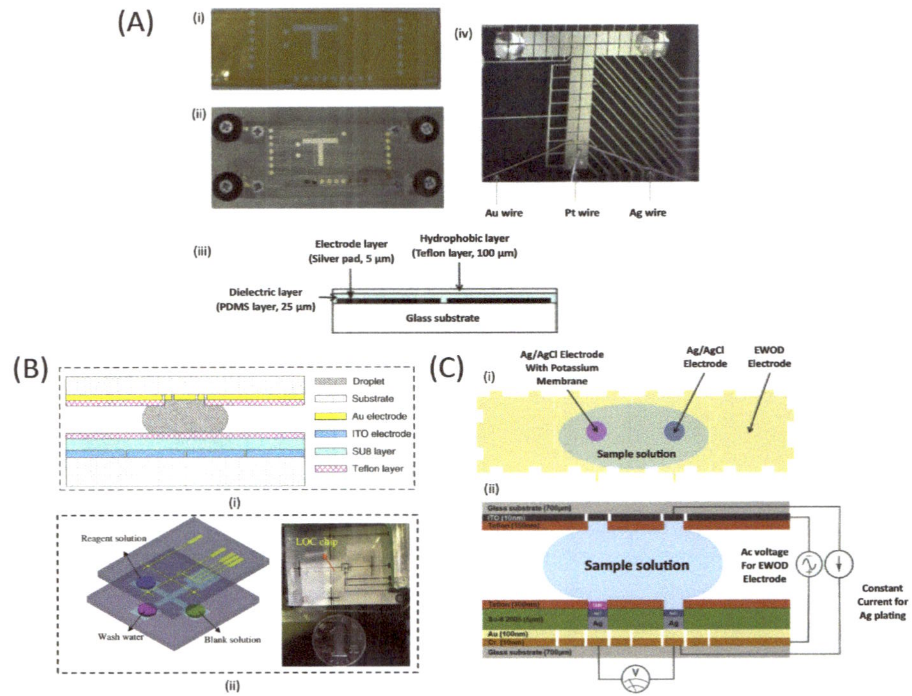

FIGURE 6.4

(A) (i) Glass substrate with photoresist pattern. (ii) Glass chip with EWOD silver electrodes. (iii) Cross-sectional structure of EWOD chip. (iv) EWOD chip with a three-electrode system. (B) (i) Schematic of DMF-based biochip. (ii) Assembly and picture of completed microfluidic chip. (C) Schematics of an electrochemical cell integrated with EWOD electrodes. *DMF*, digital microfluidics; *EWOD*, electrowetting-on-dielectric.

(A) Reproduced with permission C. Karuwan, K. Sukthang, A. Wisitsoraat, D. Phokharatkul, V. Patthanasettakul, W. Wechsatol, et al., Electrochemical detection on electrowetting-on-dielectric digital microfluidic chip, Talanta 84 (2011) 1384–1389, https://doi.org/10.1016/j.talanta.2011.03.073. Copyright 2011, Elsevier. (B) Reproduced with permission Y. Yu, J. Chen, J. Zhou, Parallel-plate lab-on-a-chip based on digital microfluidics for on-chip electrochemical analysis, J. Micromech. Microeng. 24 (2014), https://doi.org/10.1088/0960–1317/24/1/015020. Copyright 2014, IOP Publishing. (C) Reproduced with permission A. Farzbod, H. Moon, Integration of reconfigurable potentiometric electrochemical sensors into a digital microfluidic platform, Biosens. Bioelectron. 106 (2018) 37–42, https://doi.org/10.1016/j.bios.2018.01.048. Copyright 2018, Elsevier.

Similarly, the bottom plate serves as both an electrode for the DMF system and the seed layer for electrochemical sensor electrodes. This on-chip electroplating and fabrication technique allows for seamless integration of biosensors with the DMF platform. Moreover, the reconfiguration of sensors can be achieved without the need of disassembling or interrupting the device thanks to the on-demand fabrication method. This study also provides an efficient way to enhance the lifespan of reference electrodes of the sensing device. The proposed platform has been successfully demonstrated to be a useful tool in measuring potassium concentration, which shows great potential in low-cost and portable home-use sensing applications.

In summary, electrochemical sensors possess the merits of high sensitivity, rapid detection speed, excellent portability, and simplicity. Another big advantage is their ease of fabrication and integration with DMF devices, which makes them a promising tool for biological and chemical sensing applications [57,58]. In the future, more efforts should be made into integrating electrochemical sensors with DMF platforms using simpler fabrication and integration methods, which is of vital importance to facilitate large-scale integration as well as mass production and further promote the widespread use of integrated DMF systems.

6.3.2 Applications of digital microfluidic biosensors

The excellent sample manipulation abilities of DMF devices with the remarkable sensitivity of several sensing technologies built into it have made DMF biosensors a useful tool in a wide variety of applications, including cell-based applications, DNA-based applications, and protein-based applications.

6.3.2.1 Cell-based applications

Cell-based applications include cell culture, cell sorting, and purification. To date, various types of cell culture such as suspension culture [59,60], 2D cell culture on planar substrates [61,62], and 3D culture in hydrogels [63,64] have been demonstrated on a DMF platform. To enable single-cell culture and analysis, Kumar et al. have exploited a DMF platform to analyze isolated yeast cells [65]. Fig. 6.5 shows the procedures of conducting a cytotoxicity assay using a DMF platform, including pretreatment, cell incubation, and seeding manipulated through a DMF system and the subsequent detection. In this platform, five different concentration levels of antifungal drug amphotericin B have been successfully tested with real-time cell responses monitored through fluorescence microscopy. The experimental results show that a DMF platform has great potential in performing single-cell analysis. Recently, Zhai et al. have also developed a DMF platform for single-cell isolation and stable cell culturing through engineering 3D microstructures on chip [66]. In this work, innovative micropatterned arrays were constructed to efficiently trap a single cell and thus enable long-term cell culturing. After culturing breast cancer cells at different concentration levels, drug toxicity test was performed using a DMF system in comparison with the off-chip experimental

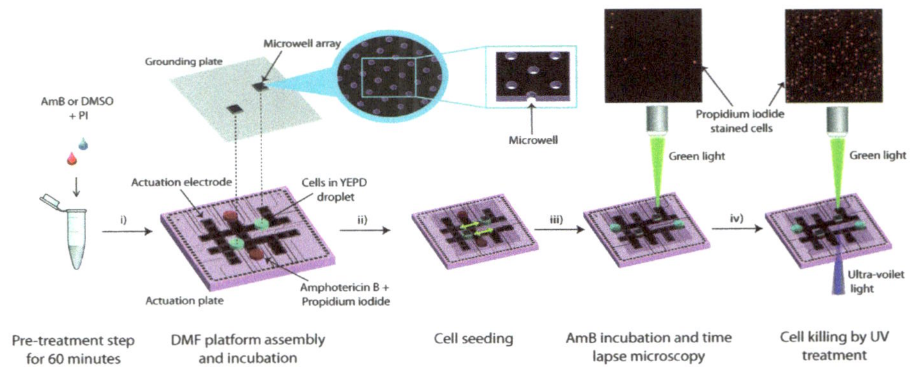

FIGURE 6.5

Flowchart of using a DMF system for conducting cytotoxicity assays on single *Saccharomyces cerevisiae* cells. *DMF*, digital microfluidics.

Reproduced with permission P.T. Kumar, K. Vriens, M. Cornaglia, M. Gijs, T. Kokalj, K. Thevissen, et al., Digital microfluidics for time-resolved cytotoxicity studies on single non-adherent yeast cells, Lab. Chip 15 (2015) 1852–1860, https://doi.org/10.1039/c4lc01469c. Copyright 2015, Royal Society Chemistry.

condition. The comparable results have proven the efficiency of single-cell trapping and culturing using the designed DMF platform, which can be a useful tool in biological applications at a single-cell level.

Many investigations have been carried out concerning cell sorting and purification, which are also of vital importance in cell-based applications. There are a variety of techniques such as utilizing electrical, optical, or magnetic forces to control the cells inside the microdroplets [1,67,68]. For example, the typical binary separation scheme was developed based on electrophoresis [69,70]. The sorting procedures normally consist of three steps, namely the separation of cells inside the mother droplet through electrophoresis, the splitting of mother droplet into daughter droplets through EWOD, and the manipulation of splitted daughter droplets individually for the subsequent operations. Nejad et al. have also studied the effect of negative dielectrophoresis trap geometry on cell sorting and manipulation based on DMF systems [71]. This study provides insight into designing and optimizing the geometry of trapping sites, which can largely increase the efficiency of cell capturing and sorting. Kumar et al. have recently integrated DMF devices with optical tweezers to enable effective single cell trapping, relocation as well as proliferation [72]. The DMF platform was composed of microwell arrays on the ground substrate. The optical tweezers can offer the ability of selective trapping and relocation of cells into the microwell array based on the expression of fluorescent proteins. Thus, this platform has been successfully demonstrated to be capable of selective retrieving and trapping of single bacteria as well as the subsequent proliferation at a single-cell level. In another study, this group has also demonstrated the use of magnetic force for manipulating cells inside the microdroplet [73]. In this system, an external magnet was placed above the top plate of the DMF device to induce the magnetic field, which was exploited to immobilize and manipulate the individual magnetic particle-conjugated protoplast droplet. An average of seven protoplasts can be manipulated and analyzed in every field of view through DMF manipulation. This integrated platform provides a useful solution to immobilize nonadherent cells and perform efficient analysis at a single-cell level.

6.3.2.2 DNA-based applications

DNA-based applications normally cover fields like DNA extraction, DNA amplification, as well as DNA sequencing. Normally, a DNA sample needs to be extracted and purified in advance of further processing and detection. Microfluidic technology has become a useful tool in performing automatic DNA extraction through integrating multiple functional units on chip. In DMF platforms, Hung et al. have developed a magnetic beads-based protocol for DNA extraction with reduced washing cycles in the room-temperature environment [74]. In this protocol, a series of droplet manipulation strategies were adopted. Broadly, two steps were included: collecting the magnetic beads into one side of the droplet through the meniscus-aid collection technique and washing out the unbound DNA from the magnetic beads. The results have also been verified through both SYBR-Green dye staining and gel electrophoresis, which confirms its validity in performing DNA extraction with minimized reagent use and reaction time.

Loop-mediated isothermal amplification (LAMP) and PCR are two most frequently adopted methods for DNA amplification. DMF platforms can serve as an important tool to implement these two methods [1,29,75]. As shown in Fig. 6.6, Wan et al. have developed an integrated DMF platform for LAMP-based pathogen nucleic acid detection [76]. They have integrated a temperature sensor as well as a heater with the control unit of a DMF platform for real-time thermal regulation and actuation signal generation. This DMF LAMP system was capable of both sample loading and detection of amplification products. Moreover, this platform has the potential of realizing fully integrated point-of-care diagnostics when further integrated with sample preparation and fluorescence detection modules. Hua et al. have exploited DMF devices for real-time multiplexed PCR with an extremely high amplification efficiency up to 94.7% [77]. In this system, the thermocycling was effected through transporting target droplets cyclically between different temperature zones at a temperature required for PCR. Moreover, by separating the reactions in space, that is, using a set of droplets containing different primer sets as reactors for parallel experiments, the DMF cartridge can be configured to perform multiplexed PCR analysis of both multiple genetic targets and multiple DNA samples in parallel. These works have proven that integrated DMF devices can be an excellent platform for DNA amplification due to its flexibility, droplet manipulation ability as well as thermocycling capability.

After DNA extraction, purification, and amplification, one extremely important step is DNA sequencing, which plays an important role in genotyping and clinical diagnostics. Many studies related to performing DNA sequencing on a DMF platform have been reported. Welch et al. have provided a groundwork for performing pyrosequencing on EWOD devices at a picoliter scale [78]. To carry out the pyrosequencing process, droplets containing ATP and luciferase solutions were mixed using DMF techniques. Afterward, the light intensity produced during the reaction was recorded to determine the number of bases added to the DNA strand. The detection limit can reach approximately 7 nM. This work has made efforts in decreasing the reagent volume and reducing the cost of DNA sequencing. Later on, Kim et al. have further improved the integration level of the DMF devices by adding sample preparation processes to the system [79]. Through integrating the distinct capillary-based modules into the DMF platform, a fully integrated device for sample preparation and analysis has been built. More recently, a DMF sequencing platform capable of real-time, automatic analysis of DNA mutations has been proposed by Zou et al. [80]. This portable system is composed of 48 electrodes, a photomultiplier tube module, and a circular magnet assembly. Experimental results showed that a DNA template with up to 53 bp can be sequenced by this system with 100% accuracy within 2 hours. These studies show that an integrated DMF platform can prove to be a powerful tool in next-generation sequencing.

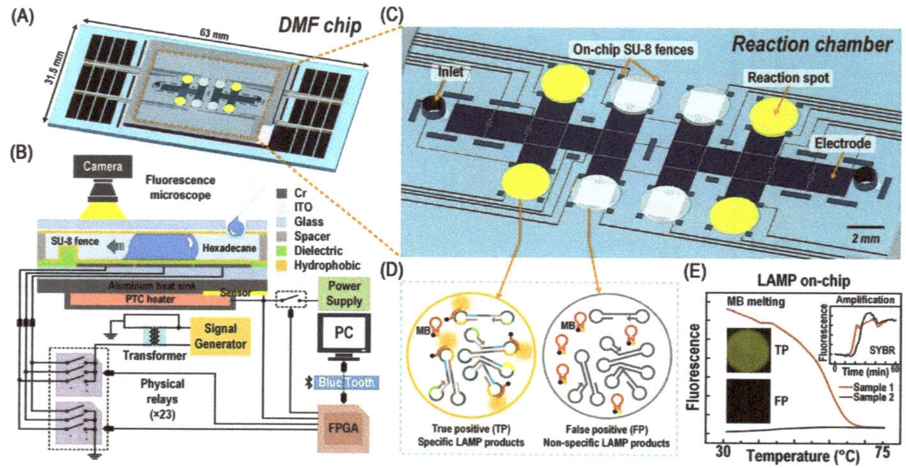

FIGURE 6.6

Overview of the DMF system for LAMP reaction and analysis. *DMF*, digital microfluidics; *LAMP*, loop-mediated isothermal amplification.

Reproduced with permission L. Wan, T. Chen, J. Gao, C. Dong, A.H. Wong, Y. Jia, et al., A digital microfluidic system for loop-mediated isothermal amplification and sequence specific pathogen detection. Sci. Rep. 7 (2017) 14586, https://doi.org/10.1038/s41598-017-14698-x. Copyright 2017, Springer Nature.

6.3.2.3 Protein-based applications

Microfluidic technology is a promising and advantageous tool in protein-based applications due to the fact that these applications usually require multiple sample processing procedures before being analyzed by mass spectrometry or any other methods. Among various microfluidic technologies, DMF techniques have been exploited in applications related to proteomics such as protein extraction [81,82], protein crystallization [83,84], and mass spectrometry [85–87] due to their advantages of efficient sample manipulation and minimized reagent use. In 2012, Luk et al. have integrated DMF devices with hydrogels to perform proteomic sample processing, including encapsulation and digestion [88]. In DMF systems, gel disc microreactors can be integrated directly with the relevant electrodes. As shown in Fig. 6.7, sample droplets containing protein can be manipulated to

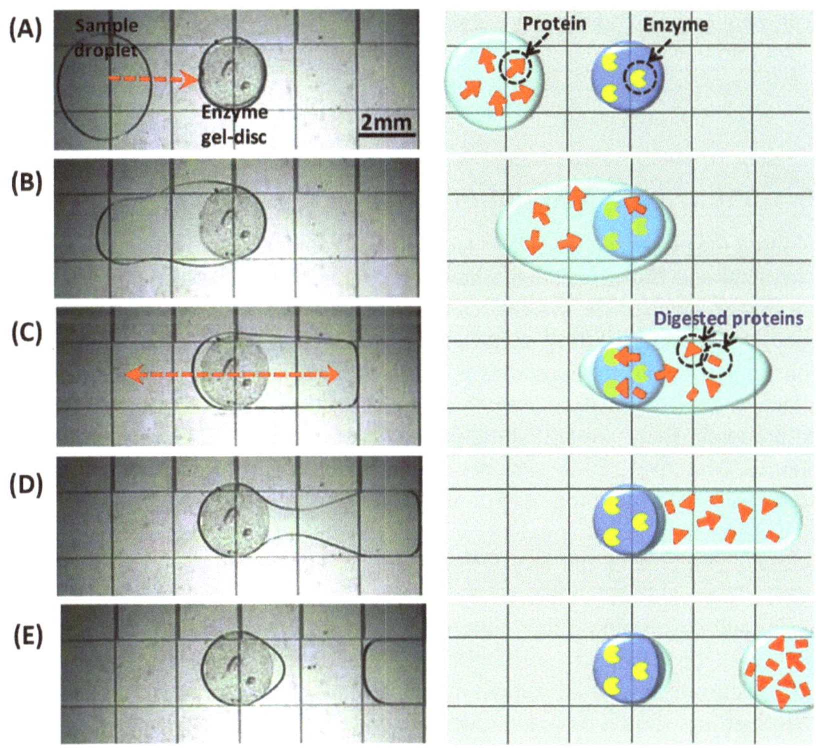

FIGURE 6.7

Images and schematic that show the operation of hydrogel proteolytic enzyme microreactors on a DMF device. *DMF*, digital microfluidics.

Reproduced with permission V.N. Luk, L.K. Fiddes, V.M. Luk, E. Kumacheva, A.R. Wheeler, Digital microfluidic hydrogel microreactors for proteomics, Proteomics 12 (2012) 1310–1318, https://doi.org/10.1002/pmic.201100608. Copyright 2012, John Wiley and Sons.

merge with the stationary gel discs for sample digestion. Afterward, the digested droplets containing digest peptides can be split from gel discs for subsequent sample processing or analysis. This platform has the potential of parallel sample processing based on multiple gel discs and thus can play an important role in proteome profiling.

In a recent study, Leipert et al. developed an efficient sample preparation workflow for proteomics based on DMF platforms, which was also compatible with the subsequent LC-M analysis [89]. By introducing a single-pot, solid-phase-enhanced sample preparation approach on-chip, the polymeric detergent removal and magnetic bead-enabled protein clean-up can be made possible, which enables the whole workflow, including cell lysis, protein extraction, reduction, alkylation as well as proteolytic digestion. Exploiting this system, they have successfully conducted proteome analysis on Jurkat T cells and identified up to 2500 proteins from approximately 500 cells. The results have shown the capability of this system in handling and analyzing a small volume of protein input, which provides insights into developing miniaturized protein-based applications.

6.4 Conclusions and outlooks

Exploiting the advantages of DMF techniques, over the years, various applications in chemical and biological fields have been presented based on DMF platforms. In particular, since DMF systems can provide a convenient way to sample transporting, mixing and splitting, significant progress has been made in sample preparation steps. Despite its numerous advantages, DMF technology still faces some challenges. To begin with, although a sophisticated electromechanical theory has been proposed, there are still some phenomena such as CAS that cannot be fully explained. Moreover, DMF systems are normally equipped with other peripheral modules and need a complicated fabrication process, which may hinder its widespread application to some extent. In addition, most of the current DMF systems only addressed part of the issues in a complete set of bioassays or biodetection process. There are still quite a few challenges facing the integration of all the modules and developing a fully automated on-chip device. For example, the compatibility of the sensing module with the hydrophobic coatings and its influence on the electrowetting-based sample manipulation should be further studied.

More efforts should be made toward addressing these challenges of DMF platforms and promoting its applications in point-of-care diagnostics in the future. First, more studies should be carried out on exploring a more comprehensive and solid theory of DMF mechanisms, which will help gain a deeper understanding of certain phenomenon and promote their practical applications. Second, it is highly desired to further lower the cost and facilitate large-scale fabrication of DMF devices to promote its widespread applications. PCB-based and paper-based DMF substrates have already been explored over the years with more accessibility in terms of cost and fabrication. These emerging substrates have provided more possibilities in DMF-based applications. More importantly, despite the current developments in the integration of various detection techniques, there is still much room for developing integrated DMF platforms coupled with more sensitive and miniaturized sensing systems. The multiplexing abilities as well as real-time monitoring capabilities should be further explored. We believe an ideal DMF platform integrated with all the necessary steps, including sample loading, manipulation, detection, and analysis, will be a reality in the near future, which will play an important role in empowering chemical and biological applications.

References

[1] H. Wang, L. Chen, L. Sun, Digital microfluidics: a promising technique for biochemical applications, Front. Mech. Eng. 12 (2017) 510–525. Available from: https://doi.org/10.1007/s11465-017-0460-z.

[2] M. Abdelgawad, A.R. Wheeler, The digital revolution: a new paradigm for microfluidics, Adv. Mater. 21 (2009) 920–925. Available from: https://doi.org/10.1002/adma.200802244.

[3] M.J. Jebrail, M.S. Bartsch, K.D. Patel, Digital microfluidics: a versatile tool for applications in chemistry, biology and medicine, Lab. Chip 12 (2012) 2452–2463. Available from: https://doi.org/10.1039/c2lc40318h.

[4] E. Samiei, M. Tabrizian, M. Hoorfar, A review of digital microfluidics as portable platforms for lab-on a-chip applications, Lab. Chip 16 (2016) 2376–2396. Available from: https://doi.org/10.1039/c6lc00387g.

[5] H.-H. Shen, S.-K. Fan, C.-J. Kim, D.-J. Yao, EWOD microfluidic systems for biomedical applications, Microfluid. Nanofluid. 16 (2014) 965–987. Available from: https://doi.org/10.1007/s10404-014-1386-y.

[6] W.C. Nelson, C.-J.C. Kim, Droplet actuation by electrowetting-on-dielectric (EWOD): a review, J. Adhes. Sci. Technol. 26 (2012) 1747–1771. Available from: https://doi.org/10.1163/156856111x599562.

[7] K. Choi, A.H. Ng, R. Fobel, A.R. Wheeler, Digital microfluidics, Annu. Rev. Anal. Chem. (Palo Alto Calif) 5 (2012) 413–440. Available from: https://doi.org/10.1146/annurev-anchem-062011-143028.

[8] A.R. Wheeler, Putting electrowetting to work, Science 322 (2008) 539–540.

[9] J. Lee, H. Moon, J. Fowler, T. Schoellhammer, C.-J. Kim, Electrowetting and electrowetting-on-dielectric for microscale liquid handling, Sens. Actuator A Phys. 95 (2002) 259–268.

[10] M.G. Pollack, R.B. Fair, A.D. Shenderov, Electrowetting-based actuation of liquid droplets for microfluidic applications, Appl. Phys. Lett. 77 (2000) 1725–1726. Available from: https://doi.org/10.1063/1.1308534.

[11] T.B. Jones, On the relationship of dielectrophoresis and electrowetting, Langmuir 18 (2002) 4437–4443. Available from: https://doi.org/10.1021/la025616b.

[12] T.B. Jones, K.L. Wang, D.J. Yao, Frequency-dependent electromechanics of aqueous liquids: electrowetting and dielectrophoresis, Langmuir 20 (2004) 2813–2818. Available from: https://doi.org/10.1021/la035982a.

[13] T.B. Jones, J.D. Fowler, Y.S. Chang, C.-J. Kim, Frequency-based relationship of electrowetting and dielectrophoretic liquid microactuation, Langmuir 19 (2003) 7646–7651.
[14] C. Sung Kwon, M. Hyejin, K. Chang-Jin, Creating, transporting, cutting, and merging liquid droplets by electrowetting-based actuation for digital microfluidic circuits, J. Microelectromech. Syst. 12 (2003) 70–80. Available from: https://doi.org/10.1109/jmems.2002.807467.
[15] M.G. Pollack, A.D. Shenderov, R.B. Fair, Electrowetting-based actuation of droplets for integrated microfluidics, Lab. Chip 2 (2002) 96–101. Available from: https://doi.org/10.1039/b110474h.
[16] J. Berthier, P. Dubois, P. Clementz, P. Claustre, C. Peponnet, Y. Fouillet, Actuation potentials and capillary forces in electrowetting based microsystems, Sens. Actuator A Phys. 134 (2007) 471–479. Available from: https://doi.org/10.1016/j.sna.2006.04.050.
[17] T. Hsieh, S. Fan, Dielectric droplet manipulations by electropolarization forces, in: Proceedings of 2008 IEEE 21st International Conference on Micro Electro Mechanical Systems, 13–17 January 2008, pp. 641–644.
[18] D. Chatterjee, B. Hetayothin, A.R. Wheeler, D.J. King, R.L. Garrell, Droplet-based microfluidics with nonaqueous solvents and solutions, Lab. Chip 6 (2006) 199–206. Available from: https://doi.org/10.1039/b515566e.
[19] V. Peykov, A. Quinn, J. Ralston, Electrowetting: a model for contact-angle saturation, Colloid Polym. Sci. 278 (2000) 789–793.
[20] K.H. Kang, How electrostatic fields change contact angle in electrowetting, Langmuir 18 (2002) 10318–10322.
[21] D. Chatterjee, H. Shepherd, R.L. Garrell, Electromechanical model for actuating liquids in a two-plate droplet microfluidic device, Lab. Chip 9 (2009) 1219–1229.
[22] D. Griffiths, Introduction to Electrodynamics, Prentice-Hall, Upper Saddle River, NJ, 1999.
[23] M. Abdelgawad, P. Park, A.R. Wheeler, Optimization of device geometry in single-plate digital microfluidics, J. Appl. Phys. 105 (2009). Available from: https://doi.org/10.1063/1.3117216.
[24] E. Baird, P. Young, K. Mohseni, Electrostatic force calculation for an EWOD-actuated droplet, Microfluid. Nanofluid. 3 (2007) 635–644. Available from: https://doi.org/10.1007/s10404-006-0147-y.
[25] C.G. Cooney, C.-Y. Chen, M.R. Emerling, A. Nadim, J.D. Sterling, Electrowetting droplet microfluidics on a single planar surface, Microfluid. Nanofluid. 2 (2006) 435–446. Available from: https://doi.org/10.1007/s10404-006-0085-8.
[26] L. Malic, D. Brassard, T. Veres, M. Tabrizian, Integration and detection of biochemical assays in digital microfluidic LOC devices, Lab. Chip 10 (2010) 418–431. Available from: https://doi.org/10.1039/b917668c.
[27] X. Zeng, K. Zhang, J. Pan, G. Chen, A.Q. Liu, S.K. Fan, et al., Chemiluminescence detector based on a single planar transparent digital microfluidic device, Lab. Chip 13 (2013) 2714–2720. Available from: https://doi.org/10.1039/c3lc50170a.
[28] U.-C. Yi, C.-J. Kim, Characterization of electrowetting actuation on addressable single-side coplanar electrodes, J. Micromech. Microeng. 16 (2006) 2053–2059. Available from: https://doi.org/10.1088/0960-1317/16/10/018.
[29] B. Coelho, B. Veigas, E. Fortunato, R. Martins, H. Aguas, R. Igreja, et al., Digital Microfluidics for nucleic acid amplification, Sensors (Basel) 17 (2017). Available from: https://doi.org/10.3390/s17071495.

[30] Y. Zhang, S. Park, K. Liu, J. Tsuan, S. Yang, T.-H. Wang, A surface topography assisted droplet manipulation platform for biomarker detection and pathogen identification, Lab. Chip 11 (2011) 398–406. Available from: https://doi.org/10.1039/C0LC00296H.

[31] A.R. Wheeler, H. Moon, C.A. Bird, R.R. Ogorzalek Loo, C.-J.C. Kim, J.A. Loo, et al., Digital microfluidics with in-line sample purification for proteomics analyses with MALDI-MS, Anal. Chem. 77 (2005) 534–540.

[32] V.K. Pamula, M.G. Pollack, R.B. Fair, Apparatuses and Methods for Manipulating Droplets on a Printed Circuit Board, Google Patents, 2013.

[33] R. Sista, Z. Hua, P. Thwar, A. Sudarsan, V. Srinivasan, A. Eckhardt, et al., Development of a digital microfluidic platform for point of care testing, Lab. Chip 8 (2008) 2091–2104.

[34] V. Jain, T.P. Raj, R. Deshmukh, R. Patrikar, Design, fabrication and characterization of low cost printed circuit board based EWOD device for digital microfluidics applications, Microsyst. Technol. 23 (2015) 389–397. Available from: https://doi.org/10.1007/s00542-015-2680-7.

[35] J. Gong, C.J. Kim, Direct-referencing two-dimensional-array digital microfluidics using multi-layer printed circuit board, J. Microelectromech. Syst. 17 (2008) 257–264. Available from: https://doi.org/10.1109/JMEMS.2007.912698.

[36] M. Gong, C.-J. Kim, Two-dimensional digital microfluidic system by multilayer printed circuit board, in: Proceedings of 18th IEEE International Conference on Micro Electro Mechanical Systems, MEMS 2005, 2005, pp. 726–729.

[37] R. Fobel, A.E. Kirby, A.H. Ng, R.R. Farnood, A.R. Wheeler, Paper microfluidics goes digital, Adv. Mater. 26 (2014) 2838–2843. Available from: https://doi.org/10.1002/adma.201305168.

[38] A.T. Jafry, H. Lee, A.P. Tenggara, H. Lim, Y. Moon, S.-H. Kim, et al., Double-sided electrohydrodynamic jet printing of two-dimensional electrode array in paper-based digital microfluidics, Sens. Actuator A Phys. 282 (2019) 831–837. Available from: https://doi.org/10.1016/j.snb.2018.11.135.

[39] S. Atabakhsh, S. Jafarabadi Ashtiani, Thermal actuation and confinement of water droplets on paper-based digital microfluidics devices, Microfluid. Nanofluid. 22 (2018). Available from: https://doi.org/10.1007/s10404-018-2060-6.

[40] Z. Liao, Y. Zhang, Y. Li, Y. Miao, S. Gao, F. Lin, et al., Microfluidic chip coupled with optical biosensors for simultaneous detection of multiple analytes: a review, Biosens. Bioelectron. 126 (2019) 697–706. Available from: https://doi.org/10.1016/j.bios.2018.11.032.

[41] L. Luan, R.D. Evans, N.M. Jokerst, R.B. Fair, Integrated optical sensor in a digital microfluidic platform, IEEE Sens. J. 8 (2008) 628–635. Available from: https://doi.org/10.1109/jsen.2008.918717.

[42] R. Evans, L. Luan, N.M. Jokerst, R.B. Fair, Optical detection heterogeneously integrated with a coplanar digital microfluidic lab-on-a-chip platform, in: Proceedings of SENSORS, IEEE, 2007, pp. 423–426.

[43] L. Malic, T. Veres, M. Tabrizian, Two-dimensional droplet-based surface plasmon resonance imaging using electrowetting-on-dielectric microfluidics, Lab. Chip 9 (2009) 473–475. Available from: https://doi.org/10.1039/b814697g.

[44] L. Malic, T. Veres, M. Tabrizian, Biochip functionalization using electrowetting-on-dielectric digital microfluidics for surface plasmon resonance imaging detection of

DNA hybridization, Biosens. Bioelectron. 24 (2009) 2218–2224. Available from: https://doi.org/10.1016/j.bios.2008.11.031.

[45] K. Choi, J.M. Mudrik, A.R. Wheeler, A guiding light: spectroscopy on digital microfluidic devices using in-plane optical fibre waveguides, Anal. Bioanal. Chem. 407 (2015) 7467–7475. Available from: https://doi.org/10.1007/s00216-015-8913-x.

[46] I. Spotts, D. Ismail, N. Jaffar, C.M. Collier, Fibre-optic sensing in digital microfluidic devices, Sens. Actuator A Phys. 280 (2018) 164–169. Available from: https://doi.org/10.1016/j.sna.2018.07.039.

[47] C. Lerma Arce, D. Witters, R. Puers, J. Lammertyn, P. Bienstman, Silicon photonic sensors incorporated in a digital microfluidic system, Anal. Bioanal. Chem. 404 (2012) 2887–2894. Available from: https://doi.org/10.1007/s00216-012-6319-6.

[48] M.W. Royal, N.M. Jokerst, R.B. Fair, Droplet-based sensing: optical microresonator sensors embedded in digital electrowetting microfluidics systems, IEEE Sens. J. 13 (2013) 4733–4742. Available from: https://doi.org/10.1109/jsen.2013.2273828.

[49] S.F. Wondimu, S. von der Ecken, R. Ahrens, W. Freude, A.E. Guber, C. Koos, Integration of digital microfluidics with whispering-gallery mode sensors for label-free detection of biomolecules, Lab. Chip 17 (2017) 1740–1748. Available from: https://doi.org/10.1039/c6lc01556e.

[50] R. Ge, R. Allen, L. Aldous, M. Bown, N. Doy, C. Hardacre, et al., Evaluation of a microfluidic device for the electrochemical determination of halide content in ionic liquids, Anal. Chem. 81 (2009) 1628–1637.

[51] A. Farzbod, H. Moon, Integration of reconfigurable potentiometric electrochemical sensors into a digital microfluidic platform, Biosens. Bioelectron. 106 (2018) 37–42. Available from: https://doi.org/10.1016/j.bios.2018.01.048.

[52] E. Samiei, G.S. Luka, H. Najjaran, M. Hoorfar, Integration of biosensors into digital microfluidics: impact of hydrophilic surface of biosensors on droplet manipulation, Biosens. Bioelectron. 81 (2016) 480–486. Available from: https://doi.org/10.1016/j.bios.2016.03.035.

[53] P. Dubois, G. Marchand, Y. Fouillet, J. Berthier, T. Douki, F. Hassine, et al., Ionic liquid droplet as e-microreactor, Anal. Chem. 78 (2006) 4909–4917.

[54] C. aruwan, K. Sukthang, A. Wisitsoraat, D. Phokharatkul, V. Patthanasettakul, W. Wechsatol, et al., Electrochemical detection on electrowetting-on-dielectric digital microfluidic chip, Talanta 84 (2011) 1384–1389. Available from: https://doi.org/10.1016/j.talanta.2011.03.073.

[55] Y. Yu, J. Chen, J. Zhou, Parallel-plate lab-on-a-chip based on digital microfluidics for on-chip electrochemical analysis, J. Micromech. Microeng. 24 (2014). Available from: https://doi.org/10.1088/0960-1317/24/1/015020.

[56] Y. Yu, J. Chen, J. Li, S. Yang, S.-K. Fan, J. Zhou, Microfabrication of a digital microfluidic platform integrated with an on-chip electrochemical cell, J. Micromech. Microeng. 23 (2013). Available from: https://doi.org/10.1088/0960-1317/23/9/095025.

[57] S. Vaddiraju, I. Tomazos, D.J. Burgess, F.C. Jain, F. Papadimitrakopoulos, Emerging synergy between nanotechnology and implantable biosensors: a review, Biosens. Bioelectron. 25 (2010) 1553–1565. Available from: https://doi.org/10.1016/j.bios.2009.12.001.

[58] J.F. Rusling, C.V. Kumar, J.S. Gutkind, V. Patel, Measurement of biomarker proteins for point-of-care early detection and monitoring of cancer, Analyst 135 (2010) 2496–2511. Available from: https://doi.org/10.1039/c0an00204f.

[59] S.C. Shih, P.C. Gach, J. Sustarich, B.A. Simmons, P.D. Adams, S. Singh, et al., A droplet-to-digital (D2D) microfluidic device for single cell assays, Lab. Chip 15 (2015) 225–236. Available from: https://doi.org/10.1039/c4lc00794h.

[60] S.H. Au, S.C. Shih, A.R. Wheeler, Integrated microbioreactor for culture and analysis of bacteria, algae and yeast, Biomed. Microdevices 13 (2011) 41–50. Available from: https://doi.org/10.1007/s10544-010-9469-3.

[61] D. Bogojevic, M.D. Chamberlain, I. Barbulovic-Nad, A.R. Wheeler, A digital microfluidic method for multiplexed cell-based apoptosis assays, Lab. Chip 12 (2012) 627–634.

[62] I.A. Eydelnant, U. Uddayasankar, M.W. Liao, A.R. Wheeler, Virtual microwells for digital microfluidic reagent dispensing and cell culture, Lab. Chip 12 (2012) 750–757.

[63] L.K. Fiddes, V.N. Luk, S.H. Au, A.H. Ng, V. Luk, E. Kumacheva, et al., Hydrogel discs for digital microfluidics, Biomicrofluidics 6 (2012) 014112.

[64] S.M. George, H. Moon, Digital microfluidic three-dimensional cell culture and chemical screening platform using alginate hydrogels, Biomicrofluidics 9 (2015) 024116. Available from: https://doi.org/10.1063/1.4918377.

[65] P.T. Kumar, K. Vriens, M. Cornaglia, M. Gijs, T. Kokalj, K. Thevissen, et al., Digital microfluidics for time-resolved cytotoxicity studies on single non-adherent yeast cells, Lab. Chip 15 (2015) 1852–1860. Available from: https://doi.org/10.1039/c4lc01469c.

[66] J. Zhai, H. Li, A.H.-H. Wong, C. Dong, S. Yi, Y. Jia, et al., A digital microfluidic system with 3D microstructures for single-cell culture, Microsyst. Nanoeng. 6 (2020). Available from: https://doi.org/10.1038/s41378-019-0109-7.

[67] J.L. He, A.T. Chen, J.H. Lee, S.K. Fan, Digital microfluidics for manipulation and analysis of a single cell, Int. J. Mol. Sci. 16 (2015) 22319–22332. Available from: https://doi.org/10.3390/ijms160922319.

[68] L. Pang, J. Ding, X.-X. Liu, S.-K. Fan, Digital microfluidics for cell manipulation, TrAC. Trends Anal. Chem. 117 (2019) 291–299. Available from: https://doi.org/10.1016/j.trac.2019.06.008.

[69] Y. Zhao, U.-C. Yi, S.K. Cho, Microparticle concentration and separation by traveling-wave dielectrophoresis (twDEP) for digital microfluidics, J. Microelectromech. Syst. 16 (2007) 1472–1481. Available from: https://doi.org/10.1109/jmems.2007.906763.

[70] S.K. Cho, Y. Zhao, C.J. Kim, Concentration and binary separation of micro particles for droplet-based digital microfluidics, Lab. Chip 7 (2007) 490–498. Available from: https://doi.org/10.1039/b615665g.

[71] H.R. Nejad, O.Z. Chowdhury, M.D. Buat, M. Hoorfar, Characterization of the geometry of negative dielectrophoresis traps for particle immobilization in digital microfluidic platforms, Lab. Chip 13 (2013) 1823–1830. Available from: https://doi.org/10.1039/c3lc41292j.

[72] P.T. Kumar, D. Decrop, S. Safdar, I. Passaris, T. Kokalj, R. Puers, et al., Digital microfluidics for single bacteria capture and selective retrieval using optical tweezers, Micromachines (Basel) 11 (2020). Available from: https://doi.org/10.3390/mi11030308.

[73] P.T. Kumar, F. Toffalini, D. Witters, S. Vermeir, F. Rolland, M.L.A.T.M. Hertog, et al., Digital microfluidic chip technology for water permeability measurements on single isolated plant protoplasts, Sens. Actuator B Phys. 199 (2014) 479–487. Available from: https://doi.org/10.1016/j.snb.2014.04.018.

[74] P.-Y. Hung, P.-S. Jiang, E.-F. Lee, S.-K. Fan, Y.-W. Lu, Genomic DNA extraction from whole blood using a digital microfluidic (DMF) platform with magnetic beads, Microsyst. Technol. 23 (2015) 313–320. Available from: https://doi.org/10.1007/s00542-015-2512-9.

[75] X. Rui, S. Song, W. Wang, J. Zhou, Applications of electrowetting-on-dielectric (EWOD) technology for droplet digital PCR, Biomicrofluidics 14 (2020) 061503. Available from: https://doi.org/10.1063/5.0021177.

[76] L. Wan, T. Chen, J. Gao, C. Dong, A.H. Wong, Y. Jia, et al., A digital microfluidic system for loop-mediated isothermal amplification and sequence specific pathogen detection, Sci. Rep. 7 (2017) 14586. Available from: https://doi.org/10.1038/s41598-017-14698-x.

[77] Z. Hua, J.L. Rouse, A.E. Eckhardt, V. Srinivasan, V.K. Pamula, W.A. Schell, et al., Multiplexed real-time polymerase chain reaction on a digital microfluidic platform, Anal. Chem. 82 (2010) 2310–2316.

[78] E.R. Welch, Y.Y. Lin, A. Madison, R.B. Fair, Picoliter DNA sequencing chemistry on an electrowetting-based digital microfluidic platform, Biotechnol. J. 6 (2011) 165–176. Available from: https://doi.org/10.1002/biot.201000324.

[79] H. Kim, M.J. Jebrail, A. Sinha, Z.W. Bent, O.D. Solberg, K.P. Williams, et al., A microfluidic DNA library preparation platform for next-generation sequencing, PLoS One 8 (2013) e68988. Available from: https://doi.org/10.1371/journal.pone.0068988.

[80] F. Zou, Q. Ruan, X. Lin, M. Zhang, Y. Song, L. Zhou, et al., Rapid, real-time chemiluminescent detection of DNA mutation based on digital microfluidics and pyrosequencing, Biosens. Bioelectron. 126 (2019) 551–557. Available from: https://doi.org/10.1016/j.bios.2018.09.092.

[81] H. Yang, J.M. Mudrik, M.J. Jebrail, A.R. Wheeler, A digital microfluidic method for in situ formation of porous polymer monoliths with application to solid-phase extraction, Anal. Chem. 83 (2011) 3824–3830. Available from: https://doi.org/10.1021/ac2002388.

[82] P.A. Wijethunga, Y.S. Nanayakkara, P. Kunchala, D.W. Armstrong, H. Moon, On-chip drop-to-drop liquid microextraction coupled with real-time concentration monitoring technique, Anal. Chem. 83 (2011) 1658–1664. Available from: https://doi.org/10.1021/ac102716s.

[83] A.P. Aijian, D. Chatterjee, R.L. Garrell, Fluorinated liquid-enabled protein handling and surfactant-aided crystallization for fully in situ digital microfluidic MALDI-MS analysis, Lab. Chip 12 (2012) 2552–2559.

[84] T. Xu, K. Chakrabarty, V.K. Pamula, Defect-tolerant design and optimization of a digital microfluidic biochip for protein crystallization, IEEE Trans. Comput. Des. Integr. Circuits Syst. 29 (2010) 552–565. Available from: https://doi.org/10.1109/tcad.2010.2042888.

[85] A.E. Kirby, A.R. Wheeler, Digital microfluidics: an emerging sample preparation platform for mass spectrometry, Anal. Chem. 85 (2013) 6178–6184. Available from: https://doi.org/10.1021/ac401150q.

[86] S.C. Shih, H. Yang, M.J. Jebrail, R. Fobel, N. McIntosh, O.Y. Al-Dirbashi, et al., Dried blood spot analysis by digital microfluidics coupled to nanoelectrospray ionization mass spectrometry, Anal. Chem. 84 (2012) 3731–3738. Available from: https://doi.org/10.1021/ac300305s.

[87] C.A. Baker, M.G. Roper, Online coupling of digital microfluidic devices with mass spectrometry detection using an eductor with electrospray ionization, Anal. Chem. 84 (2012) 2955–2960. Available from: https://doi.org/10.1021/ac300100b.
[88] V.N. Luk, L.K. Fiddes, V.M. Luk, E. Kumacheva, A.R. Wheeler, Digital microfluidic hydrogel microreactors for proteomics, Proteomics 12 (2012) 1310–1318. Available from: https://doi.org/10.1002/pmic.201100608.
[89] J. Leipert, A. Tholey, Miniaturized sample preparation on a digital microfluidics device for sensitive bottom-up microproteomics of mammalian cells using magnetic beads and mass spectrometry-compatible surfactants, Lab. Chip 19 (2019) 3490–3498. Available from: https://doi.org/10.1039/c9lc00715f.

CHAPTER

7 Emerging functional materials for microfluidic biosensors

Warakorn Limbut[1,2,3,4], Kiattisak Promsuwan[1,2,3,4], Supatinee Kongkaew[1,2,3,4], Panote Thavarungkul[1,2,5] and Wing Cheung Mak[6,7]

[1]*Center of Excellence for Trace Analysis and Biosensor, Prince of Songkla University, Hat Yai, Songkhla, Thailand*

[2]*Center of Excellence for Innovation in Chemistry, Faculty of Science, Prince of Songkla University, HatYai, Songkhla, Thailand*

[3]*Forensic Science Innovation and Service Center, Prince of Songkla University, Hat Yai, Songkhla, Thailand*

[4]*Division of Health and Applied Sciences, Faculty of Science, Prince of Songkla University, Hat Yai, Songkhla, Thailand*

[5]*Division of Physical Science, Faculty of Science, Prince of Songkla University, Hat Yai, Songkhla, Thailand*

[6]*Department of Biomedical Engineering, The Chinese University of Hong Kong, Shatin, Hong Kong*

[7]*Division of Sensor and Actuator Systems, Department of Physics, Chemistry and Biology (IFM), Biosensors and Bioelectronics Centre, Linköping University, Linköping, Sweden*

7.1 Introduction

7.1.1 Microfluidic biosensors

A biosensor is an analytical device that consists of a biological recognition element (such as enzymes, antibodies, or nucleic acids) and a physicochemical transducer. A transducer converts the signal produced by the interaction of the analyte of interest and the biological recognition element to an optical or electrical readout. Biosensors are more advantageous, cost-effective, and simple to use than other traditional laboratory-based detection techniques due to their portability, reusability, real-time response, and high specificity and selectivity. Microfluidics is a multidisciplinary technology that combines chemistry, biochemistry, engineering, physics, microtechnology, and biotechnology. Microfluidics is the science and technology of systems that process or manipulate small amounts of fluids through channels with diameters ranging from tens to hundreds of micrometers. It opens up entirely new possibilities for controlling molecule concentrations in space and time. Because of their large

Microfluidic Biosensors. DOI: https://doi.org/10.1016/B978-0-12-823846-2.00008-0

surface-to-volume ratio, microfluidic devices are highly portable, which is essential for on-site testing.

The combination of microfluidic devices and biosensors could provide an integrative platform to support sample-to-answer measurement for point-of-needs applications. Furthermore, microfluidic biosensors offer lower costs while increasing the accuracy and sensitivity of detection. An integrated microfluidic biosensor can perform full analysis, including continuous sample collection, analyte isolation, mixing, and pre-concentration. The fabrication of microfluidic biosensor structures, sample processing, and biosensing reactions are carried out within a closed and relatively stable environment, which means improved accuracy and sensitivity. Microfluidic channels (with sizes ranging from 100 nm to 500 μm) can also reduce the amount of sensing area required. With a smaller sample size, this can result in a reduction in analysis time and an increase in sensitivity, especially at lower target concentrations.

Several microfluidic devices have been developed for biosensor applications. Microfluidic biosensors could be mainly divided into three categories: enzyme-based microfluidic biosensors [1–3], microfluidic immunosensors [4–7], and DNA-based microfluidic biosensors [8].

Enzymes are proteins that catalyze chemical reactions which are used in enzyme-based microfluidic biosensors. It is a type of typical biological recognition element that is used in microfluidic biosensors because of its specific recognition and catalytic activity. Enzymes are typically immobilized onto a suitable transducer in a microfluidic device that generates a specific signal (optical signal, electrochemical signal, etc.) upon reaction with a given analyte in the microfluidic channel. Microfluidic immunosensors are based on antigen–antibody affinity interactions. In order to identify the antigen in a sample, an antibody is typically spatially in contact with the surface of a transducer as a recognition element. In addition, certain electrochemical microfluidic immunosensors can detect the formation of immunocomplexes through changes in current, resistance, refractive index, and other parameters. Lastly, for DNA-based microfluidic biosensors, complementary sequences of the target nucleic acid are immobilized in the microfluidic environment. After introducing the target nucleic acid samples into the microfluidic channel, the detection procedure is carried out in the microfluidic device. The base pair of hybridization probes and the target nucleic acid in the sample are used in the identification step. In some cases, the base-pairing procedure produces physicochemical signals that can be used to determine whether the sample contains the desired DNA. Examples of microfluidic biosensors will be discussed in Sections 7.2 and 7.3 based on different types of microfluidics and the biological recognition factors used as the sensing layer in the biosensor.

7.1.2 From conventional toward emerging functional materials

To achieve the desired need-driven analytical applications, the microfluidics/microchannel network architecture must be tailored to meet the specifications of the desired sample analysis (clinical analysis, infectious disease analysis, pathogen identification, drug analysis, DNA analysis, etc.). Therefore, the materials that are utilized in the microfluidic platform must be adequate and they must possess the necessary properties.

Silicon, glass, ceramic, and polydimethylsiloxane (PDMS) were the first materials used in microfluidic applications, but as technology advanced, advanced polymers [especially for three-dimensional (3D) printing technology], paper, polymer composites, and nanocomposites (i.e., graphene-based nanocomposites, metal–polymer nanocomposites, metal oxide-based nanocomposites) were used as emerging materials to fabricate microfluidic attachments. Materials for microfluidic devices used in research are typically focused on system usability and efficiency, whereas materials for the fabrication of microfluidic devices used for commercial purposes are typically focused on the cost of development, durability, and ease of use. As a result, the material used to manufacture microfluidic chips is critical parameter to the advancement of microfluidic technology.

Inorganic, polymeric, and biodegradable (i.e., paper) materials are the three main types of materials used in microfluidic chips. Section 7.2 focuses on a few of these products and their properties as they relate to the creation and fabrication of microfluidic platforms. Further in Section 7.3, we will discuss novel 2D nanomaterials for improved signal transduction in microfluidic biosensing. Graphene, metal chalcogenides, metal oxides, metal hydroxides, MXenes, and black phosphorous, as well as other emerging 2D nanomaterials for improved signal transduction in microfluidic biosensors will be discussed.

7.2 Materials, properties, and functions for microfluidic biosensors

7.2.1 Material for design and development of microfluidics

Inorganic materials (such as silicon, glass, and ceramic), polymeric materials [such as poly(dimethylsiloxane) (PDMS), poly(methyl methacrylate) (PMMA)], polycarbonate (PC), and combinations of these materials have been utilized for the preparation of various microfluidic devices and architectures. Photolithography and etching methods are commonly used to create microfluidic patterns from traditional materials such as glass and silicon. Nowadays, PDMS and thermoplastics are highly popular in research for the design and fabrication of microfluidics due to their availability and relative low cost. The cellulose-based paper that originated from natural wood as a porous media with the ability to wick and transport fluid is an emerging and sustainable microfluidic substrate for the creation of various innovative paper-based microfluidic devices. Microfluidics made from paper are disposable, low-cost, and environmentally friendly. Moreover, capillary action between the liquid and the porous matrix drives fluid flow along the cellulose paper without the need for external force. 3D printing, particularly fused deposition modeling (FDM) and stereolithography (SLA), is a new approach that has drawn attention to create complex 3D microfluidic architectures that cannot be made with conventional approaches.

7.2.2 Inorganic materials (silicon, glass, and ceramic)

In the early times, microfluidic devices were largely made from inorganic (non-polymeric) materials such as silicon, glass, or ceramic. Silicon was the first substrate material used for microfluidic applications due to its high chemical resistance to organic solvents and temperature stability. A silicon microfluidic platform can be created using laser ablation, chemical/physical etching, electrochemical etching, and thermal oxidation processes. For example, Fornell et al. demonstrated use of direct laser ablation to create microfluidic channels in silicon with vertical walls suitable for acoustic particle focusing via bulk acoustic waves. The acoustic resonance channel was 380-μm wide and branched into 127-μm wide side outlet channels. It is intriguing because it allows for rapid prototyping of silicon-glass microfluidic chips for acoustofluidic applications [9,10]. As a function of the silicon substrate, Kim and coworkers developed a microfluidic device for *Salmonella* Typhimurium using magnetic beads and quantum dots (QDs) [11]. The silicon microfluidic channel (width: 400 μm; height: 50 μm) was fabricated using a standard photolithography technique. The microfluidic channel consisted of two inlet and one outlet ports. The magnetic beads mixed with sample solutions were injected at inlet port 1. An antibody-conjugated QD-containing solution was injected at inlet port 2. The QD-labeled cells were captured in the detection zone using an external permanent magnet placed under the detection zone and the fluorescence signal was detected by using a custom-built fluorometer (Fig. 7.1A). Le and colleagues reported the fabrication of a filtration microfluidic chip for a particle size filtration. The Bosch deep reactive ion etching process, which combines physical and chemical etching, was used to create a microfluidic filtration chip with one inlet port and one outlet port having dimensions 28 mm $\times$ 7 mm. In this channel, there were four arrays of microposts with different distances between them: 100, 75, 50, and 25 μm, respectively. The silicon channel was 150 μm deep. To complete the filtration microfluidic chip, a silicon substrate was adhesively bonded to a mica lid. Only particles with a diameter of 25 μm or less could pass through the filter of the microfluidic chip. This chip can be used in a variety of research fields, including inkjet printing technology, particle filtration, and particle capturing [12]. Li et al. created a versatile biofunctional film on a porous silicon (pSi) chip, which is another notable function of the silicon substrate platform. The bio-microfluidic sensor was fabricated using phospholipid bilayers/pSi and an electrochemical etching process that mimics the real cell membrane. The transmembrane behavior of membrane translocating peptides can be reliably measured using real-time RIFTS. This feature leads to its practical application in the screening of peptides with membrane translocation ability. Using an optical imaging system, the microfluidic chip can perform high-

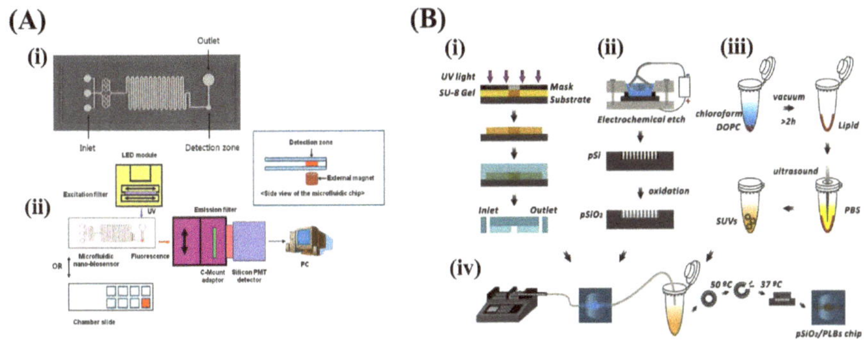

FIGURE 7.1

(A) Image of microfluidic channel. (i) Design of silicon microfluidic channel (width: 400 μm; height: 50 μm) consisted of inlet, outlet, serpentine channel, and detection zone. (ii) Schematic diagram of portable fluorometer consisted of LED unit, sampling unit, microfluidic chip, C-mount adapter, and PC. (B) The preparation of the porous silicon microfluidic sensor with supported phospholipid bilayers. (i) Patterning of the PDMS film with the microchannel (1.5 cm long, 450 μm wide, and 250 μm deep) and microwell serving as the detection zone (2 mm in diameter). (ii) Methodology for preparing the pSi sensing layer. Plasma treatment was used to seal the pSi and PDMS films together. (iii) Procedure for preparation of SUVs. (iv) The SUV suspension was injected into the microchip and self-assembled to form phospholipid bilayers on the pSi substrate. *LED*, light-emitting diode; *PC*, personal Computer; *pSi*, porous silicon; *PDMS*, polydimethylsiloxane; *SUVs*, small unilamellar vesicles.

Source: *(A) Reprinted with permission from G. Kim, et al. A microfluidic nano-biosensor for the detection of pathogenic* Salmonella, Biosens. Bioelectron. *67 (2015) 243–247. (B) Reprinted with permission from Z. Li, et al. Label-free discrimination of membrane-translocating peptides on porous silicon microfluidic biosensors,* Biomicrofluidics, *10 (2016) 064113.*

throughput screening of cell-penetrating peptides and antimicrobial peptides (Fig. 7.1B). Such a microfluidic platform was further developed for drug screening processes [13]. Biofunctionalization of silicon substrate for the immobilization of enzymes on silicon microchips platforms has been developed by Davidsson et al. The development and optimization of co-immobilized glucose oxidase (GOX) or alcohol oxidase (AOX) with horseradish peroxidase (HRP) on silicon microchips, as well as a microfluidic sequential injection analysis (SIA) system, using chemiluminescence (CL) detection is demonstrated. The flow-through micro-biosensor chips had a highly porous surface structure and was co-immobilized with enzymes GOX and HRP or AOX and HRP enabling biosensing feature. The CL detection was performed on the microchip surface and was recorded using photomultiplier tubes (PMT) mounted directly on top of the microchip unit. Chemical wet etching of silicon microchips was used to create the flow-through microchips. The GOX–HRP and AOX–HRP were immobilized on activated silicon microchip surfaces with operational stability of 8 hours and a half-life of 2–3 days [14]. Zhuo et al. created a label-free, highly selective, and sensitive S-layer protein (SLP) assay using a TiO_2-coated porous silicon (PSi)-based microfluidic biosensor. An electrochemical etching of silicon substrate was used to create PSi. A PSi-based biosensor could be easily prepared as a microfluidic device to form a microfluidic biosensing system for SLP testing. The detection of light reflection interference allowed the biosensor's sensor to be an unlabeled sensor. The noncovalently immobilized anti-SLP on the PSi surface enabled specific recognition and capturing for target molecule within the microfluidic sensing platform with a fast response time (5 min) and excellent specificity for SLP and a low detection limit at the level of pM [15]. Sun et al. developed a point-of-care (POC) system for field detection of live virus using equine respiratory infectious diseases as a model system for corresponding human diseases such as COVID-19. At the end of the reactions, specific nucleic acid sequences from five pathogens were amplified using loop-mediated isothermal amplification (LAMP) on a silicon microfluidic chip and were detected by a smartphone. The silicon microfluidic chips are highly stable, have no auto-fluorescence, and are easily manufactured, therefore are suitable for housing the LAMP assays. The chips (25 mm $\times$ 15 mm $\times$ 0.5 mm) contained 10 parallel flow channels (10 mm $\times$ 0.5 mm $\times$ 0.2 mm), each with a volume of 1 μL and sharing the same inlet. The chip surface was thermally oxidized to form a 200 nm layer of SiO_2, reducing the possibility of bare silicon interfering with the amplification process. This system correctly identified pathogen-contaminated horse nasal swab samples, with a limit of detection comparable to the traditional laboratory-based PCR test and results were obtained in 30 minutes [10].

However, due to silicon's reflective properties, which make it difficult for optical detection, glass substrates were used as an alternative material. Glass has several advanced properties such as high biocompatibility, excellent chemical tolerance, high temperature resistance, high mechanical stress resistance, moderate nonspecific adsorption, and optical clarity, making it ideal for use in capillary electrophoresis and optical sensors. Wet etching, dry etching, laser fabrication, and mechanical fabrication are all popular methods for creating glass microfluidic systems. In fabricating microfluidic channels on glass, wet etching using liquid chemicals or etchants is a well-developed process allowing the fabrication of microfluidic glass chips on a commercial scale at a reasonable cost [16]. Dry chemical etching or vapor phase etching is a chemical process that uses etchant gases to attack the glass surface or substrate [17]. Several chemical approaches have been used for fabricating glass micropatterns via dry chemical etching, i.e., CF_4/CHF_3 [18], and CF_4/Ar or CF_4/Ar [19]. Alternatively, laser ablation using CO_2, UV, and ultra-short pulse lasers have been used to fabricate microstructure on glass with better efficiency [20]. However, impurities in glass materials or trace elements in glass materials may potentially interfere with device fabrication or sensor operation [21].

As compared to glass or silicon base, ceramic has better electrical, mechanical, chemical, and temperature performance, as well as lower operating costs.

In general, low-temperature co-fired ceramic (LTCC) is used in ceramic microfluidic devices. LTCC is an aluminum oxide-based material that can be formed into a 3D device, with each sheet tested for quality control before being stacked [22]. For example, Luo et al. created a microfluidic platform with a Clark-type oxygen sensor for real-time monitoring of dissolved oxygen in a flowing microfluidic environment using LTCC technology (Fig. 7.2A) [23]. In this approach, the microfluidic device represents an improved Clark-type oxygen sensor, with the advantages of easy fabrication, flexible configuration, fast response time, incorporation of microfluidic analyte introduction, and real-time detection of dissolved oxygen. Material synthesis, cell culture, biological assays with controlled loading/feeding of reagents or analytes, and real-time monitoring of dissolved oxygen in a microfluidic environment are all possible applications. Remiszewska et al. created a microanalysis method for enzymatic urea sensor by determination of ammonium ions dependent on absorption photometry utilizing LTCC (Fig. 7.2B). This proposed LTCC microfluidic system was successful in measuring urea in cell culture medium and cell lysates [24]. Fakunle and Fritsch constructed a self-contained enzyme-linked immunosorbent assay with electrochemical detection from LTCC materials with screen-printed gold electrodes in

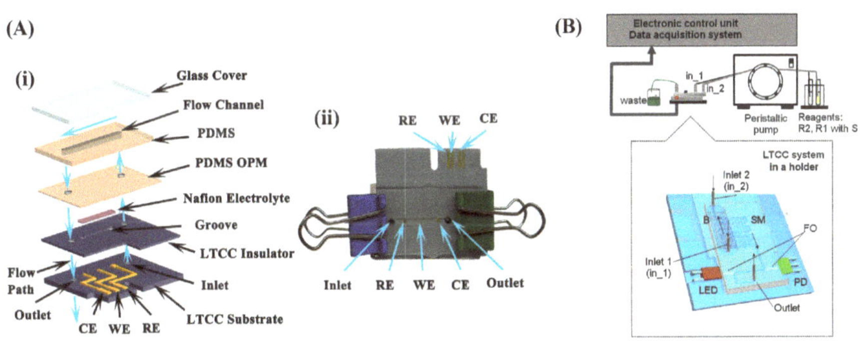

FIGURE 7.2

(A) A microfluidic platform fabricated using LTCC technology; (i) the configuration of a LTCC based Clark-type oxygen sensor consisted of three electrodes (working electrode, counter electrode and Ag/AgCl reference electrode), a solid-state proton conductive matrix (Nafion 117 membrane) and PDMS as the OPM, and used a solid-state proton conductive matrix as the electrolyte and PDMS as the OPM, (ii) Image of the bonded LTCC based microfluidic Clark-type oxygen sensor. (B) Scheme of measurements in LTCC microfluidic system based on absorption photometry. *LTCC*, Low temperature co-fired ceramic; *PDMS*, polydimethylsiloxane; *OPM*, oxygen permeable membrane.

Source: *(A) Reprinted with permission from J. Luo, et al. A low temperature co-fired ceramic based microfluidic Clark-type oxygen sensor for real-time oxygen sensing,* Sens. Actuators B: Chem., *240 (2017) 392–397. (B) Reprinted with permission from E. Remiszewska, et al. Enzymatic method of urea determination in LTCC microfluidic system based on absorption photometry,* Sens. Actuators B: Chem., *285 (2019) 375–384.*

three dimensions on all four walls. This work describes the fabrication and characterization of LTCC microchannel devices with integrated immobilized immunoassay and self-contained electrochemical detection for the detection of mouse IgG. The immunoassay components for quantifying mouse IgG were assembled onto one of the screen-printed gold surfaces, and self-contained electrochemical detection was evaluated in a three-electrode configuration with the remaining screen-printed gold elements after incubation with the enzyme precursor p-aminophenylphosphate [25]. Malecha fabricated enzymatic microreactor for urea detection, potentiometric sensor with ion selective electrodes (ISE) based array sensitive to potassium ions, and amperometric glucose sensor. Three ceramic-based microfluidic systems were constructed using the LTCC microelectronic technology. The silver/gold paste was deposited on ceramics using the screen printing technique to create printed electrodes. The individual LTCC tapes were stacked together using a lamination process under high pressure (>30 MPa). The urease enzyme was immobilized on the poured glass beads that were placed in the microreactor and detected urea based on the hydrolysis of urea. In the hydrolysis of the urea, the pH change can be detected using a micro pH sensor. Moreover, a potentiometric sensor consisting of a fluidic channel and four silver electrodes in the LTCC was used to detect potassium ions based on ISE electrode [26].

7.2.3 Polymeric Material

Polymers are alternative materials for the fabrication of microfluidic systems because they allow for more cost-effective production and easier microfabrication, have strong biocompatibility with a wide range of biological molecules, have tailorable formulations, and allow mass manufacturing processes. Elastomers, thermosets, and thermoplastics are the most popular polymers used to fabricate polymer-based microfluidic devices.

An elastomer is a material made of cross-linked polymer chains. They will expand or contract in response to external force, then return to their original shape and scale. When compared to other material components, the elastomer has a low Young's modulus and is solvent compatible. Because of their elasticity, elastomers are especially useful in microelectronics and microfluidics [27,28]. PDMS is one of the most commonly used materials for microfluidic fabrication due to its inherent properties such as biocompatibility, gas permeability (oxygen and carbon dioxide), nontoxicity and flammability, optical clarity (240–1100 nm) with low autofluorescence, and low elastic modulus (300–500 kPa). Thermosets are polymers with chains irreversibly bound together when cross-linked. Thermoset polyester (TPE) is one of the most used thermosets in microfluidics. TPE is a thermally initiated material formed by the polymerization of polyester and styrene through UV or heat. It is a hydrophobic material that requires surface modification to allow water to flow readily through microfluidic channels. One significant advantage of thermosets is that they can be used for 3D microfabrication via photopolymerization. It is also thermally stable, solvent resistant, and optically transparent. However, when heated or radiated, cross-linking occurs, resulting in a polymer network that cannot be softened or reshaped. Thermoplastics are rigid polymer fabrics with excellent longevity, mechanical processability, water absorption, and high tolerance to chemical, acidic, and essential solvents. The differences between thermoplastic polymers and thermoset or elastomer plastics may be significant, with thermoplastic polymers softening when heated to the glass transition temperature while retaining chemical and dimensional stability. Furthermore, thermoplastics are not cross-linked and can be reshaped after curing. Thermoplastics have much lower raw material and processing costs than traditional microfluidic materials like silicon and glass [29]. PMMA [1,30], PC [31,32], polystyrene (PS) [33,34], polyvinyl chloride, and polyimide [35] are common thermoplastics for microfluidic fabrication.

For all polymers, elastomer (PDMS) is one of the most commonly used soft and flexible polymers for the fabrication of microfluidic channels. Soft photolithography methods are used to create device molds, and PDMS microstructures are cast and cured on these molds. Furthermore, multiple layers can be stacked to create complex fluidic designs. Indeed, the soft lithography technique using PDMS has several advantages, including low cost, fast processing, reusability of masters, design of complex 3D systems by multilayer fabrication, excellent optical transparency, easy installation of fluidic interconnects, and applicability to a variety of biological and cellular processes due to the biocompatibility property of PDMS. PDMS substrates can be directly attached to glass by Van der Waals force or by a simple procedure, oxygen plasma, to a functionalized PDMS surface with an O–Si–O and bond by covalent connection, thus enhancing the pressure [27,36,37]. Enzyme-based microfluidic biosensors, microfluidic immunosensors, and DNA-based microfluidic biosensors have all successfully utilized PDMS as a platform. For example, Limbut et al. developed a conductimetric microfluidic bioreactor. The soft lithography technique was used to create the PDMS mold. The enzyme was immobilized in the microfluidic channel on the PDMS surface using the covalent binding method. Two gold electrodes and a laboratory-built conductimetric transducer were used to monitor the increase in solution conductivity caused by the change in charges generated by the enzyme–substrate catalytic reaction (Fig. 7.3A). The immobilized urease on the microchannel chip demonstrated good stability and repeatability. There was good agreement between the urea concentrations of human serum samples determined by the microfluidic flow injection conductimetric bioreactor system and those obtained using the Berthelot reaction [38]. Li et al. developed a nano-integrated portable enzymatic microfluidic electrochemical biochip for single-step creatinine testing. Screen-printed carbon electrodes (SPCEs), a PDMS sheet with meandering microchannels, and PTFE tubes were used to create the proposed portable microfluidic biosensing system. To improve the detection performance for creatinine, SPCEs with gold nanostructures and carbon nanotubes were combined into microfluidic devices. The microfluidic electrochemical biochip, particularly POC testing, has the

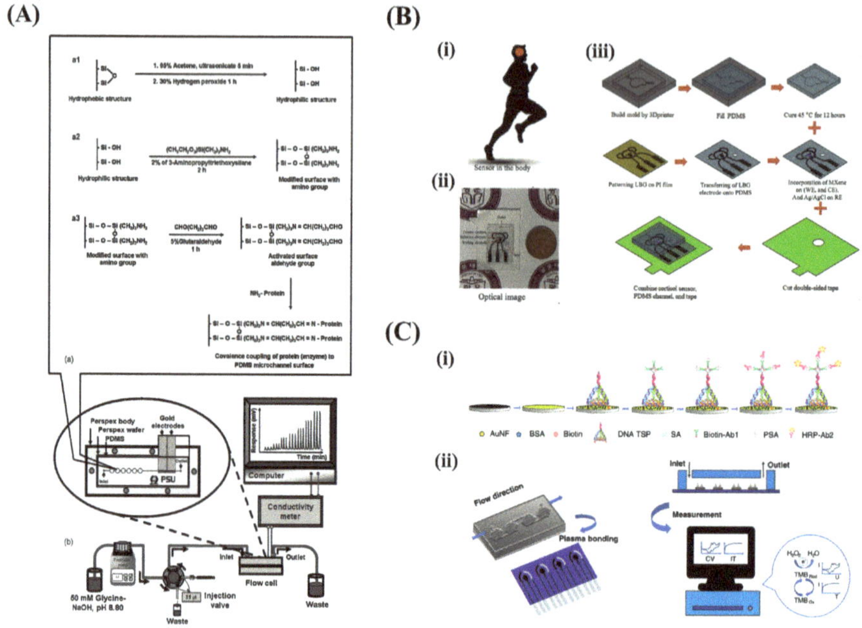

FIGURE 7.3

(A) Schematic diagram showing microfluidic flow injection conductimetric bioreactor system and immobilization of enzyme on PDMS microchannel surface by covalent binding method. (B) The wearable patch cortisol sensor for noninvasive point-of-care cortisol biomarker detection in human sweat. (i) Patch sensor attached to the body position, (ii) optical image of the wearable patch, (iii) the fabrication sequences of the wearable patch cortisol sensor integrated with a microfluidic system. (C) Schematic illustration of the μFEC system. (i) The immobilization process for PSA biosensor detection. (ii) The scheme of the μFEC system and the reaction for detection of PSA. *PDMS*, polydimethylsiloxane; *μFEC*, Microfluidic–electrochemical; *PSA*, Prostate-specific antigen.

Source: *(A) Reprinted with permission from W. Limbut, et al. Microfluidic conductimetric bioreactor, Biosens. Bioelectron. 22 (2007) 3064–3071. (B) Reprinted with permission from J. S. Nah, et al. A wearable microfluidics-integrated impedimetric immunosensor based on Ti3C2Tx MXene incorporated laser-burned graphene for noninvasive sweat cortisol detection, Sens. Actuators B: Chem. 329 (2021) 129206. (C) Reprinted with permission from D. Feng, et al. DNA tetrahedron-mediated immune-sandwich assay for rapid and sensitive detection of PSA through a microfluidic electrochemical detection system, Microsyst. Nanoeng. 7 (2021) 1–10.*

potential to become a useful tool for medical diagnosis [39]. Nah et al. developed a wearable electrochemical impedimetric immunosensor with a microfluidic channel and chamber that is based on a $Ti_3C_2T_x$ MXene-loaded laser-burned graphene (LBG) flakes 3D electrode network for noninvasive POC cortisol biomarker detection in human sweat. As a substrate, PDMS was used to create a flexible and stretchable patch sensor. The LBG was then transferred onto the PDMS substrate by removing the polyamide (PI) film. The proposed flexible microfluidic system was built with PDMS and a 3D-printed mold. The fabricated sensor was attached to the skin to collect sweat, which was then moved through the channel by natural pressure to the chamber (Fig. 7.3B) [40]. Feng et al. developed a novel microfluidic–electrochemical (μFEC) detection system. First, a μFEC system based on electrochemical biosensors, which includes screen-printed electrodes (SPEs) with modification of gold nanoflowers (Au NFs) and DNA tetrahedron structural probes (TSPs), was fabricated. Then, microfluidic chips were fabricated using DNA TSP–Au NF-modified SPEs and PDMS-designed dense meandering microchannels. The PDMS microfluidic channels were tightly coupled on the SPEs directly after plasma treatment (Fig. 7.3C). The μFEC detection system was realized with microfluidic chips integrated with the liquid automatic conveying unit and electrochemical detection platform [41]. However, because of its hydrophobicity, PDMS has the disadvantage that it can interfere with hydrophobic molecules or cause nonspecific adsorption. Other issues, such as leaching, channel deformation, and poor tolerance to organic solvents, necessitate the use of another substance or treatment when using PDMS to address these issues. Furthermore, while PDMS is widely used in academia for microfluidic fabrication, it is not suitable for mass production because it is primarily fabricated using soft lithography techniques. As a result, paper-based microfluidic devices have advantages including low cost and disposable POC diagnostics with a key advantage of being a sustainable material addressing the environmental concern [42].

7.2.4 Emerging polymer and polymer composite for 3D-printed microfluidics

The conventional methods for microfluidic fabrication include hot embossing, micromachining, micromilling, and injection molding [43,44], which are labor intensive and time-consuming. The more widely employed low-cost soft lithography allows fabrication of submicron resolution microfluidic structures, though with the downside that it involves multiple steps for fabrication, layer alignment and assembly in the case of 3D devices, and special equipped light-controlled fabrication facilities [45]. Alternative, additive manufacturing, which is more commonly known as 3D printing, is an emerging method for the fabrication of polymer-based microfluidics [46]. Myriad and freeform 3D structure designs could be done using computer-aided design (CAD) tools, allowing the manufacturing process to be autonomous, portable, and redistributable, thus making research open source [47]. These advantages of 3D-printing technology provide a multitude of potential for developing complex microfluidic devices which hasten development and implementation of microfluidics for various applications, in which it is widely used in diagnostics using lab-on-a-chip (LOC) [48–50]. 3D printing provides a simple, quick, and cheap approach to the automatable construction of these microfluidic devices, adaptable for various purposes.

The technique is limited by the printer's resolution and the materials used. Some examples of methodologies of 3D printing are multi-jet modeling (MJM) [51], FDM [43,52–58], and SLA [59–63]. The FDM and SLA have emerged as the most popular platforms for the fabrication of 3D microfluidic devices [64,65].

For FDM printed 3D microfluidics, thermo-responsive polymers are used. Melted thermoplastic filaments are extruded via the nozzle to position and pattern on the platform following the drawing template [55]. Multi-material printing is also possible using FDM [56], and external components (e.g., electrodes, membranes) could be integrated into the print during a pause process. Although theoretically many types of thermoplastic materials could be used with FDM, several materials were more favorable for 3D printing such as PDMS, poly(lactic acid) (PLA), acrylonitrile-butadiene-styrene (ABS), thermoplastic urethane (PLU), polypropylene, PS, PC, and PMMA [58,66–68] which are compatible for mass manufacturing, other than commercial proprietary polymers. For example, PMMA fulfills high optical transparency, good electrical and mechanical properties, low autofluorescence, biocompatible, and high hydrophobicity, hence making it more commonly used in capillary-driven microfluidic systems [52,55,66,69–71]. One's choice for construction materials in FDM printing depends on requirements for optical transparency, electrical and mechanical properties (elastic modulus and tensile strength), autofluorescence, biocompatibility, hydrophobicity [66,71,72]. For example, Kotz et al. demonstrated to create a microfluidic channel using the FDM method based on PMMA. The channel cross-section can be constructed down to $\sim 300\ \mu m$ and can be printed with a variety of different channel geometries and mixer structures. Biomolecules, such as biotin (5-fluorescein) conjugate and streptavidin-Cy3, with characteristic sizes of $125\ \mu m$ can be attached to the inside of the microfluidics channel by the photobleaching method. This work presented the successful biofunctionalization of 3D-printed PMMA microfluidics [55].

Although FDM is more economic compared with other 3D printing techniques, it is still difficult to print features or microchannels narrower than $100\ \mu m$. However, this can be accomplished by a careful selection of printing parameters and materials [73]. To improve the channel size limitation, the parameters of the FDM method were investigated. Bressan et al. proposed 3D-printed microfluidics devices with an integrated PMMA optical transparent window using the FDM method. The minimum production of microfluidic channels by FDM based on PMMA was $175\ \mu m$ using a design width of $300\ \mu m$. The printing parameters were optimized, such as the distance between the extruder nozzle and the PMMA slide, the printing speeds and the infill angle. They found that the actual channel width for FDM is dependent on the design width in the printer setting. The leakage of the PMMA microsystem was tested for 1 hour using a flow rate up to $2\ mL\ min^{-1}$ generated by syringe pumps. There was no observable leakage on the FDM printed microfluidic channel. Finally, the microchannel devices were applied to measure total protein concentration via a colorimetric assay [52]. Microfluidic paper-based analytical devices (PADs) coupled with FDM-3D-printed microfluidics have been studied. Puneeth et al. proposed a microchannel created on chromatography paper using a polycaprolactone (PCL) filament for the hydrophobic boundaries of a microchannel. PCL filaments are biocompatible and have a low melting point. The FDM enabled paper-based microfluidic platform was integrated with silver electrodes, which were fabricated by screen printing method, as shown in Fig. 7.4A. The microfluidic platform was applied for the determination of bovine serum albumin (BSA) and lysozyme before and after denaturation [74]. The FDM presents as an alternative way to pattern hydrophobic

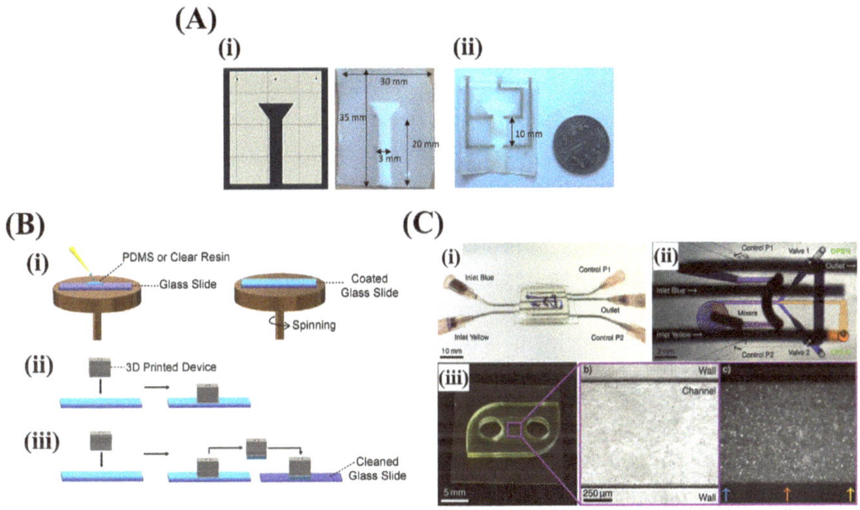

FIGURE 7.4

(A) The microfluidic design. (i) The microfluidic channel on chromatography paper fabricated by PCL. (ii) Silver tracks was screened on the microstrip. (B) Adhesive bonding of 3D-printed microfluidic channels process. (i) The PDMS on a glass slide with different interlayer materials. (ii) Direct bonding technique between 3D-printed device and coated glass slide. (iii) Transfer bonding technique between 3D-printed device and coated glass slide. (C) The microfluidic device. (i) The features of microfluidic device consisted of valve, mixing chamber, inlet, and outlet. (ii) The microfluidic device contained blue and yellow solution in the channels. (iii) The complete device for cell culture in the SLA-printed silicone hydrogel microchannels. *PCL*, Polycaprolactone; *PDMS*, polydimethylsiloxane; *SLA*, stereolithography.

Source: *(A) Reprinted with permission from S.B. Puneeth, et al. Handheld and "Turnkey" 3D printed paper-microfluidic viscometer with on-board microcontroller for smartphone based biosensing applications, Anal. Chim. Acta 1153 (2021) 338303. (B) Reprinted with permission from S. Kecili, et al. Adhesive bonding strategies to fabricate high-strength and transparent 3D printed microfluidic device, Biomicrofluidics 14 (2020) 024113. (C) Reprinted with permission from S. Zips, et al. Biocompatible, flexible, and oxygen-permeable silicone-hydrogel material for stereolithographic printing of microfluidic lab-on-a-chip and cell-culture devices, ACS Appl. Polym. Mater. 3 (2021) 243–258.*

materials on cellulose papers without the use of a wax printer. Two types of thermoplastics with special material characteristics are required for 3D-PAD fabrication, in which one choice of thermoplastic required a low melting point for the form of hydrophobic barriers on to PAD and another type of thermoplastic had a high melting point for printing the 3D feature. Jame et al. proposed the 3D-PAD fabrication using PCL (melting point 50°C–60°C) for the former, and PLA (melting point 170°C) for the latter on the paper substrate. The organic solvent resistivity was studied using ethanol, isopropanol, and acetone. The 3D-PAD was able to hold a large amount of solvent of 30 μL without leakage. However, PLA was partially soluble in acetone in 3 minutes [75].

In the SLA technique, the materials used are photosensitive polymers in liquid form and are produced layer-by-layer via patterning with a pixel-based micromirror array. The UV light or electron beam is used for the polymerization of the resin to achieve the desired structure and design. The main advantage of the SLA technique is high precision on the surface resolution. SLA systems have become highly commercialized and the most widely used 3D printing machines. The 3D-printed substrate can also be bonded to other substrates (e.g., PDMS, glass, etc.) to improve optical clarity or reinforcement by different techniques. Kecili et al. proposed collage bonding strategies for 3D microfluidics, which consisted of three layers. The microfluidic structure was first fabricated using a SLA 3D printer that can make microchannels down to 25 μm. PDMS was coated on a glass slide to be an interlayer material for 3D-printed microfluidic devices. Then, the 3D-printed structure and the glass slide were bonded by UV exposure (Fig. 7.4B). The bonding quality depends on the thickness of the interlayer materials. The proposed bonding method offers a safe integration of proteins inside microfluidic channels and a higher strength bond than traditional plasma processing. The successful fabrication of a biosensor was demonstrated by measuring the fluorescent intensity of IgG antibodies [76]. The SLA printing can be used with other methods, such as the inkjet printing method, for the fabrication of LOC devices. Zips et al. proposed the fabrication of microfluidic structures on different functional polymers (e.g., polyethylene naphthalate, PEN; polyethylene terephthalate, PET; and PDMS) without any need for additional printing steps. The hydroxyl groups were first functionalized on the surface of polymers using the oxygen plasma process. Then, the microfluidic component is covalently bonded to the polymer substrate by 3-(trimethoxysilyl)propyl methacrylate (TMSPMA) molecules during its fabrication. They showed the direct SLA printing onto functionalized substrates and subsequent bonding of two individual parts [77]. Photocurable hydrogels such as poly(ethylene glycol) diacrylates have been widely used in microfluidic fabrication and applied to cell-culture devices due to water-permeability, oxygen-permeability and cytocompatibility. However, the hydrogels are generally brittle and the swelling of hydrogels may cause the deformation of the microfluidic channels. Zips et al. demonstrated the fabrication of silicone-hydrogel hybrid materials using SLA printing for biomedical devices. A 3D-printed microfluidic device is produced layer-by-layer via patterning of a photopolymer with a pixel-based micromirror array. They used the 1 mW cm^{-2} irradiance setting to provide polymer curing while printing. The 3D-printed part was bonded onto a silanized glass slide for the production of microchannels by illumination with UV light in an inert gas atmosphere for the fabrication of optically accessible microchannels and chambers. A microfluidic device combination with horizontal and vertical channels on multiple levels, integrated valves, a circular mixing chamber, and a herringbone mixing structure is shown in Fig. 7.4C. The biocompatible material of silicone-hydrogel material was proven by the growth of the HL^{-1} cardiomyocyte cell layers in the microchannels [78].

7.2.5 Advanced polymer composites for 3D-printed microfluidics

Recent studies have shown that carbon nanomaterials reinforced by polymer filaments as raw materials for the FDM 3D printing method are popular among researchers in 3D printing applications. Carbon fibers are the most commonly used for reinforcement due to their low density and excellent thermal conductivity [79]. Different thermoplastic filaments, such as PLA, ABS, were used to incorporate conductive materials into the polymer matrix using 3D-printing technologies. Cardoso et al. demonstrated the glucose biosensor through 3D FDM printing. The platform was fabricated using PLA containing graphene (G-PLA). The pre-sensor was polished with sandpaper with deionized water and immersed in dimethylformamide (DMF) before being immobilized with glucose oxidase. Mechanical polishing and solvent immersion eliminate excess PLA and increase the availability of electron transfer of graphene in the 3D-printed electrode. The oxygenated group from carboxylic acid present at 3D-printed G-PLA provides a suitable condition for enzymatic immobilization by crosslinking with glutaraldehyde for a biosensor construction to the glucose determination in plasma. The surface treatment of the 3D-printed sensor offers greater electrochemical properties for the direct detection of nitrite and uric acid [80]. The same group (Silva and coworkers) developed a catechol biological sensor based on reduced graphene oxide/3D-printed PLA (rGO-PLA). The integrated electrochemical sensing platform composed of three electrodes: a working electrode, a counter electrode, and a pseudo-reference electrode. DMF is able to remove the excess PLA and expose the graphene structure, in which the tyrosinase enzyme was immobilized on the exposed graphene working. The rGO-PLA electrode successfully detected serotonin in synthetic urine and catechol in natural water samples [81]. The combination of nanocomposite in the filament for 3D printing is interesting since it improves the electrical conductivity and mechanical strength. Bressan et al. developed a 3D-printed microfluidic device and was then built for the synthesis of metal Au@Ag nanoparticles from the core and shell. A 3D-printed microfluidic device consisting of three inlets for the loading of reactants and an outlet to collect the synthesized nanoparticles, presenting channels with width and height of approximately 260 μm. The microfluidic channels had a width of 260 μm and the core and shell nanoparticles had a diameter of 23 nm. Silver salt on gold nanoparticle seeds was reduced by sodium borohydride in a continuous flow manner [73].

7.2.6 Paper-based microfluidics

Paper is a flexible cellulose-based material that has recently emerged as a promising microfluidic substrate to develop paper-based microfluidic devices. Paper-based microfluidics are microfluidic devices consisting of a series of hydrophilic cellulose or nitrocellulose fibers that use capillary action to transport fluid from an inlet through the porous medium to a desired outlet or region [82]. The patterning paper for paper-based microfluidic devices can be created using a variety of techniques, including paper cutting, wax printing, plasma treatment, photolithography, screen printing, and flexographic printing. Each technique aims to define hydrophilic channels and zones on a piece of paper by creating well-defined patterns of hydrophobic barriers [83]. The ability to use paper-based microfluidic devices in the field rather than in the lab is their main advantage over traditional microfluidic devices. It is small in size (approximately a few mm to 2 cm in length and width), simple to fabricate without cleanroom, lower cost, and disposable, as well as the ability to drive liquid without the use of pumps or other supporting equipment. Moreover, the development of POC diagnostic devices is the most common application of paper-based microfluidic devices, which could eliminate the need for costly and time-consuming laboratory-based analytical procedures. Dungchai and coworkers appear to have been the first to report the used of electrochemical detection for paper-based microfluidic devices to determine glucose, lactate, and uric acid in biological samples using oxidase enzyme reactions (glucose oxidase, lactate oxidase, and uricase, respectively). Microfluidic channels were fabricated on a filter paper using the photolithography technique, and electrodes were fabricated on paper-based microfluidic devices using the screen printing technique (Fig. 7.5A). In control serum samples, chronoamperometry was used to measure glucose, lactate, and uric acid. This proposed paper-based microfluidics demonstrates the successful integration of paper-based microfluidics and electrochemical detection as an easy-to-use, inexpensive, and portable alternative for POC monitoring [84].

For automated colorimetric enzyme-linked immunosorbent assays (ELISAs), Fu et al. created a paper-based microfluidic platform with thermally sensitive shape-memory polymer (Fig. 7.5B) [85]. Iyer et al. used a screen printing technique to create a color meter for complete protein detection on a paper-based microfluidic system [86]. Muthukumar et al. created a paper-based microfluidic lab on a chip for colorimetric palm oil detection in sunflower oil [87]. Using wax printed to build a hydrophobic zone on the filter paper, Hao et al. created a microfluidic paper-based system with symmetric channel distribution. Within 10 minutes, the total polyphenol content (TPC) of tea was determined simultaneously [88]. Lin et al. used a water-based photosensitive resin, polyurethane acrylate

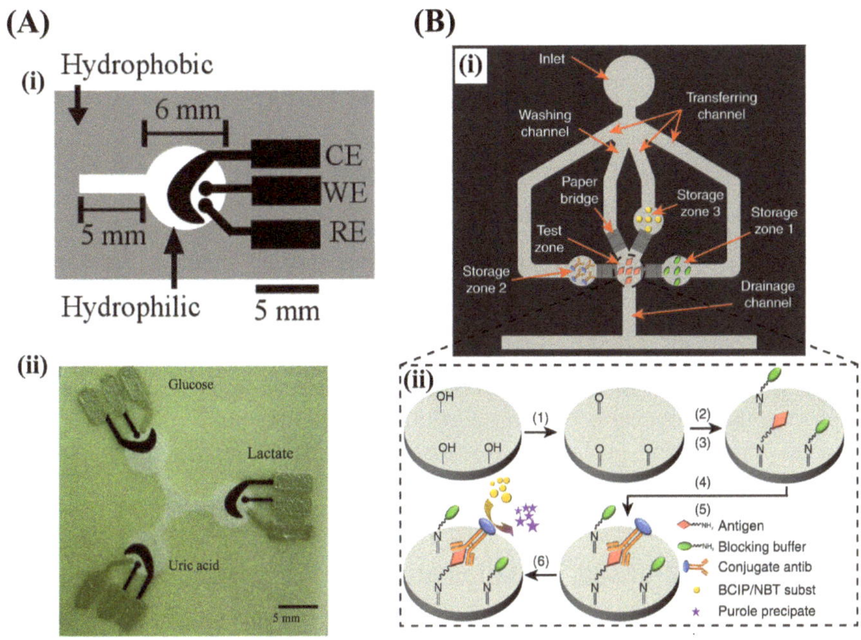

FIGURE 7.5

(A) The paper-based microfluidic devices for the determination of glucose, uric acid, and lactate. (i) The design of the electrochemical detection cell for paper-based microfluidic devices consisted of WE, RE, CE. (ii) Picture of three electrode paper-based microfluidic devices consisted of the hydrophilic area at the center of the device wicks sample into the three separate test zones where independent enzyme reactions occur (the device size of 4 cm × 4 cm). (B) The PAD design and direct ELISA protocol are shown in schematic diagrams. (i) Schematic top view of a μPAD integrating four SMP-actuated valves. (ii) The schematic protocol of a direct ELISA that can be performed in the test zone of the μPAD. A direct ELISA protocol is divided into six steps: (1) using potassium periodate (KIO_4) to modify the paper test zone surface with aldehyde groups; (2) covalently bonding antigens to the modified paper; (3) blocking the test zone surface with blocking buffer; (4) labeling antigens with enzyme-conjugated antibodies; (5) washing away unbound antibodies; and (6) adding enzyme substrate for colorimetric signal production. *WE*, working electrode; *RE*, reference electrode; *CE*, Counter electrode; *PAD*, paper-based analytical devices; *ELISA*, enzyme-linked immunosorbent assay; *μPAD*, microfluidic paper-based analytical devices; *SMP*, shape-memory polymer.

Source*: (A) Reprinted with permission from W. Dungchai, et al., Electrochemical detection for paper-based microfluidics, Anal. Chem. 81 (2009) 5821–5826. (B) Reprinted with permission from H. Fu, et al., A paper-based microfluidic platform with shape-memory-polymer-actuated fluid valves for automated multi-step immunoassays, Microsyst. Nanoeng. 5 (2019) 1-12.*

(PUA), to fabricate a microfluidic paper-based analytical system that they used for colorimetric detection of *Escherichia coli* as a model goal [89].

7.3 Emerging materials for improved signal transduction in microfluidic biosensor

Advanced nanomaterials (i.e., nano nanomaterials and 2D materials) have proved to improve the performance of microfluidics biosensors enabling better signal transduction and tailored interfacial properties with sample and transducers. Among all, graphene, metal chalcogenides, metal oxides, metal hydroxides, and MXenes, as well as black phosphorous, have all made significant impacts on improving the analytical performance of microfluidic biosensors over the last few decades. This section focuses on the applications of various emerging materials for improved signal transduction in microfluidic biosensors.

7.3.1 Graphene-based 2D nanomaterials

Graphene (Gr) is a novel 2D carbon material that has emerged as a rapidly rising star in the field of material science. Gr consists of a single layer of a sp^2-hybridized carbon lattice with excellent electrical properties (200,000 $cm^2\ V^{-1}\ s^{-1}$) [90]. Gr and its derivatives can be produced by oxidation of graphite, which has a light-yellow color form and relatively stable aqueous colloids to facilitate assembly of macroscopic structures, which are well known as graphene oxide (GO). GO is one of the derivatives Gr that has low electrical conductivity. However, GO can be improved to higher conductance by reducing the oxygen contained molecules on the GO surface via chemical, electrochemical, and thermal methods to form rGO. Graphene quantum dots (GQDs) are another type of graphene derivative with one or few-layered graphene sheets whose size is <100 nm [91]. Furthermore, GQDs are expected to have the significant unique properties of graphene because of their good biocompatibility, suitable conductivity, and low toxicity [91,92]. Thus, GQDs have gained many applications in electrochemical and optical sensing platforms and bioimaging [92]. Recently, the combination of Gr and its derivatives with other materials (polymer, metal nanoparticles, and other materials) to construct the new nanocomposite for improving the signal transduction of microfluidic electrochemical sensing and biosensing platforms. Furthermore, Gr-based microfluidic devices for integrating with an electrochemical sensor and biosensor can offer several advantages, including real-time analysis, rapid diagnosis, increased sensitivity, the use of small sample volumes, and cost savings.

Gr-functionalized polymer and metal nanoparticles with integrated microfluidic electrochemical sensors have been created and used for improving the analytical performance of biosensors. Liu et al. reported the fabrication of portable microfluidic devices based on enzyme-free glucose sensing by using the nanocomposite of chitosan grafted reduced Gr with decorated nickel nanoparticles (CS-rGO–NiNPs) modified SPE. Fig. 7.6A shows the electrodeposition processes of CS-rGO–NiNPs on the SPE surface and its glucose detection mechanism. The synthesis reaction starts with the electrostatic and hydrophobic interactions between negatively charged GO and positively charged Ni^{2+} and CS chains to form nanocomposite precursors. Protons in the acid solution around the electrode surface are consumed when a cathodic potential is applied, resulting in a relatively high localized pH. As a result, CS chains in nanocomposite precursors near the electrode become insoluble, and GO and Ni^{2+} are deposited from bulk solution together with CS chains. On the electrode surface, GO and Ni^{2+} are in situ reduced into rGO and NiNPs and deposited simultaneously with the deposition of CS. The detection reaction for glucose begins with the oxidation of $Ni(OH)_2$ and NiO species when anodic constant potentials are applied. These NiO species are then further oxidized to form strongly oxidizing Ni(III) as NiO(OH) species, which electrochemically catalyzes the oxidation of glucose into gluconolactone, liberating two hydrogen atoms. Meanwhile, NiO(OH) is reduced back into NiO species. This CS-rGO–NiNPs/SPE was integrated with a portable electrochemical microfluidic system, in which the microfluidic device was constructed using PDMS and the device holder was prepared using ABS plastic. The PDMS chamber was 7.0 mm in diameter and 0.2 mm in height, and two holes were drilled in it to accept the inlet and outlet tubes. To set up the microfluidic device, the PDMS chamber was carefully placed on top of the active area of the modified SPE sensor and fixed with the ABS holder. In the detection procedure, a sample volume of 7.7 μL was manually delivered into the detection area in the fluidic chamber via a 10 mL microsyringe that connected with the inlet tube. This electrochemical microfluidic device performed well in terms of sensing, had good operational characteristics, was simple to make, and was inexpensive. for the development of this electrochemical microfluidic device as a POC healthcare monitoring device [2]. Liu et al. reported the use of the poly(diallyl dimethylammonium) chloride functionalized on rGO (PDDA-rGO) to improve the electrostatic charges and the dispersibility of rGO, and used for simultaneous quantification of dopamine and 5-hydroxytryptamine on multi-walled carbon nanotube carbon paste electrodes with integrated microfluidic devices.

The microfluidic system is composed of glass microchannels and electrodes. Wet etching with HF solution was used to create the fluidic channels for the working electrode, reference electrode, and counter electrode. The device was then assembled face-to-face using prepared PDMS sheets with holes punched using a hollow needle and glass surfaces, then activated in a plasma machine to bond and form the device. After that, silver/silver chloride (Ag/AgCl) wire electrode, Pt wire, and MWCNT-CPE/PDDA-Gr fused-silica capillary tube were inserted into the hollow needle on the PDMS sheets to form a microfluidic electrochemical sensor device (Fig. 7.6B). The microfluidic electrochemical device showed high response for both analytes with excellent sensitivity, wide linear range, and good stability and repeatability for 20 μL sample detection [93].

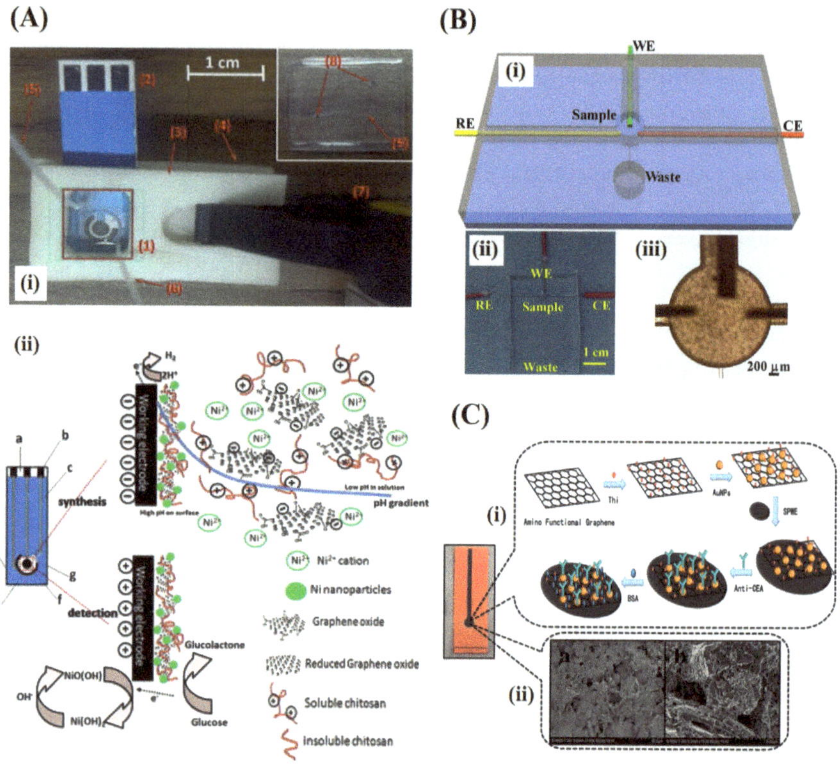

FIGURE 7.6

(A) (i) Photograph of the as-fabricated microfluidic electrochemical device with (ii) diagram of the electrodeposition processes of CS-RGO–NiNPs on SPE surface and its Glc detection mechanism. (B) (i) Schematic of the microfluidic electrochemical sensor device. (ii) A photograph of the three-electrode system coupled with a microfluidic device. (iii) Microscopic image of a microvolume detecting cell. WE, RE, and CE represent working, reference, and counter electrodes, respectively. (C) (i) The preparation and (ii) SEM images of NH_2-G/Thi/AuNPs-SPW. *CS-RGO–NiNPs*, chitosan-reduced graphene oxide–nickel nanoparticles; *SPE*, screen-printed electrode; *Glc*, glucose; *WE*, working electrode; *RE*, reference electrode; *CE*, Counter electrode; *SEM*, scanning electron microscopy; *NH_2-G/Thi/AuNPs-SPW*, amino functional graphene/thionine/gold nanoparticles nanocomposites.

Source: *(A) Reprinted with permission from J. Yang, et al., Nickel nanoparticle–chitosan-reduced graphene oxide-modified screen-printed electrodes for enzyme-free glucose sensing in portable microfluidic devices, Biosens. Bioelectron. 47 (2013) 530–538. (B) Reprinted with permission from Z. Liu,et al., Electrochemical sensor integrated microfluidic device for sensitive and simultaneous quantification of dopamine and 5-hydroxytryptamine, Sens. Actuators B: Chem. 273 (2018) 873–883. (C) Reprinted with permission from Y. Wang, et al., A novel label-free microfluidic paper-based immunosensor for highly sensitive electrochemical detection of carcinoembryonic antigen, Biosens. Bioelectron. 83 (2016) 319–326.*

Besides, a Gr-based microfluidic immunosensor has been developed, in which biomolecule (antibody, DNA, enzyme, and aptamer) can be immobilized with a functional group on graphene surface (carboxylic or amino group) by specific reactive groups, including glutaraldehyde, ethyl (dimethylamino propyl) carbodiimide/N-hydroxysuccinimide (EDC/NHS) coupling chemistry. For example, Garg et al. reported a LOC-based electrochemical detection of ferritin using amine group functionalized GO functionalized ferritin antibodies with cost on a screen-printed working electrode. Using CNC milling, a microfluidic flow cell was designed and fabricated from two layers of PMMA. Both layers of the plates and the electrode were clamped together by a magnet, and an O-ring ensured leak-free operation. A syringe pump was used to pump the sample liquid through the microfluidic chip at a flow rate of 10 μL min^{-1}. This microfluidic electrochemical ferritin sensor shows a dynamic linear range from 7.81 to 500 ng mL^{-1} and the limit of detection (LOD) of 0.413 ng mL^{-1} [4].

Singh et al. constructed a microfluidic chip integrated with an rGO for electrochemical label-free influenza virus immunosensor (H1N1). The microelectrodes and microfluidic chips were fabricated by photolithography and integrated with PDMS microchannels on a glass substrate. The antibody-H1N1 was immobilized on EDC/NHS activated rGO on the working electrode. PDMS microchannels were fabricated on a silicon wafer. Tubing was set up manually using tips and tubes, followed by connecting a syringe pump to the inlet mold for a microfluidic channel. Prepared PDMS microchannels were sealed on the glass chips using conformal contact and the sample solution was injected into the system by a syringe pump with a flow rate of 130 μL min^{-1}. This microfluidic device performed chronoamperometry and exhibited good selectivity and a low detection limit of 0.5 PFU mL^{-1} [6].

Furthermore, Wang et al. developed a microfluidic immunosensor for the detection of carcinoembryonic antigen (CEA) using amino functional graphene (NH_2-Gr) combined with thionine (Thi) and AuNPs to form nanocomposites for modifying a screen-printed working electrode (Fig. 7.6C), in which the sensor showed a linear working range of 50 pg mL^{-1} to 500 ng mL^{-1} and the LOD for CEA was 10 pg mL^{-1}. The microfluidic PAD was fabricated on two pieces that were integrated using double-sided tape. The three SPEs (working electrode, reference electrode, and counter electrode) were connected once the paper electrochemical cell was filled with solution and samples [7]. Cincotto et al. created a disposable microfluidic electrochemical paper-based device (ePAD) consisting of two spot sensors in the same working electrode for the simultaneous determination of uric acid and creatinine, which GQDs were used to modify on spot 1 for direct uric acid oxidation, and a nanocomposite consisting of GQDs, creatininase, and ruthenium. Filter paper was used to create the microfluidic channel. The ePAD was made by using a double tape cut with a cutter printer to sandwich the sheets and paper. The device was sealed using a pressing machine, with the heating system off. The sample was loaded to the detection zone of microfluidic ePAD by simple injection and was delivered by a passive process. The device showed excellent analytical performance with high selectivity for uric acid and creatinine in the human urine samples [94]. In addition, it showed a linear working range of 50 pg mL^{-1} to 500 ng mL^{-1}, and the LOD for CEA was 10 pg mL^{-1} [7]. The development of 2D graphene-based nanocomposite for the emerging 2D nanomaterials to fabricate electrochemical microfluidic devices is still important for improved signal transduction in microfluidic optical and electrochemical biosensing. With 2D graphene-based nanocomposites, it will be possible for microfluidic electrochemical devices to be applied to pre-warning and real-time detection in many fields.

7.3.2 2D transition metal nanomaterials

Due to their large surface-to-volume ratios and electron confinement, 2D, transition metal nanomaterials have outperformed common 2D materials like graphene in catalysis, sensing, and many other surface-enhanced applications [95]. Over the past few decades, significant progress has been made in the synthesis of a variety of 2D transition metal nanomaterials, which can be divided into two categories: transition metal oxides (TMOs) and transition metal chalcogenides (TMDs).

7.3.2.1 Transition metal oxides

2D TMOs consist of oxides of metals that are pre-transitioned, transitioned, or post-transitioned, which shows distinct differences in comparison to their bulk counterparts [96–99]. In TMOs, the transition metal electrons are strongly pulled by oxygen, and consequently, the structural, physical, and chemical properties are mostly determined by the strongly correlated d electrons [97]. TMOs are also highly tunable owing to the diversity of their chemical composition, crystal structure, and relative ease in inducing oxygen defects. A wide range of 2D TMOs-based transducers was used in microfluidic biosensors with improved signal transduction in electrochemical sensing and biosensing.

For example, Alizadeh et al. reported using cerium oxide nanosheets as a mimic enzyme to detect H_2O_2 in living cells and integrated with an analytical microfluidic chip, which was made from PDMS. The fabrication process is divided into three major steps: (1) the fabrication of a three-electrode setup, (2) PDMS casting, and (3) plasma bonding of PDMS over the prepared electrodes on the glass substrate. The width and depth of the microfluidic channel were both 200 μm. The H_2O_2 sample solution was injected into the chip with peristaltic pumps which were connected to the inlet of the microfluidic device. This microfluidic sensor based on cerium oxide nanosheets shows a strong response to H_2O_2 with presents a high sensitivity over a wide detection range and a low detection limit at the nanomolar level. This microfluidic chip also provides a promising platform for applications in the field of diagnostics and sensing [100]. In addition, cerium dioxide was used to couple with AuNPs and glucose oxidase as an electrochemical probe for signal amplification in μPADs constructed for sensitive detection of microRNA (miRNA) by Sun et al. CeO_2 acted as an electrocatalyst for

H_2O_2, which is the product from catalyzing glucose of GOx. The μPADs were fabricated by printing using a commercially available solid-wax printer, which was designed as a three-electrode system and modified working electrode surface by a direct drop-casting method. Sample solutions were injected directly into the detection zone of the devices (Fig. 7.7A). This microfluidic paper-based electrochemical sensing platform exhibited a wide linear range and low detection limit in femtomolar (fM) [101]. Ali et al. fabricated a uniquely structured immunoelectron made of electrospun carbon-doped titanium dioxide nanofibers ($nTiO_2$) modified on porous hierarchical graphene foam (GF) as a working electrode for label-free microfluidic immunosensor for early detection of epidermal growth factor receptor 2 (EGFR2 or ErbB2) proteins. The channel of the microfluidic device was made for PDMS using the soft lithographic method, which used PMMA as a mold. The PDMS slab containing the immunoelectrode was permanently bonded to the glass substrate containing the counter (CE) and reference (RE) electrodes by oxygen plasma treatment (Fig. 7.7B). The microfluidic device integrated with the GF–$nTiO_2$ composite-based electrode with a large specific surface area, porous structure, and 3D conductive pathways and provides remarkable detection of breast cancer biomarkers [102].

7.3.2.2 Transition metal chalcogenides

TMDs are 2D nanomaterials composed of a layer of transition metals (e.g., tungsten [W], molybdenum [Mo], vanadium [V], and niobium [Nb]) covalently bonded between two layers of chalcogen atoms (e.g., selenium [Se] or sulfur [S]). However, when compared to 2D graphene, the single sheets of TMDs are thicker and less mechanically strong, and they exhibit different electrical behavior due to their indirect bandgap. For example, tungsten disulfide (WS_2) and molybdenum disulfide (MoS_2) are semiconductors, while vanadium diselenide (VSe_2) and niobium disulfide (NbS_2) are conductors. TMDs have currently shown great promise as emerging nanomaterials for improved signal transduction in microfluidic optical and electrochemical sensors and biosensors due to their ease of synthesis, high biocompatibility, exceptional physicochemical properties, and specifically at the nanoscale, similar to graphene and other 2D materials [5]. The use of 2D TMDs nanomaterials in microfluidic optical and electrochemical sensors and biosensing has been highlighted as a way to improve signal transduction. For *S.* Typhimurium detection, Singh et al. used MoS_2 nanosheets functionalized with cetyltrimethylammonium bromide (CTAB-MoS_2-NS) conjugated with anti-*S.* Typhimurium on the indium tin oxide (ITO) microelectrode (Fig. 7.7C). The microelectrodes were integrated with a Y-shaped PDMS channel prepared by soft lithography technique forming the microfluidic biosensors. The channels of the microfluidic chip have a height and width of 200 μm and a length of 2.0 cm. The fluid flow of samples loaded into the microfluidic channels was controlled by syringe pumps [5]. TMDs have also been used for catalytic sensing applications. For example, Toh et al. have integrated four t-BuLi exfoliated TMD materials (MoS_2, $MoSe_2$, WS_2, and WSe_2) within the LOC devices by the drop-casting

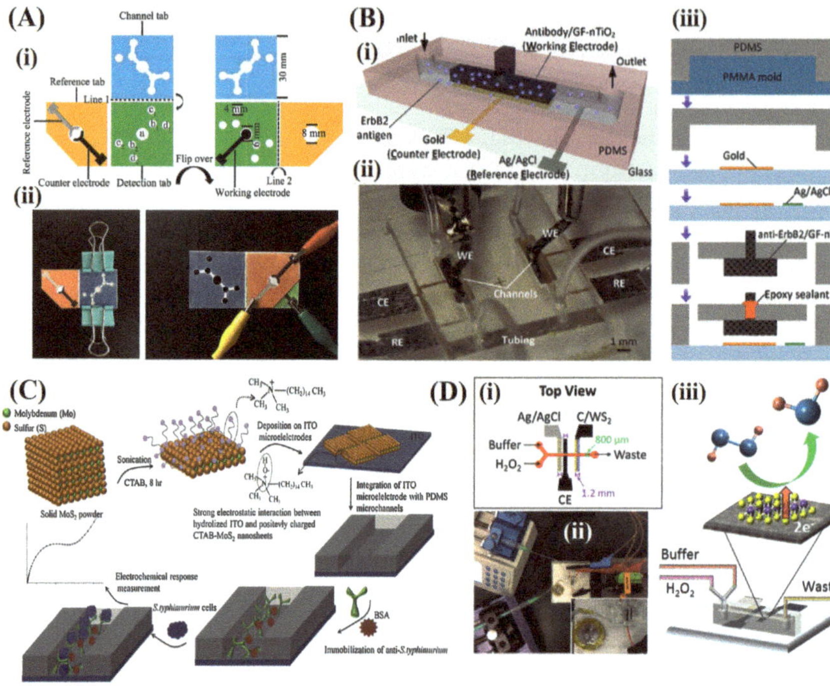

FIGURE 7.7

(A) (i) The schematic representation, size, and shape of μPADs. (ii) The fabrication of microfluidic paper-based analytical devices for detection of miRNA combining chromogenic reaction and electrochemistry. (B) (i) Schematic of the microfluidic immunosensor with 3D porous GF electrode modified with carbon-doped TiO_2 nanofibers for the detection of breast cancer biomarkers. (ii) Photo of two microfluidic immunosensors. (iii) Schematic representation of the fabrication of the microfluidic sensor. (C) Schematic representation of designing of CTAB-MoS_2-NS based microfluidic biosensor for *S* Typhimurium detection. (D) Schematic representation (not to scale) (i) and photograph (ii) of lab-on-a-chip setup for the electrocatalytic reduction of H_2O_2 with $1 coin for scale (iii). *μPAD*, microfluidic paper-based analytical devices; *miRNA*, microRNA; *GF*, Graphene foam; *CTAB-MoS$_2$-NS*, cetyltrimethyl ammonium bromide functionalized MoS_2 nanosheets.

Source: *(A) Reprinted with permission from X. Sun, et al., Ultrasensitive microfluidic paper-based electrochemical/visual biosensor based on spherical-like cerium dioxide catalyst for miR-21 detection, Biosens. Bioelectron. 105 (2018) 218–225. (B) Reprinted with permission from M. A. Ali, et al., Microfluidic immuno-biochip for detection of breast cancer biomarkers using hierarchical composite of porous graphene and titanium dioxide nanofibers, ACS Appl. Mater. Interfaces 8(32) (2016) 20570–20582. (C) Reprinted with permission from C. Singh, et al., Functionalized MoS$_2$ nanosheets assembled microfluidic immunosensor for highly sensitive detection of food pathogen, Sens. Actuators B: Chem. 259 (2018) 1090–1098. (D) Reprinted with permission from R.J. Toh, et al., Group 6 layered transition-metal dichalcogenides in lab-on-a-chip devices: 1T-phase WS2 for microfluidics non-enzymatic detection of hydrogen peroxide, Anal. Chem. 89 (2017) 4978–4985.*

method and compared their performance for H_2O_2 detection. The electrochemical LOC was made from PDMS integrated with the TMD modified electrodes. The PDMS microfluidic device with 49-μm deep channels was fabricated using photolithography techniques with SU8 2035 photoresist. The LOC was set up by using syringe pumps connected to a Y-connector at the inlet of the device (Fig. 7.7D), which injected phosphate-buffered saline (PBS) (10 mM, pH 7.4) into the microfluidic chips at a flow rate of 40 μL min^{-1} to obtain a baseline, followed by the injection of H_2O_2 (5 μL) sample into the channel at 50 μL min^{-1}. The result showed that the 1T-phase WS_2-based LOC device exhibited wide linear range (20 nM to 20 μM and 100 μM to 2 mM), a low detection limit (2.0 nM), and good selectivity and feasibility for H_2O_2 determination in human serum [103].

7.3.3 MXenes

MXenes, which include transition metal carbides, nitrides, and carbonitrides, are a new class of 2D layered materials that have emerged as graphene analogs. MXenes, which have a layered structure, metallic conductivity, and hydrophilicity due to the presence of a negative charge on the surface, were first introduced in 2011. MXenes are synthesized by a controlled acid etching of the "A" group element from "MAX" phase powders ("M" is a transition metal, "A" is an element mostly from groups 13 and 14 of a periodic table, and "X" is a carbon or a nitrogen atom) resulting in a 2D layered structure.

Recently, nanostructured MXene has emerged as a crucial 2D material to fabricate the highly sensitive electrochemical sensor and biosensor devices owing to its low framework density, large hydrophilic surface, high electrochemical activity, and good electrical conductivity. Moreover, when compared to 2D graphene, MXene has high conductivity (9880 S cm^{-1}) [104], high Young's modulus (333 ± 30 GPa) [105], and high water flux (85.4 L m^{-2} h^{-1}) with excellent salt rejection (~99.5%) properties [106]. For example, MXene-Ti_3C_2Tx-based SPE incorporated with a dialysis microfluidic chip is constructed for a direct and continuous multicomponent analysis of whole blood by Liu and coworkers. For the fabrication of the microfluidic chip, the PMMA plate was used as the chip substrate (length: 60 mm; width: 40 mm), where the channel (width: 1 mm; depth: 0.5 mm) and the chamber (diameter: 5.75 mm; depth: 1 mm) on the substrate were prepared via laser etching. The microfluidic chip consists of four layers. The top layer has a channel for setting aside blood flow, and the second layer is a dialysis membrane for allowing the molecules smaller than 1000 Da to pass. The third layer possesses a flow channel for isotonic solutions, in which blood can be dialyzed into this channel and gathered in the sensing area of the electrode located in the bottom layer. Each layer of the microfluidic chip was assembled via screw tightening. The sample solution was pumped into a microfluidic chip via the lower channel at a flow rate of 30 μL min^{-1}. This device is applied in hemodialysis, and a continuous assay of uric acid, urea, and creatinine levels in human blood is realized, which shows a new path for 2D MXene in microfluidic biomedical and sensing applications [107].

7.3.4 Black phosphorus

2D black phosphorus (2D BP) is a new 2D nanomaterial that was first discovered in 2014. Nanostructured BP (nano-BPs) include BP nanosheets, BP QDs, and BP nanoparticles [8,56]. 2D BP is a metal-free layered semiconductor that consists of P atoms that are covalently bonded with three neighboring P atoms by sp3 hybridization to form a puckered honeycomb structure and have the thickness-dependent bandgap tunable from ~0.3 eV for the bulk to 2.0 eV for a single layer [8]. 2D BP possesses a much higher surface-to-volume ratio due to the puckered lattice configuration and has more excellent biocompatibility and lower toxicity compared with other 2D materials [108] and also shows unique semiconducting characteristics of extremely high hole mobility, anisotropic conductance and higher surface-to-volume ratio led to enhances the loading capacity of functional molecules compared to other 2D materials (graphene, MoS_2) [107–109]. In recent years, some researchers have used BP in the field of electrochemical sensors and devices, which are often assembled with metal nanoparticles [110] for immobilizing biomolecule (i.e., antibodies [111], aptamers [110]). Moreover, 2D BP nanomaterials have been used to improve signal transduction in microfluidic electrochemical biosensing. For example, Chen et al. reported the use of few-layer BP nanosheets as support material for gold nanoparticles (BP-Au) and immobilized aptamer that specific to okadaic acid and modified SPCE surface-based microfluidic aptasensor for the detection of okadaic acid Fig. 7.8A and B.

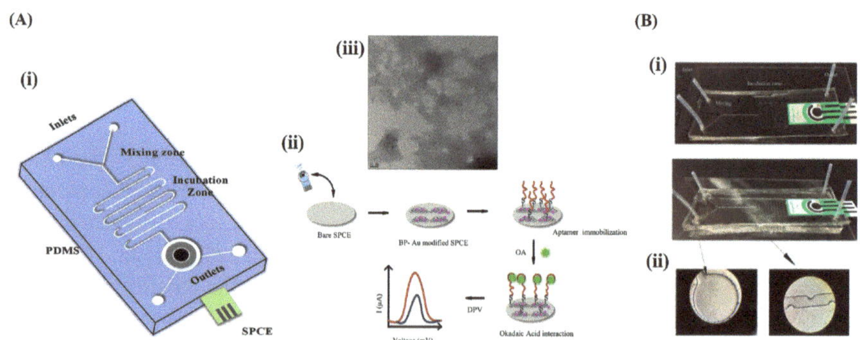

FIGURE 7.8

(A) (i) Graphical representation of the fabricated PDMS microfluidic chip with (ii) schematic representation of the process of BP-Au based aptamer sensing and (iii) SEM of BP-Au. (B) (i) Prototype of the final sensor and (ii) PDMS channels containing inlet, mixing channels, incubation chambers and modified SPCE on a glass substrate (for illustration). *PDMS*, polydimethylsiloxane; *BP*, Black phosphorus; *SEM*, scanning electron microscopy; *SPCE*, screen-printed carbon electrode.

Source: *Reprinted with permission from S. Ramalingam, et al., Phosphorene-gold nanocomposite based microfluidic aptasensor for the detection of okadaic acid, Biosens. Bioelectron. 135 (2019) 14–21.*

In this research, BP–Au nanocomposite modified SPCE was applied as a supporting material and amplifier of the signal transduction in a microfluidic electrochemical biosensing system. The fabrication of a flexible microfluidic chip was constructed by PDMS using the soft lithographic technique. The microfluidic chip was designed as a Y-channel microfluidic device comprising two inlets, an incubation channel, a sensing zone, and an outlet. The width of each channel was set at 200 μm and the sensing zone had a diameter of 4 mm. The PDMS with a channel bearing layer for Apt-BP-Au/SPCE was precisely glued onto the groove of the bottom layer and then treated with oxygen plasma for 30 seconds to seal the top and the bottom layers together. 50 μL of the sample was delivered into the microfluidic-modified SPCE by manual injecting [110].

Advanced nanomaterials are demonstrated as a promising material for future development of high-performance microfluidic electrochemical sensors and biosensors with excellent electrocatalytic property, admirable enzyme-loading capacity, good biocompatibility, and abundant surface-terminating groups. A variety of advanced nanomaterials, including graphene, metal chalcogenides, metal oxides, metal hydroxides, MXenes, and BP, have been utilized for improved signal transduction in various microfluidic electrochemical sensing and biosensing system summarized in Table 7.1.

7.4 Conclusions and outlooks

Advanced functional materials are important for the fabrication of innovative and complex functional microfluidic platforms, as well as to create high-performance electrochemical transducers for microfluidic biosensors. An overview of creating material functions for the fabrication of microfluidic units for biosensors and an overview of emerging advanced nanomaterials for improved signal transduction in microfluidic biosensors are presented in Figs. 7.9 and 7.10.

It should be noted that the three major considerations related to the design and development of microfluidic biosensors are (1) creating material functions (fabrication process for material functions); (2) modifying or changing material functions (what are the modifications and changes that can be made to material functions); and (3) using material functions (how to use material functions within your materials). These considerations are often overlooked but these are required to be realized in practical microfluidic-based biosensors.

Based on existing materials such as polymeric compounds, advanced compositing and reinforcing of polymers in tandem with adapting various fabrication techniques and surface treatments are emerging as a viable enhancement to fulfill new device design requirements for various novel applications in microfluidics. In combination with 3D printing, which allows automation of construction of microfluidic devices other than its intricate freeform design capabilities, particularly FDM and SLA, which are new approaches that have drawn dramatic

Table 7.1 Summary table of electrochemical microfluidic devices which used advanced nanomaterials-based working electrodes with their analytical performance.

Materials-based microfluidics	Materials-based WE	Sample volume (μL)	Analytes	LOD	Linearity	Real sample	Ref.
PDMS-based Chip	PDDA-RGO/MWCNT	20	Dopamine	0.016 μM	0.05–120 μM	Rat's plasma	[93]
PDMS-based Chip	CS-rGO–NiNPs	7.7	Glucose	4.1 μM	Up to 9 mM	Human urine	[2]
PMMA-based Chip	Ab-NH_2-GO	3.30	Ferritin	0.413 ng mL^{-1}	7.81–500 ng mL^{-1}	Human serum	[4]
ePAD	Ab-NH_2G/Thi/AuNPs	–	CEA	10 pg mL^{-1}	0.05–500 ng mL^{-1}	Human serum	[7]
PMMA based Chip	Ab/RGO/CA/Au	–	H1N1 virus	0.5 PFU mL^{-1}	1–104 PFU mL^{-1}	–	[6]
ePAD	GQDs	20	Uric acid	8.4 nM	0.010–3.0 μM	Human urine	[94]
PDMS-based Chip	CTAB-MoS_2	–	*Salmonella* Typhimurium	1.56 CFU mL^{-1}	10^1–10^7 CFU mL^{-1}	–	[5]
PDMS-based Chip	WS_2	–	H_2O_2	2.0 nM	0.02–2000 μM	–	[103]
PMMA-based Chip	MXene	–	Uric acid	5.0 μM	30–500 μM	Blood samples	[107]
PMMA-based Chip	Urease-MB/MXene	–	Urea	–	Up to 3 mM	Blood samples	[107]
PMMA-based Chip	MXene	–	Creatinine	1.6 μM	10–400 μM	Blood samples	[107]
PDMS-based Chip	Apta-Au BP-	–	Okadaic acid	8.0 pM	10–250 nM	Mussel extract	[110]

GQD, Graphene quantum dot; *MWCNT*, multiwall carbon nanotube; *PDDA*, poly(diallyl dimethylammonium); *PDMS*, polydimethylsiloxane; *PMMA*, poly(methyl methacrylate); *WE*, working electrode; *LOD*, limit of detection; *CEA*, carcinoembryonic antigen.

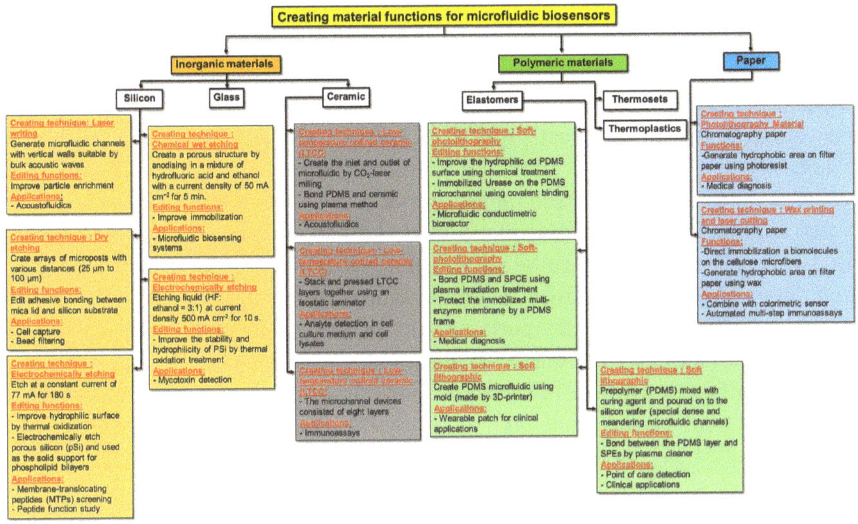

FIGURE 7.9

Overview of creating material functions for microfluidic biosensors.

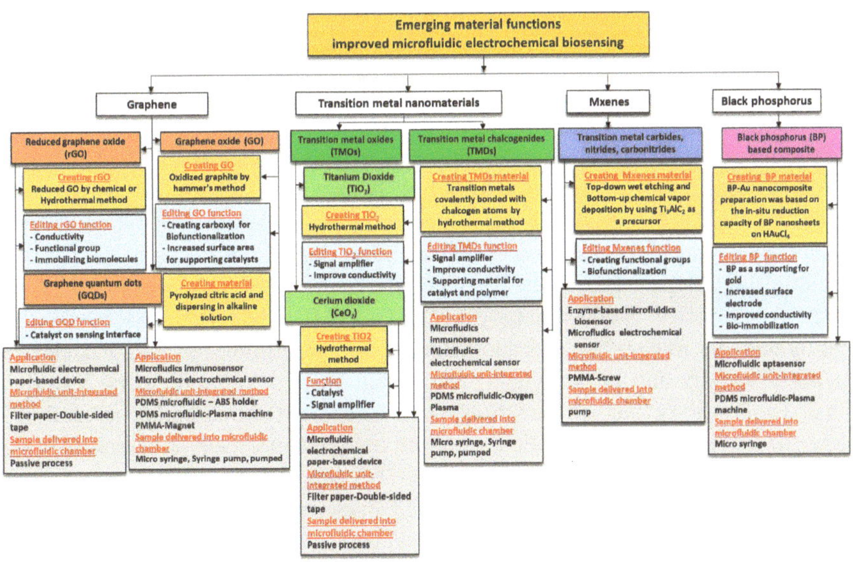

FIGURE 7.10

Overview of emerging advanced nanomaterials for improved signal transduction in microfluidic biosensors.

attention to fabricating complexed 3D microfluidics structure that is unable to be easily created in the past. Elastomers, thermosets, and thermoplastics are the most popular polymers used to fabricate microfluidic devices due to their variety of properties choosable suiting different applications, such as thermal, mechanical, and chemical stability, optical properties, and the ability to design complex 3D systems by multilayer fabrication and fluidic interconnects, and also their potential for compositing and surface modifications.

An alternative microfluidics platform is the paper-based microfluidics, utilizing a low-cost hydrophilic cellulose substrate that functions via manipulation of fluid flow guided by a defined capillary action within a patterned hydrophobic barrier, with devices that are portable and support fluid flow without external forces. Intricately designed microchannels and microstructures enable the device with capabilities to perform full analysis, including continuous sample collection, analyte isolation, mixing, and pre-concentration. Sensors can either be enzyme-based, immunoassay-based, DNA-based, or used in tandem with electrochemical detection using 2D nanomaterials for signal transduction amplification for enhanced sensitivity and specificity. Integration of paper-based microfluidics and electrochemical detection exponentially improves analytical properties, performance, and throughput as an alternative to POC monitoring.

Graphene, metal chalcogenides, metal oxides, metal hydroxides, MXenes, BP, as well as transition metals have all made significant advances in the synthesis of a variety of nanocomposite materials over the last decade and their development is still in demand. TMO properties are diverse due to their variable chemical composition, crystal structure, and relative ease in inducing oxygen defects. Transition metal dichalcogenides exhibit exceptional physicochemical properties due to their unique chalcogen-atoms-layered transition metal composition. MXenes have a large hydrophilic surface, high electrochemical activity, and electrical conductivity due to the presence of a negative charge on the surface and have also emerged as a crucial 2D material. A recent metal-free layered semiconductor material 2D BP showed better biocompatibility, lower toxicity, anisotropic conductance, and higher surface-to-volume ratio due to the puckered lattice configuration as compared to other 2D materials, hence are favored for compositing metal nanoparticles. Studies employing TMOs, TMDs, MXenes, and 2D BP thus highlighted the integration of nanomaterials in microfluidic optical and electrochemical sensors and biosensing as a potential way to improve signal transduction. The various studies mentioned had reported remarkable microfluidic biosensor devices of many applications with unique and excellent chemical, structural, optical, and physical properties using 2D nanomaterials for biosensors, most importantly with specificity, sensitivity, and linear working range ideally comparable to conventional standard techniques that require extensive process and instruments. All in all, advanced materials provide multiple functional tools for the future development of novel microfluidic biosensors enabling better microfluidics design and fabrication with new features and improved signal transduction with nanocomposite transducers.

References

[1] G. Kulsharova, N. Dimov, M.P.C. Marques, N. Szita, F. Baganz, Simplified immobilisation method for histidine-tagged enzymes in poly(methyl methacrylate) microfluidic devices, N. Biotechnol. 47 (2018) 31–38.

[2] J. Yang, J.-H. Yu, J. Rudi Strickler, W.-J. Chang, S. Gunasekaran, Nickel nanoparticle–chitosan-reduced graphene oxide-modified screen-printed electrodes for enzyme-free glucose sensing in portable microfluidic devices, Biosens. Bioelectron. 47 (2013) 530–538.

[3] Z. Wang, L. Jinlong, Z. An, M. Kimura, T. Ono, Enzyme immobilization in completely packaged freestanding SU-8 microfluidic channel by electro click chemistry for compact thermal biosensor, Process Biochemistry 79 (2019) 57–64. Available from: https://doi.org/10.1016/j.procbio.2018.12.007.

[4] M. Garg, M.G. Christensen, A. Iles, A.L. Sharma, S. Singh, N. Pamme, Microfluidic-based electrochemical immunosensing of ferritin, Biosensors 10 (8) (2020) 91.

[5] C. Singh, M.A. Ali, V. Kumar, R. Ahmad, G. Sumana, Functionalized MoS2 nanosheets assembled microfluidic immunosensor for highly sensitive detection of food pathogen, Sens. Actuators B: Chem. 259 (2018) 1090–1098.

[6] R. Singh, S. Hong, J. Jang, Label-free detection of influenza viruses using a reduced graphene oxide-based electrochemical immunosensor integrated with a microfluidic platform, Sci. Rep. 7 (1) (2017) 42771.

[7] Y. Wang, H. Xu, J. Luo, J. Liu, L. Wang, Y. Fan, et al., A novel label-free microfluidic paper-based immunosensor for highly sensitive electrochemical detection of carcinoembryonic antigen, Biosens. Bioelectron. 83 (2016) 319–326. Available from: https://doi.org/10.1016/j.bios.2016.04.062.

[8] Y. Sun, H. Jin, X. Jiang, R. Gui, Black phosphorus nanosheets adhering to thionine-doped 2D MOF as a smart aptasensor enabling accurate capture and ratiometric electrochemical detection of target microRNA, Sens. Actuators B: Chem. 309 (2020) 127777.

[9] A. Fornell, P. Söderbäck, Z. Liu, M. De Albuquerque Moreira, M. Tenje, Fabrication of silicon microfluidic chips for acoustic particle focusing using direct laser writing, Micromachines 11 (2) (2020) 113.

[10] F. Sun, A. Ganguli, J. Nguyen, R. Brisbin, K. Shanmugam, D.L. Hirschberg, et al., Smartphone-based multiplex 30-minute nucleic acid test of live virus from nasal swab extract, Lab Chip 20 (9) (2020) 1621–1627.

[11] G. Kim, J.H. Moon, C.Y. Moh, J.G. Lim, A microfluidic nano-biosensor for the detection of pathogenic Salmonella, Biosens Bioelectron 67 (2015) 243–247. Available from: https://doi.org/10.1016/j.bios.2014.08.023. 25172028.

[12] N.N. Le, K.K. Huynh, T.C.H. Phan, T.M.D. Dang, M.C. Dang, Fabrication of 25 μm-filter microfluidic chip on silicon substrate, Adv. Nat. Sci.: Nanosci. Nanotechnol. 8 (1) (2017) 015003.

[13] Z. Li, Q. Luo, J. Wu, Label-free discrimination of membrane-translocating peptides on porous silicon microfluidic biosensors, Biomicrofluidics 10 (2016) 064113. Available from: https://doi.org/10.1063/1.4971440.

[14] R. Davidsson, F. Genin, M. Bengtsson, T. Laurell, J. Emnéus, Microfluidic biosensing systems Part I. Development and optimisation of enzymatic chemiluminescent μ-biosensors based on silicon microchips, Lab Chip 4 (5) (2004) 481–487.

[15] S. Zhuo, M. Xun, M. Li, X. Kong, R. Shao, T. Zheng, et al., Rapid and label-free optical assay of S-layer protein with high sensitivity using TiO2-coated porous silicon-based microfluidic biosensor, Sens. Actuators B: Chem. 321 (2020) 128524.

[16] A.P. Nayak, M.S. Islam, V.J. Logeeswaran, Wet etching, in: B. Bhushan (Ed.), Encyclopedia of Nanotechnology, Springer Netherlands, Dordrecht, 2012, pp. 2829–2830.

[17] S.B. Inayat, A.P. Nayak, V.J. Logeeswaran, M.S. Islam, Dry etching processes, in: B. Bhushan (Ed.), Encyclopedia of Nanotechnology, Springer Netherlands, Dordrecht, 2016, pp. 1–6.

[18] P.W. Leech, Reactive ion etching of quartz and silica-based glasses in CF4/CHF3 plasmas, Vacuum 55 (3) (1999) 191–196.

[19] D.A. Zeze, R.D. Forrest, J.D. Carey, D.C. Cox, I.D. Robertson, B.L. Weiss, et al., Reactive ion etching of quartz and Pyrex for microelectronic applications, J. Appl. Phys. 92 (7) (2002) 3624–3629.

[20] T. Wang, J. Chen, T. Zhou, L. Song, Fabricating microstructures on glass for microfluidic chips by glass molding process, Micromachines 9 (6) (2018) 269.

[21] S. Prakash, M. Pinti, B. Bhushan, Theory, fabrication and applications of microfluidic and nanofluidic biosensors, Philos. Trans. R. Soc. A: Math., Phys. Eng. Sci. 370 (1967) (2012) 2269–2303.

[22] P.N. Nge, C.I. Rogers, A.T. Woolley, Advances in microfluidic materials, functions, integration, and applications, Chem. Rev. 113 (4) (2013) 2550–2583.

[23] J. Luo, T. Dziubla, R. Eitel, A low temperature co-fired ceramic based microfluidic Clark-type oxygen sensor for real-time oxygen sensing, Sens. Actuators B: Chem. 240 (2017) 392–397.

[24] E. Remiszewska, K. Malecha, J. Kruk, J. Jankowska-Śliwińska, W. Torbicz, A. Samluk, et al., Enzymatic method of urea determination in LTCC microfluidic system based on absorption photometry, Sens. Actuators B: Chem. 285 (2019) 375–384.

[25] E.S. Fakunle, I. Fritsch, Low-temperature co-fired ceramic microchannels with individually addressable screen-printed gold electrodes on four walls for self-contained electrochemical immunoassays, Anal. Bioanal. Chem. 398 (6) (2010) 2605–2615.

[26] K. Malecha, M. Dawgul, D.G. Pijanowska, L.J. Golonka, LTCC microfluidic systems for biochemical diagnosis, Biocybern. Biomed. Eng. 31 (4) (2011) 31–41. Available from: https://doi.org/10.1016/S0208-5216(11)70024-9.

[27] H. Becker, C. Gärtner, Polymer microfabrication technologies for microfluidic systems, Anal. Bioanal. Chem. 390 (1) (2008) 89–111.

[28] S. Yang, K. Jiang, Elastomer Application in Microsystem and Microfluidics, 2012.

[29] C.-W. Tsao, D.L. DeVoe, Bonding of thermoplastic polymer microfluidics, Microfluidics Nanofluidics 6 (1) (2009) 1–16.

[30] I. Bilican, M. Tahsin Guler, Assessment of PMMA and polystyrene based microfluidic chips fabricated using CO_2 laser machining, Appl. Surf. Sci. 534 (2020) 147642.

[31] U. Sarma, S.N. Joshi, Machining of micro-channels on polycarbonate by using Laser-Induced Plasma Assisted Ablation (LIPAA), Opt. Laser Technol. 128 (2020) 106257.

[32] S. Su, G. Jing, M. Zhang, B. Liu, X. Zhu, B. Wang, et al., One-step bonding and hydrophobic surface modification method for rapid fabrication of polycarbonate-based droplet microfluidic chips, Sens. Actuators B: Chem. 282 (2019) 60–68.

[33] X. Hu, X. Lin, Q. He, H. Chen, Electrochemical detection of droplet contents in polystyrene microfluidic chip with integrated micro film electrodes, J. Electroanalytical. Chem. 726 (2014) 7–14.

[34] X. Hu, Y. Dong, Q. He, H. Chen, Z. Zhu, Fabrication of a polystyrene microfluidic chip coupled to electrospray ionization mass spectrometry for protein analysis, J. Chromatogr. B 990 (2015) 96–103.

[35] Y. Lu, Q. Feng, R. Zhang, H. Lu, J. Su, Y. Cui, et al., An online pH detection system based on a microfluidic chip, Anal. Chim. Acta 1106 (2020) 71–78.

[36] A. Alrifaiy, O.A. Lindahl, K. Ramser, Polymer-based microfluidic devices for pharmacy, biology and tissue engineering, Polymers 4 (3) (2012) 1349–1398.

[37] Y.-H. Lin, S.-W. Kang, T.-Y. Wu, Fabrication of polydimethylsiloxane (PDMS) pulsating heat pipe, Appl. Therm. Eng. 29 (2) (2009) 573–580.

[38] W. Limbut, S. Loyprasert, C. Thammakhet, P. Thavarungkul, A. Tuantranont, P. Asawatreratanakul, et al., Microfluidic conductimetric bioreactor, Biosens. Bioelectron. 22 (12) (2007) 3064–3071.

[39] J. Li, Z. Li, Y. Dou, J. Su, J. Shi, Y. Zhou, et al., A nano-integrated microfluidic biochip for enzyme-based point-of-care detection of creatinine, Chem. Commun. 57 (38) (2021) 4726–4729.

[40] J.S. Nah, S.C. Barman, M.A. Zahed, M. Sharifuzzaman, H. Yoon, C. Park, et al., A wearable microfluidics-integrated impedimetric immunosensor based on Ti3C2Tx MXene incorporated laser-burned graphene for noninvasive sweat cortisol detection, Sens. Actuators B: Chem. 329 (2021) 129206.

[41] D. Feng, J. Su, Y. Xu, G. He, C. Wang, X. Wang, et al., DNA tetrahedron-mediated immune-sandwich assay for rapid and sensitive detection of PSA through a microfluidic electrochemical detection system, Microsyst. Nanoeng. 7 (1) (2021) 33.

[42] S. Arshavsky-Graham, E. Segal, Lab-on-a-Chip Devices for Point-of-Care Medical Diagnostics. Berlin, Heidelberg, Springer Berlin Heidelberg: 1–19.

[43] A.V. Nielsen, M.J. Beauchamp, G.P. Nordin, A.T. Woolley, 3D printed microfluidics, Annu. Rev. Anal. Chem. (Palo Alto Calif.) 13 (1) (2020) 45–65.

[44] S.K. Mitra, S. Chakraborty, Microfluidics and Nanofluidics Handbook (2011). Available from: https://doi.org/10.1201/b16864.

[45] X. Li, Z.T. Yu, D. Geraldo, S. Weng, N. Alve, W. Dun, et al., Desktop aligner for fabrication of multilayer microfluidic devices, Rev Sci Instrum 86 (7) (2015) 075008. Available from: https://doi.org/10.1063/1.4927197. 26233409.

[46] R. Su, J. Wen, Q. Su, M.S. Wiederoder, S.J. Koester, J.R. Uzarski, et al., 3D printed self-supporting elastomeric structures for multifunctional microfluidics, Sci Adv 6 (41) (2020). Available from: https://doi.org/10.1126/sciadv.abc9846. 33036980.

[47] S. Wüst, R. Müller, S. Hofmann, Controlled Positioning of Cells in Biomaterials-Approaches Towards 3D Tissue Printing, J Funct Biomater 2 (3) (2011) 119–154. Available from: https://doi.org/10.3390/jfb2030119. 24956301.

[48] R. Barry, D. Ivanov, Microfluidics in biotechnology, J Nanobiotechnol 2 (2004). Available from: https://doi.org/10.1186/1477-3155-2-2.

[49] L. Guo, J. Feng, Z. Fang, J. Xu, X. Lu, Application of microfluidic "lab-on-a-chip" for the detection of mycotoxins in foods, Trends in Food Science & Technology 46 (2) (2015) 252–263. Available from: https://doi.org/10.1016/j.tifs.2015.09.005.

[50] A.A. Yazdi, A. Popma, W. Wong, T. Nguyen, Y. Pan, J. Xu, 3D printing: an emerging tool for novel microfluidics and lab-on-a-chip applications, Microfluidics and

Nanofluidics 20 (2016) 50. Available from: https://link.springer.com/article/10.1007/s10404-016-1715-4.
[51] R.D. Sochol, E. Sweet, C.C. Glick, S. Venkatesh, A. Avetisyan, K.F. Ekman, et al., 3D printed microfluidic circuitry via multijet-based additive manufacturing, Lab. Chip 16 (4) (2016) 668–678.
[52] L.P. Bressan, C.B. Adamo, R.F. Quero, D.P. de Jesus, J.A.F. da Silva, A simple procedure to produce FDM-based 3D-printed microfluidic devices with an integrated PMMA optical window, Anal. Methods 11 (8) (2019) 1014–1020.
[53] S. Dul, L. Fambri, A. Pegoretti, Fused deposition modelling with ABS–graphene nanocomposites, Compos. A: Appl. Sci. Manuf. 85 (2016) 181–191.
[54] L.M. Galantucci, F. Lavecchia, G. Percoco, Experimental study aiming to enhance the surface finish of fused deposition modeled parts, CIRP Ann. 58 (1) (2009) 189–192.
[55] F. Kotz, M. Mader, N. Dellen, P. Risch, A. Kick, D. Helmer, et al., Fused deposition modeling of microfluidic chips in polymethylmethacrylate, Micromachines 11 (9) (2020) 873. Available from: https://doi.org/10.3390/mi11090873.
[56] H. Liu, K. Hu, D. Yan, R. Chen, Y. Zou, H. Liu, et al., Recent advances on black phosphorus for energy storage, catalysis, and sensor applications, Adv. Mater. 30 (32) (2018) 1800295.
[57] E.J. McCullough, V.K. Yadavalli, Surface modification of fused deposition modeling ABS to enable rapid prototyping of biomedical microdevices, J. Mater. Process. Technol. 213 (6) (2013) 947–954.
[58] M.D. Nelson, N. Ramkumar, B.K. Gale, Flexible, transparent, sub-100 μm microfluidic channels with fused deposition modeling 3D-printed thermoplastic polyurethane, J. Micromech. Microeng. 29 (9) (2019).
[59] M.P. Lee, G.J. Cooper, T. Hinkley, G.M. Gibson, M.J. Padgett, L. Cronin, Development of a 3D printer using scanning projection stereolithography, Sci. Rep. 5 (2015) 9875.
[60] Y. Pan, X. Zhao, C. Zhou, Y. Chen, Smooth surface fabrication in mask projection based stereolithography, J. Manuf. Process. 14 (4) (2012) 460–470.
[61] S.H. Park, D.Y. Yang, K.S. Lee, Two-photon stereolithography for realizing ultraprecise three-dimensional nano/microdevices, Laser Photonics Rev. 3 (1–2) (2009) 1–11.
[62] C.I. Rogers, K. Qaderi, A.T. Woolley, G.P. Nordin, 3D printed microfluidic devices with integrated valves, Biomicrofluidics 9 (1) (2015) 016501.
[63] C. Zhou, H. Ye, F. Zhang, A novel low-cost stereolithography process based on vector scanning and mask projection for high-accuracy, high-speed, high-throughput, and large-area fabrication, J. Comput. Inf. Sci. Eng. 15 (1) (2015).
[64] S. Razavi Bazaz, O. Rouhi, M.A. Raoufi, F. Ejeian, M. Asadnia, D. Jin, et al., 3D printing of inertial microfluidic devices, Sci. Rep. 10 (1) (2020) 5929.
[65] S. Waheed, J.M. Cabot, N.P. Macdonald, T. Lewis, R.M. Guijt, B. Paull, et al., 3D printed microfluidic devices: enablers and barriers, Lab Chip 16 (11) (2016) 1993–2013.
[66] Y. Chen, L. Zhang, G. Chen, Fabrication, modification, and application of poly(methyl methacrylate) microfluidic chips, Electrophoresis 29 (9) (2008) 1801–1814.
[67] P.J. Kitson, M.H. Rosnes, V. Sans, V. Dragone, L. Cronin, Configurable 3D-printed millifluidic and microfluidic 'lab on a chip' reactionware devices, Lab Chip 12 (18) (2012) 3267–3271.

[68] E. Lepowsky, S. Tasoglu, Emerging anti-fouling methods: towards reusability of 3D-printed devices for biomedical applications, Micromachines (Basel) 9 (4) (2018).

[69] Z. Chen, Y. Gao, R. Su, C. Li, J. Lin, Fabrication and characterization of poly (methyl methacrylate) microchannels by in situ polymerization with a novel metal template, Electrophoresis 24 (18) (2003) 3246–3252.

[70] F. Kotz, K. Arnold, S. Wagner, W. Bauer, N. Keller, T.M. Nargang, et al., Liquid PMMA: a high resolution polymethylmethacrylate negative photoresist as enabling material for direct printing of microfluidic chips, Adv. Eng. Mater. 20 (2) (2018).

[71] S. Qu, X. Chen, D. Chen, P. Yang, G. Chen, Poly(methyl methacrylate) CE microchips replicated from poly(dimethylsiloxane) templates for the determination of cations, Electrophoresis 27 (24) (2006) 4910–4918.

[72] R. Amin, S. Knowlton, A. Hart, B. Yenilmez, F. Ghaderinezhad, S. Katebifar, et al., 3D-printed microfluidic devices, Biofabrication 8 (2) (2016) 022001.

[73] L.P. Bressan, T.M. Lima, G.D. da Silveira, J.A.F. da Silva, Low-cost and simple FDM-based 3D-printed microfluidic device for the synthesis of metallic core–shell nanoparticles, SN Appl. Sci. 2 (5) (2020) 984.

[74] S.B. Puneeth, S. Goel, Handheld and 'Turnkey' 3D printed paper-microfluidic viscometer with on-board microcontroller for smartphone based biosensing applications, Anal. Chim. Acta 1153 (2021) 338303.

[75] J.S. Ng, M. Hashimoto, 3D-PAD: paper-based analytical devices with integrated three-dimensional features, Biosensors 11 (3) (2021) 84.

[76] S. Kecili, H.C. Tekin, Adhesive bonding strategies to fabricate high-strength and transparent 3D printed microfluidic device, Biomicrofluidics 14 (2) (2020) 024113. Available from: https://doi.org/10.1063/5.0003302. 32341724.

[77] S. Zips, O.J. Wenzel, P. Rinklin, L. Grob, K. Terkan, N.Y. Adly, et al., Direct stereolithographic 3D printing of microfluidic structures on polymer substrates for printed electronics, Advanced Materials Technologies 4 (3) (2019) 1800455. Available from: https://doi.org/10.1002/admt.201800455.

[78] S. Zips, L. Hiendlmeier, L.J.K. Weiß, H. Url, T.F. Teshima, R. Schmid, et al., Biocompatible, flexible, and oxygen-permeable silicone-hydrogel material for stereolithographic printing of microfluidic lab-on-a-chip and cell-culture devices, ACS Appl. Polym. Mater. 3 (1) (2021) 243–258.

[79] S. Saleh Alghamdi, S. John, N. Roy Choudhury, N.K. Dutta, Additive manufacturing of polymer materials: progress, promise and challenges, Polymers 13 (5) (2021) 753.

[80] R.M. Cardoso, P.R.L. Silva, A.P. Lima, D.P. Rocha, T.C. Oliveira, T.M. do Prado, et al., 3D-Printed graphene/polylactic acid electrode for bioanalysis: Biosensing of glucose and simultaneous determination of uric acid and nitrite in biological fluids, Sens. Actuators B: Chem. 307 (2020) 127621.

[81] V.A.O.P. Silva, W.S. Fernandes-Junior, D.P. Rocha, J.S. Stefano, R.A.A. Munoz, J. A. Bonacin, et al., 3D-printed reduced graphene oxide/polylactic acid electrodes: A new prototyped platform for sensing and biosensing applications, Biosens. Bioelectron. 170 (2020) 112684.

[82] M. Liu, S. Suo, J. Wu, Y. Gan, D. Ah Hanaor, C.Q. Chen, Tailoring porous media for controllable capillary flow, J. Colloid Interface Sci. 539 (2019) 379–387.

[83] S. Nishat, A.T. Jafry, A.W. Martinez, F.R. Awan, Paper-based microfluidics: simplified fabrication and assay methods, Sens. Actuators B: Chem. 336 (2021) 129681.

[84] W. Dungchai, O. Chailapakul, C.S. Henry, Electrochemical detection for paper-based microfluidics, Anal. Chem. 81 (14) (2009) 5821–5826.

[85] H. Fu, P. Song, Q. Wu, C. Zhao, P. Pan, X. Li, et al., A paper-based microfluidic platform with shape-memory-polymer-actuated fluid valves for automated multi-step immunoassays, Microsyst. Nanoeng. 5 (2019) 1–12. Available from: https://doi.org/10.1038/s41378-019-0091-0.

[86] M.S. Iyer, P.E. Resmi, C.P. Shyamlal, P.V. Suneesh, T.G. Satheesh Babu, Fabrication and testing of handheld electronic meter for colorimetric paper microfluidic devices, Materials Today: Proceedings 33 (2020) 2421–2425. Available from: https://doi.org/10.1016/j.matpr.2020.08.379.

[87] R. Muthukumar, A. Kapoor, S. Balasubramanian, V. Vaishampayan, M. Gabhane, Detection of adulteration in sunflower oil using paper-based microfluidic lab-on-a-chip devices, Materials Today: Proceedings 34 (2021) 496–501. Available from: https://doi.org/10.1016/j.matpr.2020.03.099.

[88] Z. Hao, Q. Zheng, L. Jin, S. Zhou, H. Chen, X. Liu, et al., Rapid measurement of total polyphenol content in tea by kinetic matching approach on microfluidic paper-based analytical devices., Food Chem 342 (2021) 128368. Available from: https://doi.org/10.1016/j.foodchem.2020.128368. 33158681.

[89] D. Lin, B. Li, J. Qi, X. Ji, S. Lin, W. Wang, et al., Low cost fabrication of microfluidic paper-based analytical devices with water-based polyurethane acrylate and their application for bacterial detection, Sens. Actuators B: Chem. 303 (15) (2020) 127213. Available from: https://doi.org/10.1016/j.snb.2019.127213.

[90] S. Basu, P. Bhattacharyya, Recent developments on graphene and graphene oxide based solid state gas sensors, Sens. Actuators B: Chem. 173 (2012) 1–21.

[91] P.M. Carrasco, I. García, L. Yate, R. Tena Zaera, G. Cabañero, H.J. Grande, et al., Graphene quantum dot membranes as fluorescent sensing platforms for Cr (VI) detection, Carbon 109 (2016) 658–665.

[92] M. Shehab, S. Ebrahim, M. Soliman, Graphene quantum dots prepared from glucose as optical sensor for glucose, J. Lumin. 184 (2017) 110–116.

[93] Z. Liu, M. Jin, J. Cao, R. Niu, P. Li, G. Zhou, et al., Electrochemical sensor integrated microfluidic device for sensitive and simultaneous quantification of dopamine and 5-hydroxytryptamine, Sens. Actuators B: Chem. 273 (2018) 873–883.

[94] F.H. Cincotto, E.L. Fava, F.C. Moraes, O. Fatibello-Filho, R.C. Faria, A new disposable microfluidic electrochemical paper-based device for the simultaneous determination of clinical biomarkers, Talanta 195 (2019) 62–68.

[95] J. Yang, Z. Zeng, J. Kang, S. Betzler, C. Czarnik, X. Zhang, et al., Formation of two-dimensional transition metal oxide nanosheets with nanoparticles as intermediates, Nat. Mater. 18 (9) (2019) 970–976.

[96] S.Z. Butler, S.M. Hollen, L. Cao, Y. Cui, J.A. Gupta, H.R. Gutiérrez, et al., Progress, challenges, and opportunities in two-dimensional materials beyond graphene, ACS Nano 7 (4) (2013) 2898–2926.

[97] K. Kalantar-zadeh, J.Z. Ou, T. Daeneke, A. Mitchell, T. Sasaki, M.S. Fuhrer, Two dimensional and layered transition metal oxides, Appl. Mater. Today 5 (2016) 73–89.

[98] M. Naguib, Y. Gogotsi, Synthesis of two-dimensional materials by selective extraction, Acc. Chem. Res. 48 (1) (2015) 128–135.

[99] M. Xu, T. Liang, M. Shi, H. Chen, Graphene-like two-dimensional materials, Chem. Rev. 113 (5) (2013) 3766–3798.

[100] N. Alizadeh, A. Salimi, T.-K. Sham, P. Bazylewski, G. Fanchini, Intrinsic enzyme-like activities of cerium oxide nanocomposite and its application for extracellular H_2O_2 detection using an electrochemical microfluidic device, ACS Omega 5 (21) (2020) 11883–11894.

[101] X. Sun, H. Wang, Y. Jian, F. Lan, L. Zhang, H. Liu, et al., Ultrasensitive microfluidic paper-based electrochemical/visual biosensor based on spherical-like cerium dioxide catalyst for miR-21 detection, Biosens. Bioelectron. 105 (2018) 218–225.

[102] M.A. Ali, K. Mondal, Y. Jiao, S. Oren, Z. Xu, A. Sharma, et al., Microfluidic immuno-biochip for detection of breast cancer biomarkers using hierarchical composite of porous graphene and titanium dioxide nanofibers, ACS Appl. Mater. Interfaces 8 (32) (2016) 20570–20582.

[103] R.J. Toh, C.C. Mayorga-Martinez, J. Han, Z. Sofer, M. Pumera, Group 6 layered transition-metal dichalcogenides in lab-on-a-chip devices: 1T-phase WS2 for microfluidics non-enzymatic detection of hydrogen peroxide, Anal. Chem. 89 (9) (2017) 4978–4985.

[104] C. Zhang, Y. Ma, X. Zhang, S. Abdolhosseinzadeh, H. Sheng, W. Lan, et al., Two-dimensional transition metal carbides and nitrides (MXenes): synthesis, properties, and electrochemical energy storage applications, Energy Environ. Mater. 3 (1) (2020) 29–55.

[105] M. Naguib, M. Kurtoglu, V. Presser, J. Lu, J. Niu, M. Heon, et al., Two-dimensional nanocrystals produced by exfoliation of Ti3 AlC2, Adv. Mater. 23 (37) (2011) 4248–4253.

[106] G. Liu, J. Shen, Q. Liu, G. Liu, J. Xiong, J. Yang, et al., Ultrathin two-dimensional MXene membrane for pervaporation desalination, J. Membr. Sci. 548 (2018) 548–558.

[107] J. Liu, X. Jiang, R. Zhang, Y. Zhang, L. Wu, W. Lu, et al., MXene-enabled electrochemical microfluidic biosensor: applications toward multicomponent continuous monitoring in whole blood, Adv. Funct. Mater. 29 (6) (2019) 1807326.

[108] A. Avsar, I.J. Vera-Marun, J.Y. Tan, K. Watanabe, T. Taniguchi, A.H. Castro Neto, et al., Air-stable transport in graphene-contacted, fully encapsulated ultrathin black phosphorus-based field-effect transistors, ACS Nano 9 (4) (2015) 4138–4145.

[109] J. Cai, X. Gou, B. Sun, W. Li, D. Li, J. Liu, et al., Porous graphene-black phosphorus nanocomposite modified electrode for detection of leptin, Biosens. Bioelectron. 137 (2019) 88–95.

[110] S. Ramalingam, R. Chand, C.B. Singh, A. Singh, Phosphorene-gold nanocomposite based microfluidic aptasensor for the detection of okadaic acid, Biosens. Bioelectron. 135 (2019) 14–21.

[111] L. Shijie, Z. Fuyuan, W. Junping, W. Wenjun, W. Shuo, Black phosphorus-Au nanocomposite-based fluorescence immunochromatographic sensor for high-sensitive detection of zearalenone in cereals, Nanophotonics 9 (8) (2020) 2397–2406.

CHAPTER

8 Smartphone and microfluidic systems in medical and food analysis

Aristeidis S. Tsagkaris[1], Joost L.D. Nelis[2], Katrina Campbell[2], Christopher T. Elliott[2], Jana Pulkrabova[1] and Jana Hajslova[1]

[1]Department of Food Analysis and Nutrition, Faculty of Food and Biochemical Technology, University of Chemistry and Technology Prague, Prague, Czech Republic

[2]Institute for Global Food Security, School of Biological Sciences, Queen's University, Belfast, United Kingdom

8.1 Introduction

Portable and miniaturized analysis has drawn an ever-increasing attention as it comes with unprecedented attractive features such as simplicity, rapidness, and cost-efficiency. Such methods can be clustered under the term point-of-care (POC) diagnostics including, but not limited to, lateral flow assays (LFAs), microfluidic paper-based analytical devices (μPAD), or handheld lab-on-a-chip (LOC) prototypes. Initially inspired by needs related to the medical field, e.g., testing at remote places or low-cost diagnostics for developing countries, medical POC diagnostics have been developed based on the "ASSURED" principles issued by the WHO in an effort to provide a set of working criteria for the advancement of the field. This acronym stands for Affordable, Sensitive, Specific, User-friendly, Rapid and robust, Equipment-free and Deliverable to end-users [1]. Reasonably, this breakthrough in medical analysis initiated research effort to implement the POC concept for food [2] and water [3] analysis as well. However, food POC diagnostics are at their infancy stage mostly due to the analytical challenge set by complex food matrices, which typically require exhaustive sample preparation, limiting point-of-need testing, and end-user implementation.

Regardless of the readiness difference of POC diagnostics in medical and food analytical applications, POC potential can be further enhanced by coupling them with smartphones to achieve ubiquitous monitoring [4]. Real-time one-click results due to smartphone apps, online connectivity, geo-temporal mapping, or potential end-user implementation are among the merits of smartphone utilization. POC methods with smartphone readout can provide personalized results and initi-

Microfluidic Biosensors. DOI: https://doi.org/10.1016/B978-0-12-823846-2.00002-X

ate a paradigm shift by intensifying testing for a wide variety of targets similar to what has been achieved for glucose electrochemical [5] and pregnancy colorimetric test strip methods [6]. This can be extremely valuable in both cases; in the medical field, providing rapid and accurate results is the key to control a disease and protect the socioeconomic status. This has become even more obvious during the ongoing pandemic with an ever-increasing demand for rapid on-the-spot diagnostic testing for COVID-19. In terms of food analysis, prescreening of food with a POC method will result in laboratory-based testing only for suspected samples instead of the current system of random sampling often leading to a vast majority of negative results. Therefore, rapid results from POC testing to select suspect samples, followed by confirmatory laboratory-based analysis, will lead to better utilization of available resources. Moreover, the system may introduce the importance of food testing to the general public, a process commonly forgotten or underestimated. However, there is also the potential downside to smartphone-hyphenated POCs whereby the currently developed systems may be fit for purpose for a selective group of end-users that are familiar to more complex smartphone operations but not for the less experienced smartphone users. Thus, continuous communication with all end-users on the evolving smartphone POC technology is essential for the success of this novel approach.

Nevertheless, there are challenges regarding the performance of POC diagnostics coupled to smartphones. To begin with, to achieve a truly personalized monitoring, acquired results should not be affected by the use of any smartphone model. However, this is often not the case; the majority of developed smartphone-based assays are applicable only when using certain types of smartphones. Besides, considerable controversy has been raised regarding image data acquisition and processing when using an optical smartphone readout [7,8]. Although electrochemical detection is also used (see Section 8.4.2), optical detection is more preferable and convenient so far as the smartphone camera can be used to record images or videos. In terms of data acquisition, it has not been clear so far whether it is necessary to use auxiliary attachable parts such as 3D-printed elements to standardize optical conditions or record under ambient light and correct for ambient light variation using background correction algorithms [9]. Various data processing approaches have been developed using the RGB color space (the primary color space in which smartphones capture images by default) or transforming the original signal to other color spaces (i.e., HSV or CIE-Lab) or to barcodes [10]. Apart from the readout system, the importance of developing rugged and sensitive POC diagnostics must be kept in mind. In fact, there were cases in which insufficient validation was noticed or the performance in matrix was significantly decreased [11].

In this chapter, a scientometric evaluation of the medical and food POC diagnostics with a smartphone readout is presented with the aim of outlining the advances in this field. To this end, a structured keyword search system was used to extract important information from the Scopus database. Based on this bibliometric analysis, significant approaches both in the medical and food analysis field have been comprehensively discussed. Also, expert opinion on important assets, such as the need to introduce specific guidelines on the validation of smartphone-based methods as well as the necessity to confirm the POC diagnostic results using instrumental analysis, is provided.

8.2 Scientometric analysis

It is worth to notice research trends on smartphone-based sensing via an all-time scientometric evaluation. The search was performed using the Scopus database and the following search strings were used:

- Medical applications:
  ("smartphone" AND "lateral flow" OR "lab-on-a-chip" OR "micro total analysis system" OR "microfluidic paper-based analytical device" OR "point-of-care" OR "electrochemical") AND ("medical" OR "cancer" OR "infectious disease" OR "diagnostics")
- Food applications:
  ("smartphone" AND "lateral flow" OR "lab-on-a-chip" OR "micro total analysis system" OR "microfluidic paper-based analytical device" OR "point-of-care" OR "electrochemical") AND ("food safety" OR "pesticides" OR "metals" OR "allergens" OR "pathogens")

In total, 631 hits related to medical applications and 158 hits related to food applications were found while the first archived studies were in 2006 and 2011, respectively. Obviously, medical smartphone-based diagnostics are in a more advanced stage but in both cases more breakthroughs will occur. The temporal evolution of published studies (Fig. 8.1) shows that there is a constant increase in number of publications with the exception of the period 2018–2019 for medical applications, where a slight drop was recorded. But this may be just because not all publications that came out in 2019 have been indexed in Scopus as yet.

In terms of publication type, the majority of published articles are original papers (Fig. 8.2) indicating the effort to establish this new field. In addition, the reported research originated from more than 60 countries from all over the world highlighting the global attention that smartphone-based sensing has drawn. The greatest number of studies in the medical field are related to an American affiliation, while in the case of food analysis China was the leading provider. Nevertheless, it is more important to focus on the per million capita publications rate, which is equal to the number of publications per country divided to their population in millions. This way of analyzing data shows Switzerland as being the largest country in terms of medical publications and South Korea in terms of food diagnostics (Fig. 8.2). European Union countries with traditionally strong research activity such as Germany and Spain follow with China and India just making the top 10 countries in both cases.

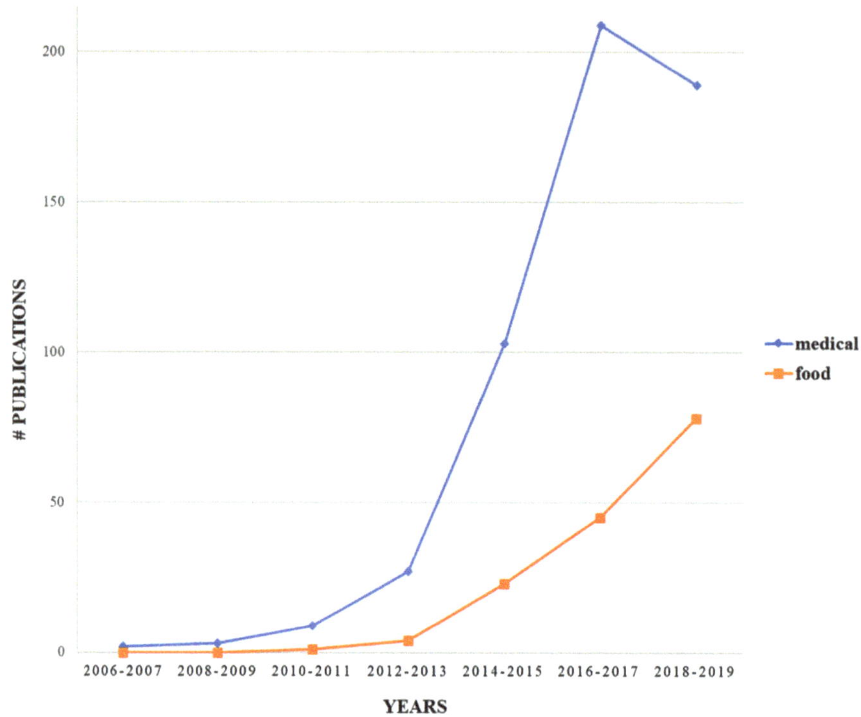

FIGURE 8.1

Temporal evolution of published work of smartphone applications in medical and food analysis (in detail, articles, conference papers, reviews, book chapters, Scopus, 9/2020).

8.3 Smartphone-based sensing technologies in the medical field

There is a large variety of smartphone-based sensing technologies available in the medical field including smartphone-hyphenated optical and electrochemical bioanalytical assays, image analysis, smartphone microscopy as well as general health monitoring and noncontact vital sign monitoring (such as respiration and pulse monitoring). Here, we will focus on novel smartphone-based bioassay technology only since hyphenation of more classical bioassays with smartphones (such as LFA, LOC, and μPADs) is more common in the food field and thus it is discussed in detail in Section 8.4. In addition, focus is on the usefulness of image analysis for skin cancer and smartphone-based microscopy for infectious diseases in remote locations because both technologies have the potential to greatly improve early detection and save lives.

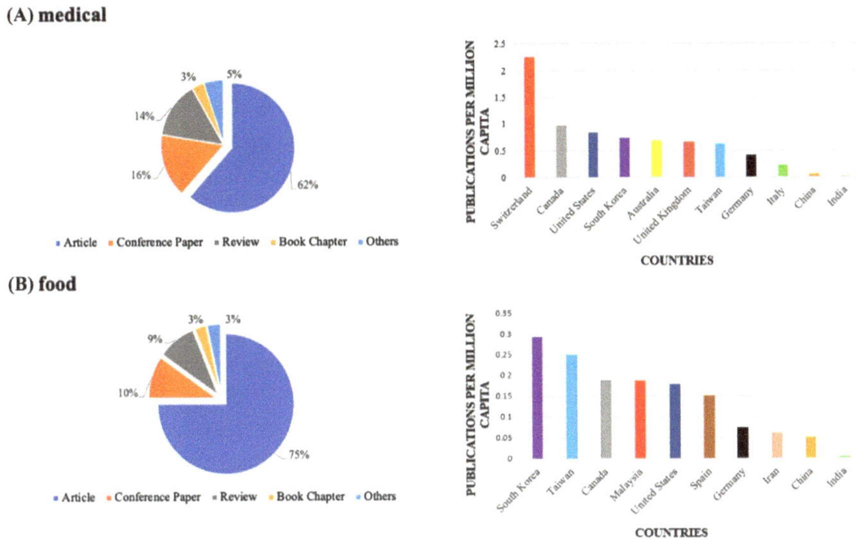

FIGURE 8.2

Published research per document type and publications per million capita of the top 10 countries of smartphone applications in (A) the medical and (B) the food field.

8.3.1 Smartphone-based tissue image analysis for early stage skin cancer detection

It was estimated by the WHO in 2003 that between 2 and 3 million skin cancers occur globally every year [12] and this number is increasing [13]. Of these, melanoma is one of the less common skin cancers (around 4% of the cases) [12]. However, due to the potential of melanoma to metastasize and spread to other sites in the body, it is responsible for the vast majority of skin cancer deaths making melanoma by far the most important type of skin cancer to detect, especially since 5-year survival rates are high (91%–95%) if it is identified early [14]. Smartphone-based detection and classification of skin cancer using image analysis may be helpful to achieve widespread early detection of melanoma. Initial strides toward this goal were made by Ramlakhan et al., who developed an automated melanoma recognition system on Android smartphones using the K-nearest neighbor classification. Although this device showed promise and had reasonable specificity (80%), the sensitivity remained quite low (60%) [15]. It is important to mention that in the medical diagnostics, specificity and sensitivity are used in a different way in comparison to that of analytical chemistry discipline. In diagnostics, sensitivity is expressed as the probability (in %) that a true positive (diseased individual) is identified correctly while specificity refers to the probability that the test correctly identifies a true negative (a disease-free individual) [16]. Thus, 40% of the individuals tested negative for melanoma with this technique when they did have melanoma.

Moreover, 20% of true melanoma-free individuals were diagnosed with melanoma using this technique. Although the 20% false positive constitutes an unnecessary follow-up examination, the 40% false negatives constitute a real issue as melanoma patients may remain unaware of the condition and not seek medical attention (which makes the system unfit for purpose). Moreover, the sample pool was very small ($n = 83$). Sensitivity could be improved by using more sophisticated and robust algorithms. In a recent study published in *Nature*, it has been shown that deep convolutional neural network (CNN) can be successfully used to correctly classify various skin cancers [17]. The developed CNN was trained using a large dataset (129,450 clinical images containing images of more than 2000 different diseases) and it was capable of diagnosing malignant melanomas versus benign nevi as well as keratinocyte carcinomas versus benign seborrheic keratosis with a similar performance as professional dermatologists. Furthermore, the potential adaptability of this system for use with mobile devices was suggested. However, the considerable computing power and energy required to run a CNN makes it very difficult to run such algorithms on a mobile phone directly. Cloud computing and image transfer via Wi-Fi may be a solution to this problem and have been used successfully to perform pixel-wise background correction for color quantification of biochemical assays under ambient light conditions [7–9]. Otherwise, the use of a library, such as a CNNdroid, that enables graphical processing unit accelerated execution of CNN on Android phones to retain functionality and limit battery use is advisable. CNNdroid is open source and available at Github [18]. Another solution is to use a Neural Compute Stick that can be used as a deep learning accelerator [19]. Such technology is interesting because it is already available in the market and can be used with existing smartphones to improve compatibility with CNN. Moreover, some of these devices, for instance Movidius, have shown potential for scientific computing using CNN [20]. This being said running deep CNN on mobile devices remains very challenging [19,21]. Thus, it may be better to opt for cloud computing instead. Finally, it is noted that the system described by Esteva et al. [17] is more complex as the dermatology apps that are currently available on the commercial market. Moreover, two independent systematic reviews (conducted to access performance of commercially available apps to detect skin cancer in 2018 [22] and 2020 [23]) concluded that these apps are often unreliable and that there is a lack of evidence regarding proper performance assessment as well as major inconsistencies in study design. Thus, further research in this area is urgently needed, as there clearly is a lack of robust clinical trials even though the technology holds great potential for rapid skin cancer diagnosis and it has been proven that it may be accurate if proper dataset and algorithm are used.

8.3.2 Improving smartphone-based infectious disease detection in remote locations

POC diagnostics using smartphone-based sensing is a very attractive platform for the detection of infectious diseases in remote locations (e.g., malaria [24–26], HIV [27–29] schistosomiasis [30–32], tuberculosis [33,34], and leprosies [35]), and the topic has been discussed in several excellent reviews [36–38]. One study that is of particular interest is by Laksanasopin et al. [27] who developed a smartphone-based dongle for rapid (within 15 minutes) duplex detection of syphilis and HIV. The device used microfluidic channels coated with antigen that was recognized by marker antibodies present in whole blood samples of HIV and/or syphilis patients flown through the channels using a hand-driven vacuum pump. Gold-labeled IgM antibodies held in a chamber in the microfluidic cassette were mixed with the blood in the flow channels where they could interact with the marker antibodies to form an antigen/marker–antibody/label complex. Then, silver reagent was added to create a silver film and the optical density was measured. This device stands out because it successfully went through a small ($n = 96$) phase 1 clinical trial in Rwanda and is currently being tested in a phase 2 trial testing the at-home use of the device [39]. Some of the major advantages of smartphone-based POCs in the studies mentioned above are the combination of worldwide smartphone availability, connectivity for rapid data transfer, and intuitive user interface with 3D-printed cost-effective microfluidics, isothermal amplification for rapid and easy-to-use nucleic acid detection, and nanotechnology for signal enhancement. The most used transducer appears to be the camera that is mainly used for colorimetric detection. Electrochemical transducers and surface plasmon resonance-based transducers are also mentioned, albeit less frequently. Evidently, the advantage of colorimetric detection is that one can simply take an image of the sample to obtain the result. However, this advantage is also a limitation. The issue is that color specification in an image is stored as RGB pixel values. These values have a maximum intensity difference of 256 point (0–255). Thus, color intensity difference from any taken image can only be measured in 256 steps, regardless of the smartphone model. Some studies suggest the conversion to alternative color spaces (mainly CieLAB, HSV) or single channels thereof to optimize color quantification. However the rationale for channel choice is still an active point of discussion and, most importantly, these conversions may not improve the overall resolution obtained because they only constitute a mathematical conversion of the raw RGB values [10]. An elegant solution to this problem was presented by the Filippini group who showed that video acquisition of one of the RGB channels (best picked based on reflectance spectroscopy measurements [40]) caused ~17-fold improvement in smartphone camera resolution while making glucose level measurements less sensitive to variations in ambient light conditions [41]. Such a concept may also be used to further improve sensitivity and time duration for fluorescence assays such as reported by Priye et al. who used isothermal PCR (reverse-transcription loop-mediated isothermal amplification or LAMP) and smartphone-based end-point fluorescence intensity quantification for the multiplex detection of Zika, chikungunya, and dengue viruses [42]. Motion-based detection is another very interesting system that circumvents the resolution issues related to color intensity measurements. The main idea is to track the movement of a target in a liquid and measure the change in speed caused by a binding event. Because the system looks at movement rather than intensity, the resolution of the system is related to the pixel density, which is ever-increasing on novel phones and commonly reaches over 10 MP per inch. This means that movement over 0.1 inch constitutes tracking over one million data points (approximately 3900-fold resolution increase compared to color intensity measurements). Recently motion-based detection has been reported by the Shafiee group for the detection of Zika virus and HIV using nanomotors [43,44] (Fig. 8.3). In the first study, platinum nanoparticles (Pt-NPs) and micropolystyrene beads are functionalized with different anti-Zika antibodies. These then form a sandwich assay if Zika virus is present. The Pt-NPs are then propelled by bubble formation caused through the Pt catalysis of H_2O_2 decomposition. Evidently free Pt-NPs move faster than those that are bound to the Zika virus/antibody/micropolystyrene complex. This change in movement is quantified using a smartphone motion tracking application and allowed a limit of detection (LOD) of 1000 viral particles mL^{-1} (Fig. 8.3A). In the second system, LAMP isothermal amplification was used to create large amplicons indicating HIV RNA presence. These amplicons could then bind to the engineered nanomotors (gold-platinum particles functionalized with ssDNA recognizing the LAMP amplicons). Catalysis of H_2O_2 decomposition was again used to propel the nanomotors and quantified using the smartphone motion tracking application. The device was then benchmarked

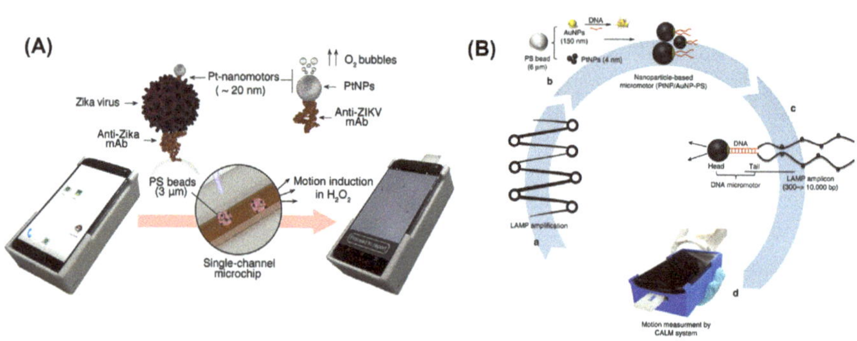

FIGURE 8.3

(A) Motion-based detection of Zika virus using a sandwhich immunoassay and Pt-nanomotors and a smartphone-based motion tracking app. (B) Motion-based detection of HIV-1 using LAMP amplification, Au/Pt nanomotors, and smartphone-based motion tracking.

Source: *(A) Reprinted (adapted) with permission from M.S. Draz, N.K. Lakshminaraasimulu, S. Krishnakumar, D. Battalapalli, A. Vasan, M.K. Kanakasabapathy, et al., Motion-based immunological detection of Zika virus using Pt-nanomotors and a cellphone, ACS Nano 12 (2018) 5709–5718. https://doi.org/10.1021/acsnano.8b01515. ©2020 American Chemical Society and (B) Panel reprinted from M.S. Draz, K.M. Kochehbyoki, A. Vasan, D. Battalapalli, A. Sreeram, M.K. Kanakasabapathy, et al., DNA engineered micromotors powered by metal nanoparticles for motion based cellphone diagnostics, Nat. Commun. 9 (2018) 4282. https://doi.org/10.1038/s41467-018-06727-8 under a Creative Commons CC BY license.*

against a reference method (quantitative real-time PCR) and 100% of the samples were correctly identified (at a clinically relevant 1000 virus particle mL^{-1} cutoff value for positive samples). Moreover, specificity and sensitivity of this device for qualitative HIV-1 detection were 99.1% and 94.6% respectively, showing the promise of the technique. It should be noted that the high concentration of H_2O_2 used (15% in [43] and 10% in [44]) may interfere with antibody–antigen binding and potentially cause DNA strand breaks. Thus, utilizing a better biocompatible method to propel the nanomotors may be beneficial.

8.3.3 Smartphone-based noninvasive diagnostics

Smartphones can also be used for noninvasive diagnostics or monitoring that can be done at a remote clinic without laboratory equipment or by the end-user at home. An interesting example of such a system is presented by Kanakasabapathy et al. [45]. The study presented an automated smartphone-based assay for POC semen analysis. A microfluidic chip fitted with a vacuum hand pump is used to load the unwashed and untreated semen sample directly from the sample cup. The chip is then inserted into a 3D-printed smartphone attachment and the semen quality is analyzed using the video-based smartphone application in less than 5 seconds. It was shown that the device is capable of evaluating semen quality (based on WHO guidelines) with $\sim$98% accuracy. This device constitutes a great stride forward since there is considerable stigma against male infertility testing which hinders many men from being tested for infertility, particularly in African countries with limited resources. Another interesting noninvasive diagnostic system is breath analysis for volatile organic compound (VOC) detection. This type of analysis has gained considerable attention in recent years due to the growing evidence that VOC patterns can contain endogenous volatile organic biomarkers indicating the presence of diseases such as lung cancer, cystic fibrosis, liver disease, coronary heart disease, and gastric cancer [46,47]. Moreover, gas detection can be very rapid and is completely noninvasive making it an ideal early detection method. Interestingly, the EU horizon 2020 project, SNIFFPHONE (https://www.sniffphone.eu/), has won the 2018 innovation award for the development of their smartphone-based POC for the detection of gastrointestinal cancer. The device (a handheld battery-powered device) uses a nanoparticle sensor array to obtain information about the VOCs present in the user's breath [48]. This information is then transmitted to the smartphone via Bluetooth and combined with other relevant user information, obtained through a questionnaire in the app, and sent to a server to perform statistical analysis. Finally, results are displayed on the smartphone. The procedure can be performed in a matter of minutes. Results of the clinical trials are pending. If successful, this device may be a major breakthrough in cancer diagnosis due to the unprecedented ease of use and portability of the device.

8.4 Smartphone-based sensing technologies in the food analysis field

Smartphone-based sensing can be either optical or electrochemical depending the hyphenated bioanalytical assay. As in the case of the medical field, smartphone-based food optical assays have been widely used due to their facility related to optical signal, e.g., a color development. However, colored food matrices set an additional challenge in optical smartphone-based sensing indicating the necessity to use sample preparation protocols. In the case of electrochemical assays, although they are not affected by colored food matrices, they have mostly exploited smartphones as an interface rather the actual sensor. In the following paragraphs, the most used assay types are discussed based on the smartphone-based detection type.

8.4.1 Optical detection

8.4.1.1 Lateral flow assays

LFAs are probably the most used POC diagnostic in the food analysis field. In principle, an LFA consists of various zones of polymeric strips on which substances can be immobilized and react with an analyte [49] (Fig. 8.4). Antibodies are commonly exploited as biorecognition elements and depending on implemented assay formats may be immobilized on two different spots, the test and the control line. The control line is used as an internal quality control necessary to verify that the LFA is functional. Two different formats of LFAs can be distinguished, namely competitive and sandwich formats. Both assays have labeled antibodies stored in the probe pad. The label can vary but is often gold nanosphere, latex microbead, or carbon black (CB). These labels are attached to the detection antibody (which interacts with the target analyte) and the control antibody (which will bind to the control line as a positive control). Importantly, liquid samples or extracts containing the analyte move through this lateral device due to

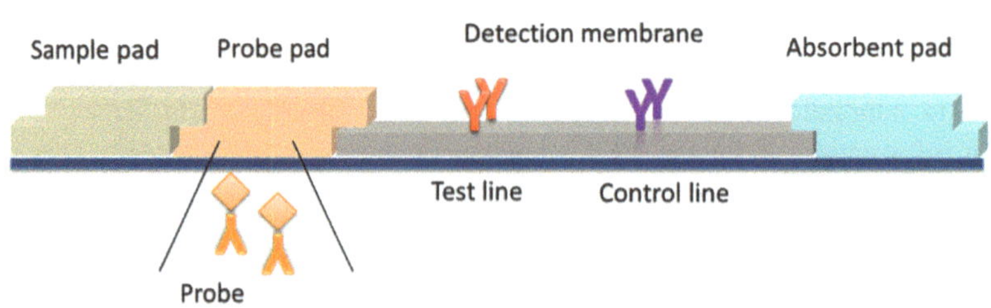

FIGURE 8.4

Illustration of a typical LFA formats. *LFA*, Lateral flow assay.

Source: *Panel reprinted from L. Anfossi, F. Di Nardo, S. Cavalera, C. Giovannoli, C. Baggiani, Multiplex lateral flow immunoassay: an overview of strategies towards high-throughput point-of-need testing, Biosensors 9 (2019) 2 (Ref. [50]) under a Creative Commons CC BY license.*

capillary forces from the sample pad through the probe pad, test lines, and finally absorbent pad. Competitive assays are used for low-molecular-weight analytes and a positive result is related to the absence of color on the test line due to the binding of the labeled detection antibody binding sites to the target molecule in the sample instead of the analyte–protein conjugates immobilized on the test line. In terms of larger molecules, for example, allergens, the sandwich format is mainly used and the analyte is immobilized between two complementary antibodies (the labeled detection antibody that bound to the analyte during the sample run and an immobilized antibody on the test line). These antibodies recognize different epitopes on the target molecule which is why a sandwich assay is not recommended for small molecules. Finally, the labeled control antibody (often a polyclonal goat anti-mouse antibody) binds with mouse immunoglobulin G (IgG) antibody that is immobilized on the control line as a positive control. It is important to notice that result evaluation is mainly qualitative and made through virtual inspection. However, the utilization of smartphones is bringing about a shift toward objective semiquantification by quantifying color intensity without the need of bulky or alternative commercial readers as used by Neogen in their LFIA testing. Such commercial readers have the advantage of working well for their respective LFA tests but the disadvantage of being incompatible with any other LFAs while the smartphone can potentially quantify any LFA. Although there is controversy that smartphones do not improve the quantification capabilities of a LFA [51], even commercial LFAs for mycotoxins screening are available in the market using a smartphone reader (https://food.r-biopharm.com/products/rida-smart-app-mycotoxin/, last accessed 04/09/2020).

Based on the scientometric analysis results, several interesting LFAs can be critically distinguished. Recently, a smartphone-based LFA was proposed for the detection of *Escherichia coli* O157:H7 in ground beef and spinach matrices [52]. A 3D-printed sample holder was attached to a smartphone to eliminate ambient light interferences. To focus on the performance of the smartphone as an analytical detector, the authors used commercial LFAs to avoid any performance variance related to the assay chemistry. The smartphone-based reader achieved a short analytical range, i.e., 10^4–10^5 CFU/mL. Another POC diagnostic for pathogen detection was reported [53], combining nucleic acid extraction, isothermal amplification, and colorimetric detection. In contrast to the conventional pathogen assays, which are time consuming, this hybrid LFA could attain results within 1 hour. In another study, the interphone variation when using LFAs was evaluated [7]. That is a comparison commonly missing from the published literature, but it is a necessary step toward technology commercialization [10]. In this study of Nelis et al. [7], LFAs for gluten, bovine milk, and cheese detection were utilized and it was concluded that camera calibration, white balance adjustment, and exposure time adjustment should be considered for LFA quantification. Last but not least, a critical comparison among different multiplex immunoassay formats (passive flow-, active flow-through, and LFAs) for nut (hazelnut and peanut) allergen screening was performed and the LFA was coupled to a smartphone readout [54]. Assay validation was performed and a detection capability (CCβ) of the smartphone assay of 1 ppm of total hazelnut protein and 10 ppm total peanut protein by testing 20 different blank and spiked samples. Although this CCβ is quite sensitive, it should be noted that visual detection was more sensitive in this study, highlighting the need to further optimize smartphone-based sensitivity. It should be noticed that insufficient validation of screening methods was reported earlier [11], and this study is a good example on how to validate screening assays in food analysis.

8.4.1.2 Paper-based assays

The main reason behind the ever-increasing interest for paper-based assays is that they provide the facility due to the use of a very common substrate to build the assay. The assays also have optimal portability, cost-effectiveness, and eco-friendliness. Finally, the paper substrate can also function as a filter for crude matrix samples. Paper is a cellulose membrane on which it is possible to immobilize a wide range of (bio)receptors [55] such as antibodies, enzymes, for example acetylcholinesterase [56], or aptamers that are able to selectively and sensitively detect analytes. To achieve that, various immobilization strategies [57], e.g., physical adsorption or covalent binding can be used, which are related to different advantages and drawbacks. Considering smartphone-based related applications, a paper-based method based on a ferrozine colorimetric assay was able to detect iron species in food supplements [58]. It combined a simple workflow including the use of a smartphone app for result semiquantification (Fig. 8.5A) to the evaluation of important performance characteristics, such as LOD or linear range. This method was applied to commonly fortified samples, including wheat flour, infant formula, and nixtamalized corn flour. The device was also successfully benchmarked toward atomic emission spectroscopy, which is a traditional method for elemental determination. In another proof-of-concept study, paper strips were embedded with graphene quantum dots for phenol and polyphenol smartphone-based screening. An application that has potential in food and environmental sample screening [59]. Fluorescent quenching was the detection principle, and a LED light connected to the smartphone was used as the excitation lighting source. In addition, a 3D-printed box was necessary to eliminate the assay from ambient light interferences (Fig. 8.5B). Comparing this platform to colorimetric smartphone sensing, this was a more sophisticated in terms of fabrication which may limit on-site applicability, but the fluorescent detection can provide better detectability. It is worthwhile to point out that the application of wax-printing to functionalize cellulosic platforms in order to better confine the immobilization area and reduce the amount of biorecognition element needed for bioaffinity platforms [61]. In this way, a wax-defined cellulosic platform able to detect starch, a common milk adulterant, was developed and coupled to a smartphone read-out [60] (Fig. 8.5C). Potassium iodide (KI) and iodine were used as chemical reagents for the colorimetric detection of starch in milk. The device attained a linear response ($R^2 = 0.99$) at the range of 0%–12% v/w starch. In conclusion, paper-based assays are versatile analytical

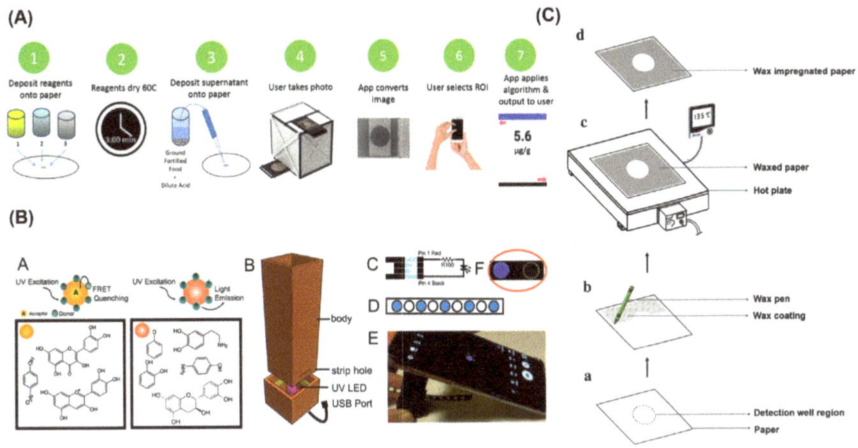

FIGURE 8.5

Selected paper-based assays with smartphone read-out in the food analysis field. (A) The developed workflow for iron species determination in food supplement. (B) UV excitable labels and UV-LEDs were used in a small and portable box that can be attached to the smartphone camera to measure polyphenols. (C) Wax-printing application on a cellulose surface for milk fraud detection.

Source: *(A) Panel reprinted from A.W. Waller, D.J. Rigsby, M. Gaytán-Martínez, J.E. Andrade, Development of a paper-based sensor compatible with a mobile phone for the detection of common iron formulas used in fortified foods within resource-limited settings, Nutrients 11 (2019) 1673, under a Creative Commons CC BY license. (B) Panel reprinted from R. Álvarez-Dlduk, J. Orozco, A. Merkoçi, Paper strip-embedded graphene quantum dots: a screening device with a smartphone readout, Sci. Rep. 7 (2017). https://doi.org/10.1038/s41598-017-01134-3, under a Creative Commons CC BY license. (C) Panel reprinted from A.K. Govindarajalu, M. Ponnuchamy, B. Sivasamy, M.V. Prabhu, A. Kapoor, A cellulosic paper-based sensor for detection of starch contamination in milk, Bull. Mater. Sci. 42 (2019) 255. https://doi.org/10.1007/s12034-019-1958-2, an open-access publication.*

platforms that are able to monitor for various analytes and within different food analytical concepts (both food safety and food fraud).

8.4.1.3 Microfluidic chips and LOC technology

Microfluidics is a relatively new field which was established in 2006 following the initial publication on the topic by G.M. Whitesides in *Nature* [62]. Microfluidics refers to the manipulation of fluids in channels with dimensions of tens of micrometers. Fluidic behavior under this microlevel confined regions significantly differs from fluidic behavior in macroscale. In this context, although no strict criteria have been proposed to define microfluidic systems, the length and internal size of the channels is considered of critical importance [63]. Microfluidic channels are combined to LOC devices to develop fully portable and autonomous analytical platforms. LOC systems are able to mimic different apparatus such as reactors and pumps to carry out injection, filtration, dilution, and detection in a reduced portion eliminating handling errors and enhancing robustness while retaining the analysis cost low [64]. A pump-free microfluidic rapid mixer combined with a paper-based channel was developed for organophosphate smartphone-based screening [65]. The detection is based on cholinesterase recognition that was immobilized on the cellulose part of the device. The device consisted of multiple layers combining laminated transparency film, double-sided adhesive tape, and an overlapping inlet ending in a paper-based reaction area. Besides developing a practical device, this study proposed also a smart and cost-efficient sample preparation protocol. In detail, adhesive tape was stuck to the vegetable surface, peeled off, and dipped into a water–methanol solution achieving a LOD around 0.20 μM for malathion depending on the tested matrix. On the downside, such a sample preparation can determine only residues found on peels. Another interesting development was a microfluidic biosensor for online, rapid, and sensitive detection of *Salmonella* Typhimurium combining immunomagnetic separation and fluorescence [66]. The attained LOD was 58 CFU/mL within 2 hours showing great potential for on-site monitoring of foodborne pathogens. Of importance is that the assay was video recorded, in contrast to the common approach of photographing in smartphone diagnostics. Video recording captures a dynamic response that can inform about the measurement quality. It should be kept in mind that smartphone assays are vulnerable to minor variations (assay positioning, sample injection, etc.) during measurements and video signals can be exploited to evaluate if a measurement is valid. This was the case in a hybrid paper-LOC injector system for carbofuran screening in apple extracts [40]. The assay was video-recorded enabling acceptance/rejection of measurements based on a principle components analysis system. Last but not least, in contrast to other paper-based assays in which cellulose strips are used only as a colorimetric sensor, in this case the paper strips acted as a permeable element to asymmetrically restrict the backward flow. Besides traditional food safety issues, a lab-on-a-disk analytical platform with a smartphone read-out was developed to screen for genetically modified (GM) papaya [67]. The discrimination of GM samples is based on a specific GM marker (P-35S, Cauliflower mosaic virus 35S promoter) whilst non-GM papaya was confirmed through papain (non-GM marker). This method provides sample-to-answer results within 20 minutes (significantly faster than the traditional polymerase chain reaction method) with a LOD of 10 pg/μL DNA using a sample of 0.02 μL papaya juice. In conclusion, it is clear that microfluidic platforms can revolutionize the current analytical scheme considering their low-cost and integrated sample handling. In addition, their hyphenation to smartphones further automates workflows as the user can get real time results and immediately communicate them.

8.4.2 Electrochemical detection

Optical smartphone-based detection using the built-in camera is a straightforward approach, but it may be limited by camera resolution and ambient lighting conditions

[68]. Moreover, food extracts are often colored and exhibit a certain level of turbidity. This can cause interference with colorimetric measurements [9,10]. In this regard, electrochemical techniques may be advantageous because measurements remain largely unaffected by color or turbidity [69]. Moreover, electrochemical sensors can produce better quantitative and more sensitive results as some colorimetric equivalents (especially LFAs) [70]. Another advantage of electrochemical detection is that it is independent of the smartphone model/capabilities, thus, avoiding the common issue of interphone variation [7,10], except perhaps if the potentiostat is integrated into the smartphone model [71]. The most common way to hyphenate a smartphone with the potentiostat is through Bluetooth or by directly inserting the device in a USB or micro-USB port. Commercial examples of such technology are the EmStat and the Sensit smart of Palmsens (https://www.palmsens.com/, last accessed 06/10/2020). Thus, data treatment, connectivity, and intuitive user interface through application use are the main advancements of smartphone technology for electrochemical detection. Notably, detection on these devices is achieved through the use of screen-printed electrodes (SPEs), which are small, disposable electrodes that can be functionalized to detect a variety of targets [72]. These electrodes are cost-efficient (especially the carbon SPE type), disposable, and small making them well adapted for POC sensing. However, the sensitivity of carbon SPEs is reduced compared to the classic glassy carbon electrode. To improve SPE sensitivity nanomaterials are often used to increase the electroactive surface area and obtain better electron transfer [73]. One example of nanomaterial modified SPEs with increased sensitivity is a reported assay using Au nanoparticle and phosphorene modified SPEs for the quantification of the marine toxin okadaic acid (LOD 0.8 ng/mL in buffer) by differential pulse voltammetry showing such portable assays are capable of very sensitive detection as well [74]. Nonetheless, it has also been noted that the use of nanomaterials is often expensive and not suitable for high-throughput production due to the complexity of the required surface modification protocols [75]. Interestingly, CB, which is a by-product created by aromatic hydrocarbon combustion in oil furnaces, has recently shown promise for the development of more cost-efficient nanomaterial modified SPEs. Indeed, CB only costs about ~1 €/kg, improves sensitivity, is resistant to fouling, has stable dispersions (~2 weeks), and no pre-treatments are required [76–79]. Moreover, it was shown that CB can equal or even outperform graphene oxide, reduced graphene oxide, and carbon nanotubes SPEs in terms of electrocatalytic activity and background currents [77]. In another study, it was shown that SPE modification with CB causes a ~10-fold improvement in the LOD for the detection of domoic acid in shellfish (from 4 ng/mL for carbon SPEs to 0.4 ng/mL for CB-SPEs) and substantially reduced the background current in a chronoamperometry based immunoassay using CB-SPEs and magnetic beads [80]. Thus, CB appears to be a very promising material for electrochemical POC sensing. Finally, another very promising development in the electrochemical field is the use of microfluidic paper-based electrochemical devices (μPEDs), especially origami based μPEDs that are pre-loaded with all the necessary reagents to perform the measurements. These devices are optimized to allow simple use by an end-user since no reagents need added.

Instead, the sample is simply dropped on a section of the paper electrode. The reagents are mixed by folding the sections containing the reagents onto each other. Moreover, since the electrode is paper-based it also allows automatic capillary flow and sample filtration. Such devices were recently developed by the Arduini group and include the detection of bisphenol A in bottled water [81], multiplex detection of the pesticides paraoxon, atrazine and 2,4-dichlorophenoxyacetic acid in river water [82], and ethanol quantification in beer [83]. Notably, CB or a CB Prussian blue nanocomposite was utilized to increase the electrochemical performance of these paper-based SPEs, and a detailed protocol describing the manufacturing of these devices has recently been published in *Nature Protocols* [69]. Fig. 8.6 shows an example of such a device for the detection of pesticides.

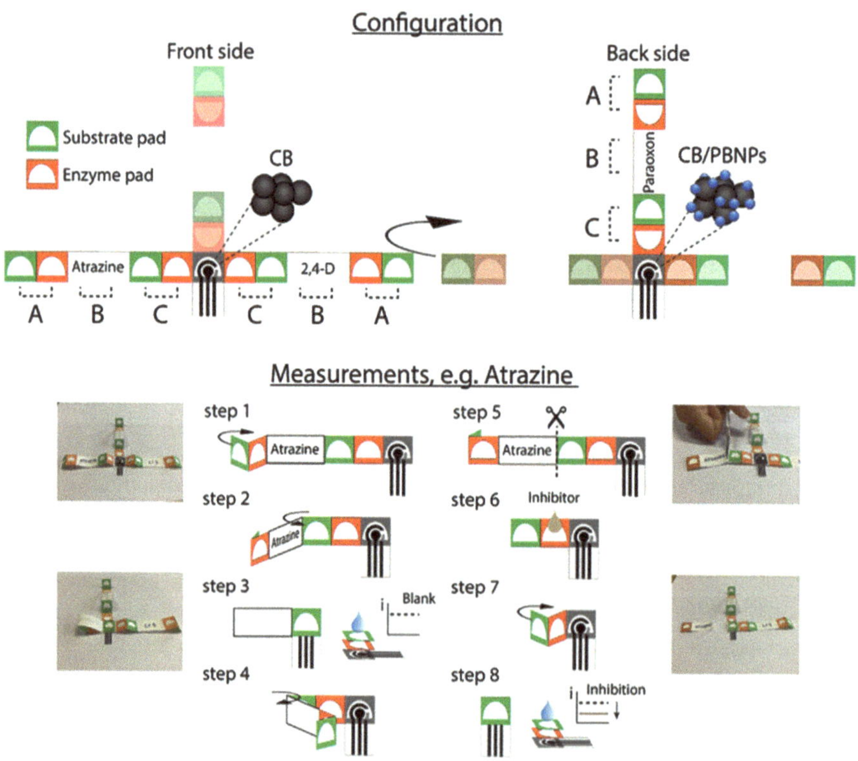

FIGURE 8.6

An overview of a reagent-free paper-based origami device for multiplex pesticide detection.

Source: *This figure was reprinted with permission from F. Arduini, S. Cinti, V. Caratelli, L. Amendola, G. Palleschi, D. Moscone, Origami multiple paper-based electrochemical biosensors for pesticide detection, Biosens. Bioelectron. 126 (2019) 346–354. https://doi.org/10.1016/j.bios.2018.10.014.*

8.5 Validation requirements and benchmarking

The unprecedented characteristics introduced by the use of smartphones as analytical detectors, it is still of utmost importance to evaluate the performance of such novel analytical platforms. However, there is an absence of guidelines on validation of smartphone-based diagnostics, thus questioning whether such equipment actually provides reliable and robust results. Available guidelines on *in-vitro* diagnostics (Directive 98/79/EC, in the medical field) or qualitative and semiquantitative methods [11] in the food safety field can be considered a helpful tool. Nevertheless, large differences between smartphone-based diagnostics and classical chemical assays do exist, for example on-site applicability or potential end-user involvement; therefore, it is necessary to establish tailor-made validation guidelines for smartphone-based methods. In this way, detection capability (CCβ) and interphone result variation are two key aspects that smartphone-based studies should evaluate to demonstrate their applicability. CCβ calculation requires a high number of tests in the tested matrix, specifically at least 20 blank and 20 contaminated [at a screening target concentration (STC)] different samples need to be evaluated. In terms of food safety assays, this STC should be always related to the legislative limits (if they exist) of a contaminant and preferably to be set at 50% of the limit. Otherwise, higher number of samples need to be tested, for example, 60 blank and 60 contaminated samples when the STC is set equal to the legislative limit. Regarding interphone result variation, this is a smartphone inherent problem related to different optical systems and processing algorithms, depending on the manufacturer. Thus, smartphone hyphenated assays need to be tested with a statistically significant number of different smartphones (minimum three different brands) to fully express their potential in chemical analysis. In fact, as we recently showed [10], the evaluation of interphone result variation is a practice commonly missing in the published studies and more effort needs to be devoted to this issue. Undoubtedly, it is also necessary to benchmark smartphone-based methods toward instrumental methods. Although for food analysis this is a regulatory follow-up requirement (Decision 2002/657/EC) for unequivocal identification and quantification, cross-comparison method performance is not always reported in the literature. Within this context, liquid chromatography tandem mass spectrometry (LC–MS/MS) and gas chromatography tandem mass spectrometry (GC–MS/MS) are commonly used as they combine attractive characteristics such as great detectability, robustness or wide linear ranges. LC–MS/MS is more used as it can measure a wider number of compounds, in contrast to GC–MS/MS which can only detect thermostable volatile analytes. It is important to mention that the price of such instrumentation is constantly reduced enabling more laboratories to implement and apply high-end analytical methods. Moreover, for pathogen detection PCR is a very important confirmatory analytical tool as this has been even more evident during the COVID-19 pandemic [84]. PCR is a sensitive technique used for the amplification of a target specific DNA sequence, starting by a small quantity of a sample or tissue. All in all, validation and benchmarking to accredited standards following examples as outlined in Commission Regulation (EU) No 519/2014 and AOAC International guidelines for the validation of qualitative binary chemistry methods [85] are indispensable factors to prove that smartphone-based analysis can attain reliable and rugged results and therefore increase its applicability.

8.6 Conclusion

Smartphone-based sensing is an evolving field providing new opportunities for medical and food analysis. A significant difference in terms of technology readiness between medical and food smartphone-based diagnostics has been noted, with medical analytical applications being further advanced. Nevertheless, successful clinical phase 1–3 trials are a key aspect to demonstrate the applicability of smartphone assays for medical purposes and get approval for their use. However, large and successful phase 3 trials are often still absent and further validation and development is needed to advance the field and increase sensitivity and selectivity, especially for image analysis where use of algorithms that are more advanced and proper datasets are needed. In the case of food analysis, smartphone-based studies are predominantly at the proof-of-concept stage and still face a considerable amount of challenges. Amongst them, coupling food sample preparation to smartphone assays is a substantial bottleneck as such protocols are commonly extensive limiting their application by nonexperts or on-site uses. To overcome this challenge, more effort should be paid on the development of micro-total analysis systems (μTAS). μTAS provide integrated sample preparation, a highly desired characteristic for in-situ and end-user friendly methods. To date, there is a lack of such systems, especially in the case of solid food matrices. To develop μTAS, the use of 3D printing technology is highly recommended as cost-efficient, lightweight, and practical solutions with short fabrication time can be developed. Insufficient validation of smartphone-based platforms was also noticed with most of the studies not evaluating the method performance in food matrices but in buffer solutions. In addition, benchmarking toward reference methods should be always performed, otherwise a critical comparison of the newly developed technology is not feasible. In terms of result interphone variation, this is a critical step toward commercialization of smartphone-based methods. Ideally, smartphone-based analysis should enable the implementation of individuals without laboratory skills and chemistry background. Considering that a myriad of smartphones are available, a universal approach for optical detection needs to be considered. Standardization of image data capturing and processing will enable greater opportunities while video analysis may increase resolution, reduce background interference, and provide an internal assay quality control. In this way, this technology will be transferred to consumers for at-home analysis resulting in more intense testing. Realistically, smartphone-based analysis has not reached this point yet, but still such a scenario can no longer be considered utopia. A vital aspect for acceptability for public end-user or citizen science engagement will be effective continuous communication on the evolving technology with the evolving generations.

Acknowledgment

This project has received funding from the European Union's Horizon 2020 research and innovation program under the Marie Sklodowska-Curie grant agreement No 720325, FoodSmartphone.

References

[1] J.H. Soh, H.-M. Chan, J.Y. Ying, Strategies for developing sensitive and specific nanoparticle-based lateral flow assays as point-of-care diagnostic device, Nano Today (2020) 100831.

[2] M.Y.-C. Wu, M.-Y. Hsu, S.-J. Chen, D.-K. Hwang, T.-H. Yen, C.-M. Cheng, Point-of-care detection devices for food safety monitoring: proactive disease prevention, Trends Biotechnol. 35 (2017) 288–300. Available from: https://doi.org/10.1016/j.tibtech.2016.12.005.

[3] S. Kumar, M. Nehra, J. Mehta, N. Dilbaghi, G. Marrazza, A. Kaushik, Point-of-care strategies for detection of waterborne pathogens, Sensors 19 (2019). Available from: https://doi.org/10.3390/s19204476.

[4] J. Nelis, C. Elliott, K. Campbell, The Smartphone's Guide to the Galaxy": in situ analysis in space, Biosensors 8 (2018) 96.

[5] M. Adeel, M.M. Rahman, I. Caligiuri, V. Canzonieri, F. Rizzolio, S. Daniele, Recent advances of electrochemical and optical enzyme-free glucose sensors operating at physiological conditions, Biosens. Bioelectron. 165 (2020). Available from: https://doi.org/10.1016/j.bios.2020.112331.

[6] S. Bu, K. Wang, C. Ju, Y. Han, Z. Li, P. Du, et al., A pregnancy test strip for detection of pathogenic bacteria by using concanavalin A-human chorionic gonadotropin-$Cu_3(PO_4)_2$ hybrid nanoflowers, magnetic separation, and smartphone readout, Microchim. Acta 185 (2018). Available from: https://doi.org/10.1007/s00604-018-2968-2.

[7] J.L.D. Nelis, L. Bura, Y. Zhao, K.M. Burkin, K. Rafferty, C.T. Elliott, et al., The efficiency of color space channels to quantify color and color intensity change in liquids, pH strips, and lateral flow assays with smartphones, Sensors 19 (2019) 5104.

[8] J.L.D. Nelis, Y. Zhao, L. Bura, K. Rafferty, C.T. Elliott, K. Campbell, A randomized combined channel approach for the quantification of color- and intensity-based assays with smartphones, Anal. Chem. (2020). Available from: https://doi.org/10.1021/acs.analchem.0c01099.

[9] Y. Zhao, S.Y. Choi, J.L.D. Nelis, H. Zhou, C. Cao, K. Campbell, et al., Smartphone modulated colorimetric reader with color subtraction, 2019 IEEE Sensors. Accept. Publ. (n.d.) 1–4.

[10] J.L.D. Nelis, A.S. Tsagkaris, M.J. Dillon, J. Hajslova, C.T. Elliott, Smartphone-based optical assays in the food safety field, TrAC. Trends Anal. Chem. 129 (2020) 115934. Available from: https://doi.org/10.1016/j.trac.2020.115934.
[11] A.S. Tsagkaris, J.L.D. Nelis, G.M.S. Ross, S. Jafari, J. Guercetti, K. Kopper, et al., Critical assessment of recent trends related to screening and confirmatory analytical methods for selected food contaminants and allergens, TrAC. – Trends Anal. Chem. 121 (2019). Available from: https://doi.org/10.1016/j.trac.2019.115688.
[12] S. Petry, Intersun: the global UV project: a guide and compendium, World Heal. Organ. (2003) 54. Available from: https://doi.org/10.1016/j.neulet.2013.04.050.
[13] J. Ferlay, M. Colombet, I. Soerjomataram, T. Dyba, G. Randi, M. Bettio, et al., Cancer incidence and mortality patterns in Europe: estimates for 40 countries and 25 major cancers in 2018, Eur. J. Cancer. 103 (2018) 356–387. Available from: https://doi.org/10.1016/j.ejca.2018.07.005.
[14] C.M. Balch, J.E. Gershenwald, S.-J. Soong, J.F. Thompson, M.B. Atkins, D.R. Byrd, et al., Final version of 2009 AJCC melanoma staging and classification, J. Clin. Oncol. 27 (2009) 6199–6206. Available from: https://doi.org/10.1200/JCO.2009.23.4799.
[15] K. Ramlakhan, Y. Shang, A mobile automated skin lesion classification system, Proc. - Int. Conf. Tools Artif. Intell. ICTAI. 23 (2011) 138–141. Available from: https://doi.org/10.1109/ICTAI.2011.29.
[16] R. Parikh, A. Mathai, S. Parikh, G.C. Sekhar, R. Thomas, Understanding and using sensitivity, specificity and predictive values, Indian J. Ophthalmol. 56 (2008) 45.
[17] A. Esteva, B. Kuprel, R.A. Novoa, J. Ko, S.M. Swetter, H.M. Blau, et al., Dermatologist-level classification of skin cancer with deep neural networks, Nature 542 (2017) 115–118. Available from: https://doi.org/10.1038/nature21056.
[18] S.S.L. Oskouei, H. Golestani, M. Hashemi, S. Ghiasi, CNNdroid: GPU-Accelerated Execution of Trained Deep Convolutional Neural Networks on Android, ACM. ISBM. (2015) 1201–1205. https://doi.org/10.1145/2964284.2973801.
[19] N.D. Lane, S. Bhattacharya, A. Mathur, P. Georgiev, C. Forlivesi, F. Kawsar, Squeezing deep learning into mobile and embedded devices, IEEE Pervasive Comput. 16 (2017) 82–88. Available from: https://doi.org/10.1109/MPRV.2017.2940968.
[20] M.H. Gregg, D. Ionica, The Movidius Myriad architecture's potential for scientific computing, IEEE (2009) 6–14. Available from: https://doi.org/10.1109/MM.2015.4.
[21] S. Rallapalli, H. Qiu, A.J. Bency, S. Karthikeyan, R. Govindan, R. Urgaonkar, Are very deep neural networks feasible on mobile devices?, in: Usc Conf. Proc., 2015.
[22] C. Rat, S. Hild, J. Rault Sérandour, A. Gaultier, G. Quereux, B. Dreno, et al., Use of smartphones for early detection of melanoma: systematic review, J. Med. Internet Res. 20 (2018) e135. Available from: https://doi.org/10.2196/jmir.9392.
[23] K. Freeman, J. Dinnes, N. Chuchu, Y. Takwoingi, S.E. Bayliss, R.N. Matin, et al., Algorithm based smartphone apps to assess risk of skin cancer in adults: systematic review of diagnostic accuracy studies, BMJ 368 (2020) m127. Available from: https://doi.org/10.1136/bmj.m127.
[24] C. Dallet, S. Kareem, I. Kale, Real time blood image processing application for malaria diagnosis using mobile phones, Int. Symp. Circuits Syst. (ISCAS), IEEE. 11 (2014) 2405–2408. https://doi.org/10.1109/ISCAS.2014.6865657.

[25] P.B. Lillehoj, M.-C. Huang, C.-M. Ho, A handheld, cell phone-based electrochemical biodetector, 26th Int. Conf. Micro Electro Mech. Syst. (MEMS), IEEE. (2013) 53–56. https://doi.org/10.1109/MEMSYS.2013.6474174.

[26] C.C. Stemple, S.V. Angus, T.S. Park, J.-Y. Yoon, Smartphone-based optofluidic lab-on-a-chip for detecting pathogens from blood, J. Lab. Autom. 19 (2014) 35–41. Available from: https://doi.org/10.1177/2211068213498241.

[27] T. Laksanasopin, T.W. Guo, S. Nayak, A.A. Sridhara, S. Xie, O.O. Olowookere, et al., A smartphone dongle for diagnosis of infectious diseases at the point of care, Sci. Transl. Med. 7 (2015) 273re1. Available from: https://doi.org/10.1126/scitranslmed.aaa0056.

[28] M.G. Mauk, C. Liu, M. Sadik, H.H. Bau, Microfluidic devices for nucleic acid (NA) isolation, isothermal NA amplification, and real-time detection, Methods Mol. Biol. 1256 (2015) 15–40. Available from: https://doi.org/10.1007/978-1-4939-2172-0_2.

[29] P.A. Sandoz, A.F. Coskun, A.J. Chung, W.M. Weaver, O. Adeyiga, D. Khodadadi, et al., Digital readout platform for water-in-oil droplet immunoassays running on a cell-phone for point of care viral load sensing, Proc. 16th Int. Conf. Miniaturized Syst. Chem. Life Sci. MicroTAS 2012. (2012) 338–340. http://www.rsc.org/images/loc/2012/pdf/M.1.18.pdf.

[30] J.T. Coulibaly, M. Ouattara, M.V. D'Ambrosio, D.A. Fletcher, J. Keiser, J. Utzinger, et al., Accuracy of mobile phone and handheld light microscopy for the diagnosis of schistosomiasis and intestinal protozoa infections in Côte d'Ivoire, PLoS Negl. Trop. Dis. 10 (2016) e0005550. Available from: https://doi.org/10.1371/journal.pntd.0004768.

[31] R.K.D. Ephraim, J.S. Cybulski, E. Duah, M. Prakash, M.V. D'Ambrosio, D.A. Fletcher, et al., Diagnosis of *Schistosoma haematobium* infection with a mobile phone-mounted Foldscope and a reversed-lens CellScope in Ghana, Am. J. Trop. Med. Hyg. 92 (2015) 1253–1256. Available from: https://doi.org/10.4269/ajtmh.14-0741.

[32] S.D. Holmen, E.F. Kjetland, M. Taylor, E. Kleppa, K. Lillebø, S.G. Gundersen, et al., Colourimetric image analysis as a diagnostic tool in female genital schistosomiasis, Med. Eng. Phys. 37 (2015) 309–314. Available from: https://doi.org/10.1016/j.medengphy.2014.12.007.

[33] B. Veigas, J.M. Jacob, M.N. Costa, D.S. Santos, M. Viveiros, J. Inácio, et al., Gold on paper-paper platform for Au-nanoprobe TB detection, Lab. Chip 12 (2012) 4802–4808. Available from: https://doi.org/10.1039/c2lc40739f.

[34] B. Veigas, E. Fortunato, P.V. Baptista, Mobile based gold nanoprobe TB diagnostics for point-of-need, Methods Mol. Biol. 1256 (2015) 41–56. Available from: https://doi.org/10.1007/978-1-4939-2172-0_3.

[35] M.S. Duthie, M.F. Balagon, A. Maghanoy, F.M. Orcullo, M. Cang, R.F. Dias, et al., Rapid quantitative serological test for detection of infection with *Mycobacterium leprae*, the causative agent of leprosy, J. Clin. Microbiol. 52 (2014) 613–619. Available from: https://doi.org/10.1128/JCM.02085-13.

[36] M. Zarei, Portable biosensing devices for point-of-care diagnostics: Recent developments and applications, TrAC. Trends Anal. Chem. 91 (2017) 26–41. Available from: https://doi.org/10.1016/j.trac.2017.04.001.

[37] D.S.Y. Ong, M. Poljak, Smartphones as mobile microbiological laboratories, Clin. Microbiol. Infect. 26 (2020) 421–424. Available from: https://doi.org/10.1016/j.cmi.2019.09.026.

[38] M. Zarei, Infectious pathogens meet point-of-care diagnostics, Biosens. Bioelectron. 106 (2018) 193–203. Available from: https://doi.org/10.1016/j.bios.2018.02.007.

[39] I.C. Balan, Use of Novel Mobile Technology to Screen Sexual Partners for HIV and STIs, (2016).

[40] A.S. Tsagkaris, J. Pulkrabova, J. Hajslova, D. Filippini, A hybrid lab-on-a-chip injector system for autonomous carbofuran screening, Sensors 19 (2019). Available from: https://doi.org/10.3390/s19245579.

[41] G. Comina, A. Suska, D. Filippini, Autonomous chemical sensing interface for universal cell phone readout, Angew. Chem. Int. (Ed.) 54 (2015) 8708–8712.

[42] A. Priye, S.W. Bird, Y.K. Light, C.S. Ball, O.A. Negrete, R.J. Meagher, A smartphone-based diagnostic platform for rapid detection of Zika, chikungunya, and dengue viruses, Sci. Rep. 7 (2017) 1–11. Available from: https://doi.org/10.1038/srep44778.

[43] M.S. Draz, N.K. Lakshminaraasimulu, S. Krishnakumar, D. Battalapalli, A. Vasan, M.K. Kanakasabapathy, et al., Motion-based immunological detection of Zika virus using Pt-nanomotors and a cellphone, ACS Nano 12 (2018) 5709–5718. Available from: https://doi.org/10.1021/acsnano.8b01515.

[44] M.S. Draz, K.M. Kochehbyoki, A. Vasan, D. Battalapalli, A. Sreeram, M.K. Kanakasabapathy, et al., DNA engineered micromotors powered by metal nanoparticles for motion based cellphone diagnostics, Nat. Commun. 9 (2018) 4282. Available from: https://doi.org/10.1038/s41467-018-06727-8.

[45] M.K. Kanakasabapathy, M. Sadasivam, A. Singh, C. Preston, P. Thirumalaraju, M. Venkataraman, et al., An automated smartphone-based diagnostic assay for point-of-care semen analysis, Sci. Transl. Med. 9 (2017). Available from: https://doi.org/10.1126/scitranslmed.aai7863.

[46] H. Amal, M. Leja, Y.Y. Broza, U. Tisch, K. Funka, I. Liepniece-Karele, et al., Geographical variation in the exhaled volatile organic compounds, J. Breath. Res. 7 (2013) 47102. Available from: https://doi.org/10.1088/1752-7155/7/4/047102.

[47] W. Miekisch, J.K. Schubert, G.F.E. Noeldge-Schomburg, Diagnostic potential of breath analysis—focus on volatile organic compounds, Clin. Chim. Acta 347 (2004) 25–39. Available from: https://doi.org/10.1016/j.cccn.2004.04.023.

[48] C. Jaeschke, M. Padilla, E. Turppa, I. Polaka, O. Gonzalez, K. Richardson, et al., Overview on SNIFFPHONE: a portable device for disease diagnosis, in: 2019 IEEE Int. Symp. Olfaction Electron. Nose, 2019: pp. 1–2. https://doi.org/10.1109/ISOEN.2019.8823212.

[49] K.M. Koczula, A. Gallotta, Lateral flow assays, Essays Biochem. 60 (2016) 111–120. Available from: https://doi.org/10.1042/EBC20150012.

[50] L. Anfossi, F. Di Nardo, S. Cavalera, C. Giovannoli, C. Baggiani, Multiplex lateral flow immunoassay: an overview of strategies towards high-throughput point-of-need testing, Biosensors 9 (2019) 2.

[51] G.M.S. Ross, M.G.E.G. Bremer, J.H. Wichers, A. Van Amerongen, M.W.F. Nielen, Rapid antibody selection using surface plasmon resonance for high-speed and sensitive hazelnut lateral flow prototypes, Biosensors 8 (2018). Available from: https://doi.org/10.3390/bios8040130.

[52] Y. Jung, Y. Heo, J.J. Lee, A. Deering, E. Bae, Smartphone-based lateral flow imaging system for detection of food-borne bacteria *E. coli* O157:H7, J. Microbiol. Methods. 168 (2020). Available from: https://doi.org/10.1016/j.mimet.2019.105800.
[53] J.R. Choi, J. Hu, R. Tang, Y. Gong, S. Feng, H. Ren, et al., An integrated paper-based sample-to-answer biosensor for nucleic acid testing at the point of care, Lab. Chip 16 (2016) 611−621. Available from: https://doi.org/10.1039/c5lc01388g.
[54] G. Ross, G.I.J. Salentijn, M.W.F. Nielen, A critical comparison between flow-through and lateral flow immunoassay formats for visual and smartphone-based multiplex allergen detection, Biosensors. 9 (2019) 143.
[55] C.I.L. Justino, A.C. Freitas, R. Pereira, A.C. Duarte, T.A.P.R. Santos, Recent developments in recognition elements for chemical sensors and biosensors, TrAC. Trends Anal. Chem. 68 (2015) 2−17.
[56] A.S. Tsagkaris, L. Uttl, J. Pulkrabova, J. Hajslova, Screening of carbamate and organophosphate pesticides in food matrices using an affordable and simple spectrophotometric acetylcholinesterase assay, Appl. Sci. (2020).
[57] E.W. Nery, L.T. Kubota, Evaluation of enzyme immobilization methods for paper-based devices—a glucose oxidase study, J. Pharm. Biomed. Anal. 117 (2016) 551−559. Available from: https://doi.org/10.1016/j.jpba.2015.08.041.
[58] A.W. Waller, D.J. Rigsby, M. Gaytán-Martínez, J.E. Andrade, Development of a paper-based sensor compatible with a mobile phone for the detection of common iron formulas used in fortified foods within resource-limited settings, Nutrients 11 (2019) 1673.
[59] R. Álvarez-DIduk, J. Orozco, A. Merkoçi, Paper strip-embedded graphene quantum dots: a screening device with a smartphone readout, Sci. Rep. 7 (2017). Available from: https://doi.org/10.1038/s41598-017-01134-3.
[60] A.K. Govindarajalu, M. Ponnuchamy, B. Sivasamy, M.V. Prabhu, A. Kapoor, A cellulosic paper-based sensor for detection of starch contamination in milk, Bull. Mater. Sci. 42 (2019) 255. Available from: https://doi.org/10.1007/s12034-019-1958-2.
[61] A.S. Tsagkaris, D. Migliorelli, L. Uttl, D. Filippini, J. Pulkrabova, J. Hajslova, A microfluidic paper-based analytical device (μPAD) with smartphone readout for chlorpyrifos-oxon screening in human serum, Talanta (2020) 121535. Available from: https://doi.org/10.1016/j.talanta.2020.121535.
[62] G.M. Whitesides, The origins and the future of microfluidics, Nature 442 (2006) 368−373.
[63] B. Rapp, Microfluidics: modeling, mechanics and mathematics, ELSEVIER, 2016.
[64] A. Suska & D. Filippini, Autonomous lab-on-a-chip generic architecture for disposables with integrated actuation, submitted, Submitted. (2019).
[65] I. Jang, D.B. Carrão, R.F. Menger, A.R. Moraes de Oliveira, C.S. Henry, Pump-free microfluidic rapid mixer combined with a paper-based channel, ACS Sens. 5 (2020) 2230−2238. Available from: https://doi.org/10.1021/acssensors.0c00937.
[66] S. Wang, L. Zheng, G. Cai, N. Liu, M. Liao, Y. Li, et al., A microfluidic biosensor for online and sensitive detection of *Salmonella* Typhimurium using fluorescence labeling and smartphone video processing, Biosens. Bioelectron. 140 (2019). Available from: https://doi.org/10.1016/j.bios.2019.111333.
[67] J.F.-C. Loo, G.W.-C. But, H.-C. Kwok, P.-M. Lau, S.-K. Kong, H.-P. Ho, et al., A rapid sample-to-answer analytical detection of genetically modified papaya using loop-mediated isothermal amplification assay on lab-on-a-disc for field use,

Food Chem. 274 (2019) 822–830. Available from: https://doi.org/10.1016/j.foodchem. 2018.09.049.

[68] D. Zhang, Q. Liu, Biosensors and bioelectronics on smartphone for portable biochemical detection, Biosens. Bioelectron. 75 (2016) 273–284. Available from: https://doi.org/10.1016/j.bios.2015.08.037.

[69] S. Cinti, D. Moscone, F. Arduini, Preparation of paper-based devices for reagentless electrochemical (bio)sensor strips, Nat. Protoc. 14 (2019) 2437–2451. Available from: https://doi.org/10.1038/s41596-019-0186-y.

[70] J.L.D. Nelis, A.S. Tsagkaris, Y. Zhao, J. Lou-Franco, P. Nolan, H. Zhou, et al., The end user sensor tree: an end-user friendly sensor database, Biosens. Bioelectron. 130 (2019) 245–253. Available from: https://doi.org/10.1016/j.bios.2019.01.055.

[71] J. Guo, Uric acid monitoring with a smartphone as the electrochemical analyzer, Anal. Chem. 88 (2016) 11986–11989. Available from: https://doi.org/10.1021/acs.analchem.6b04345.

[72] O.D. Renedo, M.A. Alonso-Lomillo, M.J.A. Martínez, Recent developments in the field of screen-printed electrodes and their related applications, Talanta 73 (2007) 202–219. Available from: https://doi.org/10.1016/j.talanta.2007.03.050.

[73] L. Reverté, B. Prieto-Simón, M. Campàs, New advances in electrochemical biosensors for the detection of toxins: nanomaterials, magnetic beads and microfluidics systems. A review, Anal. Chim. Acta 908 (2016) 8–21. Available from: https://doi.org/10.1016/j.aca.2015.11.050.

[74] S. Ramalingam, R. Chand, C.B. Singh, A. Singh, Phosphorene-gold nanocomposite based microfluidic aptasensor for the detection of okadaic acid, Biosens. Bioelectron. 135 (2019) 14–21. Available from: https://doi.org/10.1016/j.bios.2019.03.056.

[75] F. Ricci, G. Adornetto, G. Palleschi, A review of experimental aspects of electrochemical immunosensors, Electrochim. Acta 84 (2012) 74–83. Available from: https://doi.org/10.1016/j.electacta.2012.06.033.

[76] F. Arduini, S. Cinti, V. Mazzaracchio, V. Scognamiglio, A. Amine, D. Moscone, Carbon black as an outstanding and affordable nanomaterial for electrochemical (bio)sensor design, Biosens. Bioelectron. 156 (2020) 112033. Available from: https://doi.org/10.1016/j.bios.2020.112033.

[77] S. Cinti, F. Arduini, M. Carbone, L. Sansone, I. Cacciotti, D. Moscone, et al., Screen-printed electrodes modified with carbon nanomaterials: a comparison among carbon black, carbon nanotubes and graphene, Electroanalysis 27 (2015) 2230–2238. Available from: https://doi.org/10.1002/elan.201500168.

[78] F. Arduini, F. Di Nardo, A. Amine, L. Micheli, G. Palleschi, D. Moscone, Carbon black-modified screen-printed electrodes as electroanalytical tools, Electroanalysis 24 (2012) 743–751. Available from: https://doi.org/10.1002/elan.201100561.

[79] D. Talarico, F. Arduini, A. Constantino, M. Del Carlo, D. Compagnone, D. Moscone, et al., Carbon black as successful screen-printed electrode modifier for phenolic compound detection, Electrochem. Commun. 60 (2015) 78–82. Available from: https://doi.org/10.1016/j.elecom.2015.08.010.

[80] J.L.D. Nelis, D. Migliorelli, S. Jafari, S. Generelli, J. Lou-Franco, J.P. Salvador, et al., The benefits of carbon black, gold and magnetic nanomaterials for point-of-harvest electrochemical quantification of domoic acid, Microchim. Acta 187 (2020). Available from: https://doi.org/10.1007/s00604-020-4150-x.

[81] D. Jemmeli, E. Marcoccio, D. Moscone, C. Dridi, F. Arduini, Highly sensitive paper-based electrochemical sensor for reagent free detection of bisphenol A, Talanta 216 (2020) 120924. Available from: https://doi.org/10.1016/j.talanta.2020.120924.

[82] F. Arduini, S. Cinti, V. Caratelli, L. Amendola, G. Palleschi, D. Moscone, Origami multiple paper-based electrochemical biosensors for pesticide detection, Biosens. Bioelectron. 126 (2019) 346–354. Available from: https://doi.org/10.1016/j.bios.2018.10.014.

[83] S. Cinti, M. Basso, D. Moscone, F. Arduini, A paper-based nanomodified electrochemical biosensor for ethanol detection in beers, Anal. Chim. Acta 960 (2017) 123–130. Available from: https://doi.org/10.1016/j.aca.2017.01.010.

[84] R. Funari, K.-Y. Chu, A.Q. Shen, Detection of antibodies against SARS-CoV-2 spike protein by gold nanospikes in an opto-microfluidic chip, Biosens. Bioelectron. 169 (2020) 112578. Available from: https://doi.org/10.1016/j.bios.2020.112578.

[85] I. AOAC, Guidelines for validation of qualitative binary chemistry methods, J. AOAC Int. 97 (2014) 1492.

CHAPTER

CMOS-based microanalysis systems

9

Tugba Kilic[1], Sara S. Ghoreishizadeh[2] and Sandro Carrara[3]

[1]*Center for Systems Biology, Massachusetts General Hospital, Harvard Medical School, Boston, MA, United States*

[2]*Department of Electrical and Electronic Engineering, University College London, London, United Kingdom*

[3]*Integrated Circuits Laboratory, School of Engineering, Institute of Microengineering, École Polytechnique Fédérale de Lausanne (EPFL), Neuchâtel, Switzerland*

9.1 Introduction

In the field of science and technology, Complementary Metal-Oxide Semiconductor (CMOS)-based microanalysis involves a large number of different devices, technologies, methods, and approaches, which is pretty evident from the scientific literature. For example, we can find the use of CMOS-based image sensors in investigations on pathological tissues [1], in implantable microimagers [2], in micro-electro-mechanical systems (MEMS) [3], in electrophoretic systems [4], in CMOS chips for electrical sensing of microfluidic transport [5], in CMOS chemical microsensors [6], in fluorescence integrated sensors [7], in amperometric sensors [8], in CMOS electronics and memristive biosensors [9], and are sometimes also coupled with microfluidics [10]. As a proof of expansion of the field, we can just consider that a handbook published some years ago by Cambridge University Press about bioelectronics contains 44 chapters consisting of more than 500 pages [11], while a new handbook on sensors in press now will be published with more than 900 pages [12]. Therefore, it is quite easy to understand that an exhaustive overview of the findings published so far through using the state-of-the-art technology cannot be produced in a single book chapter because it will require a full book, at least.

On the other hand, the aim of this chapter is to focus on analytical systems at the micron scale with integrated electronics, namely CMOS-enabled biosensors, and try our best to summarize the most advanced state of the art, mainly considering CMOS systems with integrated biosensors and microfluidics for applications to detection of relevant analytes in biomedical applications. CMOS-enabled biosensors represent an evolution in the field of biosensors where biosensing is carried out by a device integrated with an electronic reader, also providing system autonomy [13]. To some extent, the application of the new concept of the

Microfluidic Biosensors. DOI: https://doi.org/10.1016/B978-0-12-823846-2.00006-7

Internet-of-Things (IoT) in the field of biosensors has given further push to the development of CMOS-enabled biosensors [14]. The CMOS (complementary metal-oxide silicon) process is being widely used in industrial processes through which the modern integrated electronics are successfully built in enormous quantities at low cost per unit. The CMOS process is behind empowering modern humanity to carry in their pockets, typically embedded in smartphones, a computing capability much larger than that possible in 40 seconds using electronics computers larger than entire rooms (e.g., the ENIAC [15]) or in 80 seconds using electronics computers still larger than a bulky desk (e.g., the Cray [16]). The name of the technology itself explains how it works: the complementary metal and oxide layers are built on silicon in order to obtain a series of transistors that are directly connected by using the same process in the very same structure, while realizing extremely complex architectures. In the CMOS process, a single bulk of silicon is used to create all the differently doped silicon regions forming the *drains* and the *gates* of many transistors; the oxide is used on top to create the required *gates* as well as to passivate the electrical connections to avoid shortcuts; the metal layers are finally used to create the multiple connections among different transistors. The technology enables accomodating up to 50 million transistors in a single processing unit [17]. The CMOS process is very reliable and all modern electronics are built based on this process. Some of the integrated circuits (IC) are also very useful, widely used, and robust and are still manufactured (e.g., the very famous IC μA 741 [18]). Of course, such a powerful, low-cost, and successful fabrication process has impacted the industry of consumer electronics enormously, making it possible to manufacture electronic devices in a range of affordable prices. Also, it is expected that the CMOS industry will make possible manufacturing IoT devices at affordable prices, with a current forecast of about 24 billion of IoT devices by 2030 [19]. In this context, the electronics industry will definitely be benefitted alongside biosensors and microanalytical systems due to the potential of the CMOS process in developing new IoT devices for use as autonomous biosensors for medical applications at home, in intensive care units, and in surgery rooms as well [20]. This has motivated us to introduce CMOS-based microanalysis systems in this chapter: emergence of a new generation of biomedical IoTs.

The chapter consists of five subsections: (1) microfluidics-based device fabrication, (2) detection methods for different kinds of analytes, (3) electrical readouts with CMOS-embedded electronics, and (4) applications. In the first two sections, we aim to provide a detailed update about microfluidic biosensors, by also highlighting the advantages and limitations of each subclass in order to give insights for better future designs. We specifically explain the fabrication methods, i.e., photolithography, e-beam lithograpy, wax (2-D) printing, 3-D printing, soft lithography, and nanoimprinting and surface functionalization strategies with various bioreceptors like enzymes, antibodies, and aptamers, and discuss how to target a wide variety of analytes such as cell metabolites (e.g., glucose and lactate), protein disease biomarkers (e.g., cancer markers or neurotransmitters), ions, and electrolytes (e.g., Na, Li, K) via several detection methods including but not limited to potentiometry, amperometry, conductometry, ion-sensitive field-effect transistor (ISFET) and metal oxide semiconductor field effect transistor (MOSFET). Due to the fact that the research field is very broad as mentioned above, readers will be directed to more specific reviews, books, and additional references for further reading. Then, the state-of-the-art biosensors' instrumentation is reviewed as well, with a focus on high-specification systems that implement the above-mentioned common electrochemical techniques, such as amperometry, potentiometry, and impedance spectroscopy. So, the emphasis is more on electrochemical sensing here. Recent developments in both printed circuit board (PCB) and application-specific IC (ASIC) sensor instrumentation are highlighted and related design strategies, challenges, opportunities, and trends are discussed in depth by considering that small size, low power consumption, and high reliability are of utmost importance. Finally, we provide a brief overview of the monolithically integrated amperometric/potentiometric biosensors and PCB/ASIC electronics and discuss the challenges and future direction of this new promising area of research. Applications to point-of-care (POC) diagnostics, food safety, and healthcare monitoring with wearable and implantable devices are then introduced as part of such a huge plethora of technological approaches been applied today, by also referring to the more recent trends in literature to go further beyond the present state-of-the-art.

9.2 Fabrication techniques for microfluidic biosensors

In this section, we aim to provide a detailed update on microfluidic biosensors highlighting the advantages and limitations of each subclass to give insights to researchers for better future designs. We specifically explain the fabrication methods, surface functionalization strategies with various bioreceptor molecules to target a wide variety of analytes, and the detection modalities. Due to the fact that this research field is very broad, readers will be directed to more specific reviews, books, and additional references for further reading.

Considering the advantages of microfluidics, microfluidic biosensors or lab-on-a-chip devices can scale down the routine wet lab practices to a miniaturized format. While those biosensors provide a versatile usage, the success of the analysis heavily relies on the microfluidic design. Depending on materials to critical design dimensions, each fabrication technique has its own advantages and drawbacks. Table 9.1 gives a comparative analysis of current microfluidic fabrication techniques that are briefly described in the following sections.

Paper-based microfluidics has attracted the attention of researchers over the past 10 years due to its specific advantages such as low sample volume requirement, ease of use, portability, high sensitivity, and no necessity to well-equipped laboratory equipment and well-trained manpower. These characteristics have

Table 9.1 List of current microfluidic fabrication techniques.

Technique	Materials Used	Device Resolution	Advantages	Drawbacks	Reference
Photolithography	• Photoresists (PRs) • SU8	<100 nm	• Parallel processing • High resolution • Mass production	Diffraction-limited	[21]
E-beam lithography	• PMMA • SU8 • PEG	<1 nm	High resolution	• Low throughput • Complex protocols • Expensive	[22]
Wax (2-D) printing	• Filter paper • Chromotography papa	<100 μm	• Simple fabrication • Low cost • Fast • Mass production	Sensitivity Loss of sample	[23]
3-D printing	• LED light source • Curable liquid polymers • Thermoplastic materials	<100 μm		• Biocompability • Surface quality • Optical transparency • Low resolution	[24]
Soft lithography	• PDMS • SU8	30 nm to 100 μm	Gas permeableNonspecific adsorption	• Low cost • Versatile • Simple • Biocompatible	[25]
Nanoimprinting	• Thermoplastic polymer • Photopolymer	<10 nm		• Low cost	

PDMS, polydimethylsiloxane; *PEG*, polyethylene glycol; *PMMA*, polymethyl methacrylate.

made paper platforms a promising alternative for a variety of applications such as clinical diagnosis and quantitative analysis of chemical and biological substances. Among the wide range of fabrication methods for microfluidic paper-based analytical devices (μPADs), the wax printing method is suitable for high-throughput production and requires only a commercial printer and a heating source to fabricate complex 2- or 3-D structures for multipurpose systems. μPADs can be used by anyone for in situ diagnosis and analysis; therefore, wax-printed μPADs are promising especially in resource-limited environments where people cannot get access to sensitive and fast diagnosis of their serious health problems and where food, water, and related products are not able to be screened for toxic elements. This review paper is focused on the applications of paper-based microfluidic devices fabricated by the wax-printing technique and used for international health. Besides presenting the current limitations and advantages, the future directions of this technology, including the commercial aspects, are discussed. As a conclusion, the wax-printing technology continues to overcome the current limitations and to be one of the promising fabrication techniques. In the near future, with increasing interest of the industrial sector in the paper-based technology, the wax-printed paper-based platforms are expected to emerge to cater to the requirements of the healthcare industry.

9.2.1 Photolithography

Photolithography by its literal meaning is image formation by using light. It is a technique that has been in use for more than two centuries and relies on the usage of photosensitive materials, i.e photoresists. Upon exposure to light by direct writing or through a photomask, spin-coated photoresist either becomes hard and insoluble (negative photoresist) or stays soluble (positive photoresist). Photolithography is one of the most widely used techniques for the fabrication of microdevices such as FETs and ICs thanks to its adaptability for parallel processing and high resolution, which is affected only by the wavelength of the light [21]. Currently, the highest resolution (<100 nm) photolithography can be achieved via using deep ultraviolet light (248 and 193 nm). Created channels could be enclosed within flat substrates through various bonding strategies such as anodic, fusion, or plasma bonding [22]. Microfluidic biosensors based on photolithography are generally made of silicon, glass [23], polydimethylsiloxane (PDMS) [22,24,25], and polycarbonate [26]. PDMS has attracted a lot of attention due to its biocompatibility, low cost, and suitability for various applications.

9.2.2 Electron beam lithograpy

Electron beam (e-beam) lithography (EBL) is similar to scanning electron microscopy (SEM) since it also utilizes an electron gun, but instead of collecting the secondary electrons to image the surface, EBL uses beams to create nanopatterns (<1 nm) on a substrate [27]. Like photolithography, e-beam lithography requires resists such as polymethyl methacrylate (PMMA), SU8, and polyethylene glycol (PEG), which are cross-linked to become soluble (positive) or less soluble (negative) upon electron struck. E-beam lithography is suitable for developing nanobiosensors that require a minimum sample input, e.g., those used in various diagnostic platforms such as nanowire [28] and nanoplasmonic [29].

9.2.3 Wax (2-D) printing

Wax (2-D) printing is one of the fabrication techniques to design μPADs [30]. It is based on creating hydrophobic patterns on a hydrophilic paper (cellulose fibers) by using wax. The whole process can be completed in two steps: (1) wax printing and (2) wax melting. Wax printing can be achieved by various techniques such as screen printing [31], inkjet printing [32], and ink stamping [33]. Wax printing techniques offer many advantages. Since commercial printers can be employed, mass production can be achieved, and the unit cost of each device is much lower than other techniques. Thanks to its simple fabrication, wax printing has a widespread use in low-source settings. Diagnostic μPADs fabricated via wax printing have been used for chemical analysis [34,35], blood typing [36], pH [37], nucleic acid biomarkers [38], and ions (Ni, Co, Fe etc.) [39,40].

9.2.4 3D printing

Three-dimensional (3D) printing is an additive manufacturing method for creating 3D objects layer by layer. Microfluidic devices can be developed via 3-D printing based on stereolithography [41], multijet modeling [42], fused deposition modeling [43], and selective laser sintering [44]. Stereolithograpy is a 3D printing technique that utilizes a photo resin and an LED light source to form an assembly-free 3D polymeric microfluidic device [45]. Multijet modeling, on the other hand, is used to create complex shapes by using single or multi-inkjet heads that dispense various curable liquid photopolymers [46]. As an alternative method to using photopolymers, fused deposition modeling relies on low-cost thermoplastic materials such as acrylonitrile butadiene styrene, poly-lactic acid, and polyethylene terephthalate to create 3D microfluidic devices via an extrusion process in which a nozzle head prints the material that hardens upon temperature decrease. Every 3D printing technique has its own advantages and challenges. For instance, stereolithography can render higher resolution, whereas resins used are cytotoxic in contrast to biocompatible plastics that could be used in other 3D printing techniques.

9.2.5 Soft lithography

Soft lithography is a nonlithographic technique that is widely used as a low-cost alternative for micro- and nano-fabrication. It comprises various patterning methods such as micro-contact printing, micro-molding in capillary, micro-transfer molding, and replica molding to create an elastomeric stamp that is used to create patterns on soft substrates like room-temperature vulcanizing silicones and PDMS.[29] PDMS is the most preferred material to fabricate fluidic channels thanks to its low cost, and robust mechanical and optical properties. Until now, a wide range of PDMS-based microfluidic devices have been developed for 3D cell culture applications [47]; organs-on-chips [48]; and sensing/biosensing optical [49] and electrochemical [50–54] signals, and photonics [55]. Despite many advantages, nonspecific adsorption of proteins is a challenge for PDMS devices. Therefore, several groups have developed strategies for eliminating nonspecific protein adsorption [56].

9.2.6 Nanoimprint lithography

Nanoimprint lithography (NIL) is a technique used to create nanoscale patterns in a low-cost and high-throughput fashion. Similar to the hot embossing technique, it also uses a hot press heating. Basically, a thermoplastic polymer coated on a silicon wafer is placed into the press and a stamp or a template, and is pushed towards the polymer. Decrease in thermoplastic polymer temperature enables the stamp lift-off and the patterns are created [57]. In another version of NIL, instead of temperature, UV can be used to harden the liquid photopolymer. Therefore, in contrast to NIL, UV–NIL does not require high temperatures and can be performed at room temperature. By NIL, it is possible to design sub-10 nm and high aspect ratio ($>50:1$) patterns [58]. NIL has been used for designing nanoelectromechanical systems, and localized surface plasmon resonance (LSPR) [59], photonics crystal [60], and electrochemical [61] biosensors.

9.3 Detection methods for different analytes

Development of biosensors begins with the selection of the target analyte that can be a marker either for a disease, cancer, infectious diseases, or for the cell fate and metabolism. Hence, analytes can have different molecular structures and can be available at physiological concentrations. We subgrouped analytes into four main categories: ions and electrolytes, metabolites, proteins, and nucleic acids, and summarized the most important ones belonging to each category, citing the related literature.

9.3.1 Ions and electrolytes

Detection of ions and electrolytes has been of great interest since they can be used as markers for electrolyte disorders and organ dysfunction—any deviation from the physiological ranges for sodium (Na^+), potassium (K^+), calcium (Ca^{2+}), and chloride (Cl^-) could be an indication of electrolyte disorders, or about the viability of living organisms, and quality of water soil and air. Moreover, ion detection/monitoring is necessary for the screening of drugs designed for ion channels. Up to now, various instrumental methods based on spectroscopy and chromatography have been developed. However, chemical and biochemical settings require specific detection schemes, i.e. on-site, online, biocompatible, sensitive, and selective monitoring. In order to address these needs, a wide variety of novel sensing techniques have been developed. These techniques can be subgrouped into five categories based on the mode of detection: (1) fluorescence, (2) absorbance, (3) colorimetry, (4) SERS (Surface Enhanced Raman Scattering) and chemiluminescence, and (5) electrochemical.

9.3.1.1 Fluorescence-based detection of ions and electrolytes

Fluorescence-based detection has been a method of choice for many microfluidic devices designed for ion detection thanks to available fluorescent molecular probes specific to ions, types of detectors, exceptions sources, and filters. Fluorescent probes provide high sensitivity and temporal resolution and hence facilitate microfluidic design in various formats and materials. These probes have been introduced in microfluidic chips in an organo-aqueous phase, which will be mixed with the sample stream to detect K+ [62], Pb^{2+} [63], Cd^{2+} [64], Hg^{2+} [65], and Cs^{+} [66]. In addition to the fluorescent probes, quantum dots have also been utilized for detecting ions. Qi et al. developed paper-based 3D origami microfluidics for multiplexed Hg^{2+} and Cu^{2+} ions in water samples. They functionalized glass fiber paper with quantum dots that emit luminescent energy upon ion imprinting [65]. Yetisen et al. developed another paper-based microfluidic device for Na^{+}, K^{+}, and Ca^{2+} detection in tear to be applied for the detection of early-stage dry eye [67]. Another application that could be useful for POC settings has been demonstrated by Ghaderinezhad et al. The team developed a paper-based sensor coupled with a smartphone-enabled platform for detecting the range of Na^{+}, K^{+}, Ca^{2+}, Cl^{-}, and NO^{2-} in urine [68].

9.3.1.2 Absorbance-based detection of ions and electrolytes

Although not so common as fluorescent detection, absorption spectroscopy has also been used for ion sensing since the setup requirement is more cost efficient compared to the fluorescent detection. Typically, an LED light source and a photodiode detector are required for the measurement setup used in a variety of probes. Gold nanoparticles (AuNPs), for instance, have a strong SPR band and therefore highly sensitive to any changes occurring on their surface. Pedro et al. used ionophore-modified AuNPs for continuous monitoring of Hg^{2+} ions [69]. In another work, NO^2 was detected with 20 nM sensitivity by using a PMMA-based microfluidic device using the Griess method, which relies on the formation of an Azo dye [70].

9.3.1.3 Colorimetric detection of ions and electrolytes

Like absorption spectroscopy, colorimetric detection relies on changes in the absorption. However, in colorimetric detection, visible changes in color enable researchers to develop platforms such as paper-based microfluidic devices that do not allow transmission but could be used for reflection-based color analysis, which is easy to be carried out via smartphones in resource-limited settings. Up to know, various articles reviewed the colorimetric μPADs [71,72]. These devices in general rely on detecting reagents immobilized on the cellulose membrane. For instance, Sudkate et al. developed a μPAD for the detection of Cu^{2+} using the catalytic etching of silver nanoparticles (AgNPs), which change color in contact with the analyte [73]. In a more recent article, scientists presented for the first time a colorimetric μPAD for the detection of Li^{+} from blood. The μPAD has two paper-based elements responsible for blood cell separation and colorimetric detection [74].

9.3.1.4 SERS and chemiluminescence-based detection of ions and electrolytes

SERS and chemiluminescence techniques have rarely been used for ion detection although very common in use for protein and nucleic acid detection. SERS is a technique that relies on a local physical phenomenon called SERS effect or plasmon resonance seen in the holes or gaps of a rough metal surface. Therefore, gold or silver surfaces or their nanoparticles have been used to develop biosensors. Qi et al. used glutathione 4-mercaptopyridine (4-MPY) modified AgNPs to enhance the SERS effect and detect As(III) ions that cause aggregation of AgNPs due to the affinity of As(III) ions to glutathione. The assay enabled a linear detection range of 3–200 ppb and a limit of detection (LOD) of 0.67 ppb [75]. Wang et al., on the other hand, used AuNPs to develop a droplet-based microfluidic device to monitor Hg^{2+} in water based on SERS. The device showed rapid and reproducible detection through signal changes of rhodamine B [76].

Chemiluminescence (CL) is, as its name implies, the generation of radiation via a chemical reaction. CL-based-detection modalities therefore rely only on phototransducers since there is no need for light sources to do measurements. Hence, CL technique has been utilized for microfluidic designs. Inpota et al. demonstrated the use of CL for the detection of CO^{2+} via using polyethyleneimine and polyethyleneimine-functionalized poly (methyl methacrylate) nanoparticles [77].

Another group used the CL technique to create a disposable μPAD device fabricated by wax printing to detect Cr(III) upon the luminol oxidation by hydrogen peroxide in the presence of Cr(III). The μPAD demonstrated a fast (<1 minute analysis time) and sensitive (LOD: 0.02 ppm) detection of Cr(III) [78].

9.3.1.5 Electrochemical detection of ions and electrolytes

Like fluorescent-based detection, electrochemistry plays a major role in ion/electrolyte sensing platforms. As one of the first commercialized detection components, ion selective electrodes (ISEs) have been in use since 1970s in a variety of both gas-electrolyte and clinical chemistry analyzers. Park et al. used a Ca^{2+} ISE to monitor cellular activities of HepG2/C3As cultured in a microchannel [79].

Similarly, K^+-selective nano-petal electrodes were developed consisting of PVC (polyvinyl chloride) K^+-selective membranes to monitor cell viability of HepG2/C3A cells upon osmotic shock [80]. Among other techniques that will be discussed in following sections, anodic stripping voltammetry enables the most sensitive detection of heavy metals thanks to its preconcentration step. Jothimuthu et al. used anodic stripping voltammetry to develop a lab-on-a-chip device for the detection of Pb^{2+}, Cd^{2+}, Zn^{2+}, Ti^+, In^{3+}, Cu^{2+}, and Sn^{2+} ions. The device consists of a bismuth working electrode, a silver/silver chloride (Ag/AgCl) reference electrode, and a gold auxiliary electrode that is bonded to a PDMS chip [81]. ISEs have also been used to perform noninvasive monitoring through human sweat. Gao et al. developed a flexible and fully integrated sensor array for monitoring not only electrolytes (Na^+ and K^+) but also metabolites (glucose, lactate) [82]. Similarly, another group designed a battery-free, wireless, and epidermal ISE-based detection system that relies on stretchable electrode arrays to detect Na^+, K^+, and H^+ as well as glucose [83].

ISE can also be applied to electrochemical paper-based devices (EPADs). In their work, Lan and colleagues showed that wax printers can be used for simple microfludic designs to define a PVC-based ion selective membrane area from a sample zone. The ion-sensing EPAD provided a simple, portable, and inexpensive alternative for Na^+, Ca^{2+}, and K^+ sensing [84]. Ion-sensitive field-effect transistors (ISFETs) were first introduced in 1970s after the serial development of metal–oxide semiconductor field-effect transistors (MOSFETs). ISFETs enable us to study the interface between oxides or metals and liquids. ISFETs monitor changes at the interface as a function of the bias at the reference electrode (equivalent of gate at MOSFETs) with respect to the changes in ion concentrations. Due to their versatility, ISFETs have been adapted for use in various microfluidic designs. Cordero et al. introduced a fully on-chip-integrated wearable sweat analysis platform based on ISFET sensors for the detection of pH, Na^+, and K^+ [85]. Silicon nanowire based ISFETs have also drawn a great deal of attention in the past decade thanks to their performance. Various groups have used memristive ISFETs for ion/electrolyte detection [86–88].

9.3.2 Metabolites (glucose, lactate, etc.)

Metabolites, namely glucose and lactate, are another class of biomarkers that need to be monitored for diabetes patients to understand their physiological state as well as the dosage of the drug administered. Physiological limits of blood glucose is 3.6–5.8 mM in blood [89]. In addition to diabetes, lactate monitoring is also required for monitoring sepsis and muscle fatigue. The best case scenario from the clinical point of view would be monitoring these metabolites in a continuous fashion and administration of required medications with a feedback-controlled mechanism. Up to now, a wide variety of approaches have been developed based on the detection mechanism used: electrochemical [90], optical [91], and detection media; *in vivo*—intravenous (blood) and subcutaneous or *in vitro*—physiological liquids are analyzed in the lab or on the skin from sweat [92], saliva [93], and tears [94].

9.3.3 Proteins

As the final product of the gene expression, proteins are associated with biological functions. Hence, they are considered as valuable diagnostic markers. As a result of this, protein biosensors have a significant importance in the POC diagnostics.

Optical techniques that are based on the interaction between light and biorecognition element, e.g. antibodies, dominate the microfluidic protein biosensors field. SPR technique is widely used for protein detection and could be used in microfludic setups. It is very sensitive to the changes occurring at the electrode–solution interface. Commercially available bench-top SPR systems require prism couplers for microfluidics integration, whereas fiber-optic-based or waveguide-based SPR systems could be miniaturized in a microfluidic format. In addition to planar gold electrodes, recent efforts have been focused on nano-sized arrays in various geometries to create nanoplasmonics. Nanoplasmonic devices can be in the form of nano hole arrays [95], nano wells, nanopores, nanoslits, or nanodiscs and make multiplexed detection of analytes possible. In addition, for nanoplasmonic devices, there is no need of illumination at a steep angle to initiate SPR, but a simple illumination and normal incident light could be sufficient [96].

9.4 Biosensor CMOS instrumentation

In this section, an overview of the instrumentation systems used in electrochemical biosensors is given with a focus on state-of-the-art high-specification portable systems that implement common electrochemical techniques, including amperometry and impedance spectroscopy. We then provide an overview of the monolithically integrated amperometric/potentiometric biosensors on CMOS and discuss the challenges and future direction of this new promising area of research.

9.4.1 Amperometric sensor instrumentation

Electrical current measurement is one of the most important techniques used in nanobiosensors such as ion channels, solid-state nanopores, and redox-based sensors. It also plays an important role in electrochemical impedance spectroscopy (EIS), a technique commonly used for studying a large variety of electrochemical sensors and phenomena.

Amperometric sensors produce a current at the output that is typically the result of a redox reaction that either directly involves the analyte of interest or is indirectly affected by the presence of the target analyte. The oxidase-based glucose sensors, which are used in most of the commercially available continuous glucose measurement devices, are an example of the former sensor type, and label-free amperometric DNA sensors, which incorporate probe redox species, are examples of the latter. In both sensors, the transduction mechanism is actuated through applying a specific voltage profile, which may be step-wise (chronoamperometry), triangular (cyclic voltammetry), or small-amplitude pulses superimposed on a triangular waveform (differential pulse voltammetry), among others. This leads to the generation of an electric current that is proportional to the concentration of the analyte of interest.

The generated sensor current normally flows between two electrodes: the working and the counter-electrodes of the amperometric sensor. An amperometric instrumentation connects to both electrodes and closes the loop for the sensor current. In this process, the instrumentation circuit either produces an amplified copy of the sensor current [97] using a current conveyor and current mirrors or converts the sensor current into a voltage right away through a resistor or a switch-capacitor based transimpedance amplifier (TIA) [8,98]. A switch-capacitor based TIA (SC-TIA) works in two phases. In the first phase, the input current, I_{IN}, is integrated into the integrator capacitor, while in the second phase, the capacitor is discharged through either a switch or a reference current, I_{REF}. The SC-TIA has the benefit of low-pass filtering the electrochemical noise as well and can also easily be adjusted, for example, through changing the clock frequency of the switch or the reset current, to measure a wide range of input currents. The TIA is typically followed by an analog-to-digital converter (ADC), while the current mirrors can directly drive a current to a frequency converter [97].

Portable amperometric instrumentation circuits have been either designed based on an assembly of off-the-shelf components on a PCB [99] or custom-designed for the specific application (ASIC). Although a PCB-based technique has the advantage of being capable of rapid production at low cost, and easy testing, only the ASIC devices allow millimeter-size sensor nodes, scalability (increasing the number of on-chip sensing sites), ultra-low noise, and low-power consumption.

In the quest for a more compact and low-power instrumentation circuit, several design techniques have been introduced and demonstrated mostly relying on resource sharing. For example, the switch-capacitor TIA can also act as an integral part of a dual-slope ADC. This effectively converts the input current into a pulse, whose duration is proportional to the ratio between I_{IN} and I_{REF} [100]. Here a counter can be used to measure/digitize the duration of the pulse either at the sensor node or later at the receiver/processing side [100].

The digital data may then be transferred from the sensor node to the point of display/processing using wires or wireless telemetry techniques depending on the application. Most portable sensing devices utilize standard commercially available wireless data transfer techniques such as Bluetooth or Near Field Communication (NFC). Custom-designed on-chip wireless units have also been designed [100] for more compact and fully integrated sensor nodes.

Novel ASIC-based instrumentation devices cannot only be custom-designed to match the exact criteria of a specific sensor/application but also address some major nonidealities and challenges in electrochemical sensing and offer completely new sensing applications. For example, differential readout circuits have been proposed and demonstrated to remove the relatively large background current of amperometric sensors from the measurements [8,101]. Compact in-pixel processing circuits have been developed to allow parallel recording of quantal [102] release events and high-throughput drug screening [103]. A self-powered wireless physiochemical sensing system was developed in [104] for monitoring glucose or lactate in bodily fluids without the need for a battery.

9.4.2 Electrochemical impedance spectroscopy

EIS is a powerful measurement technique to characterize and understand the interfacial behavior of a liquid–solid interface. Such investigations are important in numerous applications and fields of research, including electrochemical synthesis and deposition of materials, such as metals and conducting polymers, investigation of molecule–surface interactions, and research on sensors, batteries, and fuel cells [105]. Electrochemical impedance measurements in electrochemical sensing applications are generally performed by applying a small (mV range) alternating ac voltage through a pair of electrodes and measuring the subsequently generated current signal via the same pair. It is necessary to generate an accurate voltage amplitude independent of load variations with minimum phase delay throughout the operational bandwidth so that any measured current amplitude and phase is only attributed to the device under test [106].

Several approaches are used to perform EIS. They can be broadly classified into two categories: fast Fourier transform (FFT)-based EIS and frequency response analyzer (FRA)-based method. The most important advantage of FFT is that the frequency response at all desired frequencies can be computed at the same time. However, the FFT-based method is very hardware intensive, and although attempts have been made to integrate FFT into ASIC for portable applications, it has not been possible in some applications due to the required accuracy, preprocessing, and prestorage of the input data set [107]. The FRA-based method generates results for one frequency point at a time. An FRA-based system normally includes a quadrature signal generator, an amperometric readout, a multiplier, and an integrator. The quadrature signal generator produces a pair of sinusoid signals at the desired frequencies that have the same amplitude but a 90 degrees phase shift. The multipliers shift the readout circuit response down to DC. At the end, integrators remove the interference and produce two constant DC signals that represent the real and imaginary portions of the admittance of the biosensors in the potentiostat system [107]. Due to its relative simplicity of implementation, the FRA-based method can be implemented as compact EIS circuits on AISC [108–110].

9.4.3 Monolithic integration of biosensors

The monolithic integration of sensing electrodes with the instrumentation circuit on a single substrate presents several unique advantages. The integration improves the signal-to-noise ratio of the measurement, through removing the long interconnects between the sensor and instrumentation, provides scalability, and allows miniature sensor nodes/pixels with small (sub microwatt) power consumption [98]. Fully integrated CMOS biosensors have been implemented based on various sensing mechanisms, including the impedimetric [111], capacitive [112,113], electrochemical [114,115], ISFET [116], nanowire [117], magnetic [118], optical [119], and micromechanical [119,120] techniques, to provide real-time and high-throughput detection.

Some sensing electrodes can be fabricated using standard semiconductor technology. A good example of an integrated sensor and instrumentation is ISFETs. As described earlier in this chapter, ISFETs are normal transistors fabricated in standard CMOS technology, which can be used as a pH sensor because the intrinsic insulation layer used in the standard CMOS processes (silicon nitride) is pH-sensitive and can therefore act as the recognition element of a pH sensor. ISFETs are most-sought-after sensors as they do not require any postprocessing of CMOS chips and their transduction mechanism is inherently potentiometric.

Amperometric transduction can be implemented on ICs fabricated in semiconductor technology. Here the exposed aluminum layer (AKA pad) in a CMOS IC can be used as the basis on which a sensing electrode can be developed. Such development might also be needed when the recognition element of the sensor (either potentiometric or amperometric) can only be attached/deposited on a conductive electrode. However, Al is a poor electrode material for electrochemical analysis because of its reactivity, and it readily oxidizes upon exposure to air [121]. Therefore, the exposed Al of the electrodes is deposited with a noble metal such as gold (Au) since it is inert and chemically stable in an electrolyte solution.

The deposition of gold on Al can be performed using methods such as (1) physical vapor deposition (PVD) [122], (2) electroless nickel immersion gold (ENIG) deposition [121], which involves deoxidization of the Al surface, followed by zincation, Ni deposition, and Au deposition; and (3) electroplating through applying a specific current or voltage to the electrode. Such postprocessing may be done at either the wafer level [114] or the single-die level [122]. PVD metal deposition becomes expensive for thick layers and is generally limited to only 1—2 μm. The electroless plating is simple to carry out, as no electrical connection is needed to the electrodes being plated. However, electrodeposition provides precise control over the layer thickness and can be used to achieve high thicknesses with higher purity while not requiring cleanroom facilities.

Other metallic, nanostructure, and probe molecules may be deposited on top of the Au to achieve desired sensitivity and selectivity to the target analyte. For example, in [123], the ENIG gold covered electrode was electroplated to form gold nanostructures on which the DNA probe molecules were immobilized.

Alternatively, capacitance variation of the electrical double layer on the gold-plated electrode surface has been exploited to detect biorecognition events at the sensing electrode/electrolyte interface [124]. All the aforementioned operations were done by creating special Au electrodes developed directly on chip.

9.5 Application areas

In this section, we aim at closing the chapter by briefly summarizing the several examples that have been proposed by the literature over the last 10 years or so as the most advanced cases of CMOS-based microanalysis system. In the context of this book, we consider analytical microsystem systems with integrated electronics used for biosensing. Of the wide range of application areas, as already discussed in the Introduction, we consider the three main branches that correspond to POC diagnostics, food safety, and healthcare as well as wellbeing continuous monitoring. All these application areas are still witnessing emerging technologies, approaches, and ideas being incorporated, including the widespread use of electrochemical biosensors for glycemia tests (both POC and implantable or wearable for continuous monitoring) or use of flow microfluidics with on/off optical reading dedicated to pregnancy tests.

Similarly, an array of approaches are available for monitoring a long series of metabolites, ions, biomarkers, and exogenous therapeutic compounds, in order to provide powerful solutions in the above-mentioned areas of application. The following subsections discuss, through examples, several options available for expanding the technologies to fulfill a plethora of possible new monitoring needs and related habits or diseases.

9.5.1 POC diagnostics

An example of POC diagnostics is "*rapid tests*," which have recently proven to be extremely useful with large success in dealing with the COVID-19 pandemic The rapid tests for the COVID infection are based on exactly the same principle of the pregnancy tests: a microfluidic structure extracts the serum form the blood by capillary forces and then a predesigned series of antibody-functionalized regions are used to distinguish the positive/negative test by the control (see Fig. 9.1). The related technology is very easy to use and it costs almost nothing because they involve only a piece of plastic with very little amount of biological material (antibodies, which are currently produced in large amount by biotech or pharma companies). However, they are beyond the scope of our discussion because they do not require any CMOS electronics. These are basically on/off tests, where only the appearance of a line observable by necked eyes is sufficient to readout the test result. However, the integration of such a simple lateral flow technology with cheap CMOS electronics may help conduct a quantitative

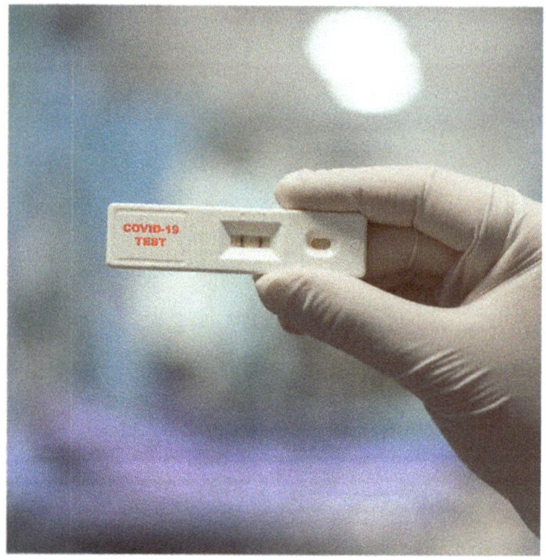

FIGURE 9.1

Covid Rapid Test based on the same lateral flow principle used in pregnancy tests, like hCG urine Pregnancy test by Goldenmed.

analysis, as well as chemiluminescent immunoassays, which has been demonstrated several years ago in literature [125]. This is of key importance for our discussion here because it demonstrates that we could have a quantitative analytical determination by using similar optical methods of a plethora of biological markers at the patients' bed, e.g. cancer markers [125], growth factors [126], insulin [127], and COVID immune-reaction [128], just to mention few.

Similarly, many different applications might be targeted using other electrochemical methods, for example, using the same approaches as used for the detection of glucose for diabetic patients in detecting other human endogenous metabolites like lactate [129], cholesterol [130], or glutamate [131], and exogenous metabolites as well, like anticancer compounds [132], antiinflammatory drugs [133], and anesthetics [134]. Besides, the coupling of optical methods with CMOS electronics proves advantageous for the electrochemical methods as well, not only because we can obtain quantitative analytical measurements of the target human molecules at the patient's bed, but also because it might address simultaneous other related needs. For example, it has been demonstrated in the case of anesthesiology the possibility to develop monitoring devices directly connected to the Internet to exploit the enormous potential of the Cloud technology [135] or to address some specific detection issues with new kinds of solutions provided by machine learning as edge-computation at the patient's bed [136]. The application possibilities are immense in principle. We can envisage developing specific devices for a plethora of molecules related to both diseases and therapy, which will help address the needs of many more patients than possible with today's technology. The scientific literature demonstrates an enormous potential here, which is difficult to cover in detail in a single book chapter. An entire book dedicated to only this topic would be required here. Therefore, we mention here a very few examples to simply give an idea to the readers of the enormous potential of this technology. Even complex fluorescence optical biodetections are carried out using extremely miniaturized microscopes with a future possibility of their portability [7]. Electrochemical detection of DNA by amperometric methods [137], by simply using pH measures [138] or by capacitive detection [112], has also been demonstrated. In some cases, the proposed integrated biosensors are so specialized, for example, for measuring pathophysiological levels of cytokines [139], glucose and lactate [140], or potassium [80] in early diagnosis of sepsis or for measuring the level of anesthesiology with portable devices [136]. Some of those published proposals for new POC devices.

A few examples out of many POC devices already introduced into the market and are based on the electrochemical detection of several human molecules relevant for different diagnostics. A good review of POC instruments already introduced into the market that deal with the requirements for genomics and related DNA detection is, for example, available in *Biosensors and Bioelectronics* by M. Trotter and co-authors [141].

9.5.2 Food safety

Another important field of application of the CMOS-based microanalysis system is ensuring food safety. In this field, it is very important to precisely identify and quantify several components of the industrial food products, ranging from food sophistication to food quality or from nutritional check to food safety. For example, for people who might manifest allergies due to lactose or other components of the milk, it might be very useful for those people to have access to easy-to-use technologies for checking the presence of milk allergens in prepared foods. Again, the scientific literature proposes a series of technical solutions aimed at fulfilling these requirements. For example, the literature proposes solutions based on electrochemical immune-sensing to look for particular allergens in milk [142], toxins in groundnut [143], and residues of antibiotics in food [144], just to mention a few. Optical methods help as well in fluoroquinolone detection in animal-derived foods [145], and toxins in almonds [146] or in maize [147], for example. The research in this field has made enormous progress in bringing those methods to our smartphones [148]. Also, there are numerous devices available in the market facilitating quality check of pH, food color, and microbiology (e.g., bacterial contaminations). Fig. 9.2 briefly summarizes some of these technologies already available in the market.

9.5.3 Healthcare monitoring

The extreme miniaturization of CMOS-based microanalysis systems has proven to be extremely useful in the field of implantable and wearable biosensors for the

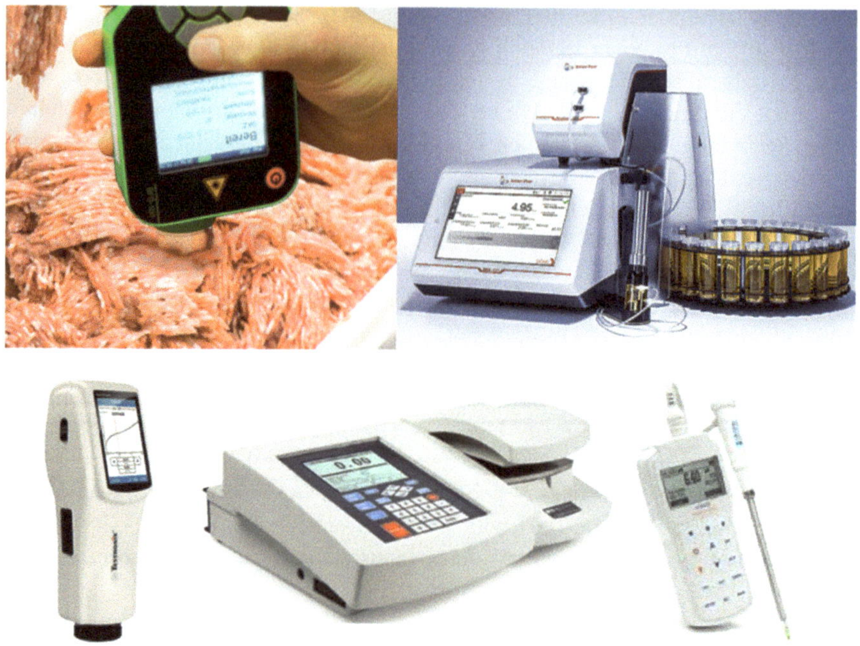

FIGURE 9.2

A short series of instruments already introduced into the market for food quality testing (from top-left, clockwise): freshdetect BFD-100, portable measurement device by FreshDetect; beer analyzing system by Anton-Paar; cheese pH meter by Hanna Instruments; food testing equipment by Saturn group; and TP800 portable spectrophotometer by Testronic offered for color control in food production.

continuous monitoring of the human healthcare. In Section 9.5.1, we have already seen that the devices available in the market are mainly dedicated to continuous monitoring of glucose. The reason being the market around diabetics is quite significant and, hence, has attracted many investments in the past. Examples include FreeStyle by Abbot, which is a wearable device for sampling the patient glucose under their skin (typically in the interstitial tissues), and facilitating continuous monitoring of the patient glycemia. The field of healthcare diagnostics is still expanding with similar technologies being introduced on a continuous basis for the detection of other metabolites and biomarkers. For example, fully integrated implantable devices have already been demonstrated to be successful not only for glucose monitoring but also for monitoring therapeutic compounds, e.g. antipyretic and analgesics [149]. The continuous monitoring of other drugs, like antiinflammatory ones [150], has been demonstrated as well. Wearable devices and sensors have been successfully demonstrated for real-time monitoring of many compounds in different contexts, e.g., lactate, glucose and ions for sport assessment [82], lithium for therapeutic monitoring of bi-polar patients [151], nicotine for discrimination of light and heavy smokers [152], and stress level [153]. At the same time, implantable CMOS-based systems have been demonstrated to be successful in many other medical applications, not limited to brain recording [154] and stimulation [155]. Some of the options already offered by the scientific literature opening the scope for the expansion of the present state-of-the art implantable devices for monitoring purposes in the field of human healthcare.

9.5.4 Trends on POC: beyond wearable and implantable devices

Before concluding this chapter, we overlook the future prospects of CMOS-based microanalysis systems. An interesting perspective that has emerged over the last 20 years of research in the field of miniaturized CMOS-based microanalysis systems has been published recently by one of the authors of this chapter [156]. In that paper, the historical perspective of the first publication about the innovative concept of *Smart Dust*, originally proposed by the University of California, Berkeley, in 1999 [157], has been expanded to influence today's research in the new fascinating fields of Internet-of-Things in biomedical [158], leading to the new emerging concepts of *Neural Dust* [159] and *Body Dust* [160]. The main idea behind the two concepts is that the CMOS integration might be developed to render fully functional CMOS devices in sizes of dust grains of few tens of hundredth of microns in order to offer the possibility of fully functional systems freely floating in the bloodstream and capable of neurostimulation or microanalysis. The concepts that, on one hand, might look too ambitious for a serious research and development plan, while, in the other hand, have been already demonstrated to be feasible experimentally, at least in the case of the stimulation of the sciatic nerve in rats [161].

The systems developed so far are available in mm-sizes [161]. On the other hand, the theoretical possibility of developing a CMOS circuit for a glucometer smaller than 10 μm^2 [162], a data transmitting circuit smaller than 50 μm^2 [163], and an inductive receiving coils for power transmission in diameters of 100 μm^2 [164] has already been demonstrated with the present state-of-the-art in CMOS technology. Therefore, the future of CMOS-based microanalysis systems undoubtedly focuses on downscaling the available technologies with the most modern CMOS technology nodes to the nanoscale, in order to achieve devices within sizes of tens of microns and, therefore, to obtain systems for microanalysis so miniaturized to pave the way for their spread in human tissues like dust moieties.

References

[1] B. Kumar, Design of CMOS based image sensor with cantilever mechanism for smart pathological detection with opto-mechanical system.

[2] D.C. Ng, T. Tokuda, S. Shiosaka, Y. Tano, J. Ohta, Implantable microimagers, Sensors 8 (2008) 3183.

[3] X. Li, D.-W. Lee, Integrated microcantilevers for high-resolution sensing and probing, Meas. Sci. Technol. 23 (2011) 022001.

[4] P.L. Urban, D.M. Goodall, E.T. Bergström, N.C. Bruce, Electrophoretically mediated microanalysis of a nicotinamide adenine dinucleotide-dependent enzyme and its facile multiplexing using an active pixel sensor UV detector, J. Chromatogr. A 1162 (2007) 132.

[5] B.C. Jacquot, C. Lee, Y.N. Shen, E.C. Kan, Time-resolved charge transport sensing by chemoreceptive neuron MOS transistors $({\rm C}\nu {\rm MOS}) $ with microfluidic channels, IEEE Sens. J. 7 (2007) 1429.

[6] A. Hierlemann, et al., Application-specific sensor systems based on CMOS chemical microsensors, Sens. Actuators B: Chem. 70 (2000) 2.

[7] D. Măriuţa, et al., Miniaturization of fluorescence sensing in optofluidic devices, Microfluidics Nanofluidics 24 (2020) 1.

[8] S.S. Ghoreishizadeh, I. Taurino, G. De Micheli, S. Carrara, P. Georgiou, A differential electrochemical readout ASIC with heterogeneous integration of bio-nano sensors for amperometric sensing, IEEE Trans. Biomed. Circuits Syst. 11 (2017) 1148.

[9] I. Tzouvadaki, et al., Multi-panel, on-single-chip memristive biosensing, IEEE Sens. J. 19 (2019) 5769.

[10] A. Vallero, et al., Memristive biosensors integration with microfluidic platform, IEEE Trans. Circuits Syst. I: Regul. Papers 63 (2016) 2120.

[11] S. Carrara, K. Iniewski, Handbook of Bioelectronics: Directly Interfacing Electronics and Biological Systems, Cambridge University Press, 2015.

[12] G.G.a.S. Carrara, Handbook of Sensors, 992, IEEE/Wiley Publisher, 2022.

[13] S.-Y. Lu, et al., A review of CMOS electrochemical readout interface designs for biomedical assays, IEEE Sens. J. 21 (2021).

[14] M. Mayer, A.J. Baeumner, A megatrend challenging analytical chemistry: biosensor and chemosensor concepts ready for the internet of things, Chem. Rev. 119 (2019) 7996.

[15] The first known programmable. <https://en.wikipedia.org/wiki/ENIAC> (last accessed 11.01.22).

[16] The Cray 2 supercomputer by Cary Inc. <https://en.wikipedia.org/wiki/Cray> (last accessed 11.01.22).

[17] S. Manipatruni, D.E. Nikonov, I.A. Young, Beyond CMOS computing with spin and polarization, Nat. Phys. 14 (2018) 338.

[18] The famous Integrated Circuit μ. 741. <https://en.wikipedia.org/wiki/Operational_amplifier> (last accessed 11.01.22).

[19] Transforma Insights research, Global IoT market to grow to 24.1 billion devices in 2030, generating $1.5 trillion annual revenue. <https://transformainsights.com/news/iot-market-24-billion-usd15-trillion-revenue-2030>, May 19, 2020.

[20] F. Stradolini et al., in 2018 IEEE international symposium on circuits and systems (ISCAS). IEEE, 2018, pp. 1–5.

[21] D. Andrews, G. Scholes, G. Wiederrecht, Comprehensive Nanoscience and Technology, Academic Press, 2010.

[22] D.C. Duffy, J.C. McDonald, O.J. Schueller, G.M. Whitesides, Rapid prototyping of microfluidic systems in poly (dimethylsiloxane), Anal. Chem. 70 (1998) 4974.

[23] C. Iliescu, H. Taylor, M. Avram, J. Miao, S. Franssila, A practical guide for the fabrication of microfluidic devices using glass and silicon, Biomicrofluidics 6 (2012) 016505.
[24] N. Wongkaew, P. He, V. Kurth, W. Surareungchai, A.J. Baeumner, Multi-channel PMMA microfluidic biosensor with integrated IDUAs for electrochemical detection, Anal. Bioanal. Chem. 405 (2013) 5965.
[25] Z. Li, et al., Pen-on-paper strategy for point-of-care testing: rapid prototyping of fully written microfluidic biosensor, Biosens. Bioelectron. 98 (2017) 478.
[26] H. Becker, L.E. Locascio, Polymer microfluidic devices, Talanta 56 (2002) 267.
[27] M. Altissimo, E-beam lithography for micro-/nanofabrication, Biomicrofluidics 4 (2010) 026503.
[28] Y. Im, et al., Investigation of a single Pd nanowire for use as a hydrogen sensor, Small 2 (2006) 356.
[29] A.A. Yanik, M. Huang, A. Artar, T.-Y. Chang, H. Altug, Integrated nanoplasmonic-nanofluidic biosensors with targeted delivery of analytes, Appl. Phys. Lett. 96 (2010) 021101.
[30] Y. Yang, et al., Paper-based microfluidic devices: emerging themes and applications, Anal. Chem. 89 (2017) 71.
[31] Y. Sameenoi, P.N. Nongkai, S. Nouanthavong, C.S. Henry, D. Nacapricha, One-step polymer screen-printing for microfluidic paper-based analytical device (μPAD) fabrication, Analyst 139 (2014) 6580.
[32] X. Li, J. Tian, G. Garnier, W. Shen, Fabrication of paper-based microfluidic sensors by printing, Colloids Surf. B: Biointerfaces 76 (2010) 564.
[33] P. de Tarso Garcia, T.M.G. Cardoso, C.D. Garcia, E. Carrilho, W.K.T. Coltro, A handheld stamping process to fabricate microfluidic paper-based analytical devices with chemically modified surface for clinical assays, RSC Adv. 4 (2014) 37637.
[34] M. Ueland, et al., Capillary-driven microfluidic paper-based analytical devices for lab on a chip screening of explosive residues in soil, J. Chromatogr. A 1436 (2016) 28.
[35] M.G. Caglayan, S. Sheykhi, L. Mosca, P. Anzenbacher, Fluorescent zinc and copper complexes for detection of adrafinil in paper-based microfluidic devices, Chem. Commun. 52 (2016) 8279.
[36] J. Noiphung, et al., A novel paper-based assay for the simultaneous determination of Rh typing and forward and reverse ABO blood groups, Biosens. Bioelectron. 67 (2015) 485.
[37] S.A. Nogueira, L.R. Sousa, N.K. Silva, P.H. Rodrigues, W.K. Coltro, Monitoring acid–base titrations on wax printed paper microzones using a smartphone, Micromachines 8 (2017) 139.
[38] K. Kaarj, P. Akarapipad, J.-Y. Yoon, Simpler, faster, and sensitive Zika virus assay using smartphone detection of loop-mediated isothermal amplification on paper microfluidic chips, Sci. Rep. 8 (2018) 1.
[39] W. Dungchai, O. Chailapakul, C.S. Henry, A low-cost, simple, and rapid fabrication method for paper-based microfluidics using wax screen-printing, Analyst 136 (2011) 77.
[40] Y. Guan, B. Sun, Detection and extraction of heavy metal ions using paper-based analytical devices fabricated via atom stamp printing, Microsyst. Nanoeng. 6 (2020) 1.
[41] A.P. Kuo, et al., High-precision stereolithography of biomicrofluidic devices, Adv. Mater. Technol. 4 (2019) 1800395.
[42] R. Sochol, et al., 3D printed microfluidic circuitry via multijet-based additive manufacturing, Lab Chip 16 (2016) 668.

[43] G. Gaal, et al., Simplified fabrication of integrated microfluidic devices using fused deposition modeling 3D printing, Sens. Actuators B: Chem. 242 (2017) 35.
[44] S. Waheed, et al., 3D printed microfluidic devices: enablers and barriers, Lab Chip 16 (2016) 1993.
[45] T. Kanai, M. Tsuchiya, Microfluidic devices fabricated using stereolithography for preparation of monodisperse double emulsions, Chem. Eng. J. 290 (2016) 400.
[46] K.B. Anderson, S.Y. Lockwood, R.S. Martin, D.M. Spence, A 3D printed fluidic device that enables integrated features, Anal. Chem. 85 (2013) 5622.
[47] S. Halldorsson, E. Lucumi, R. Gómez-Sjöberg, R.M. Fleming, Advantages and challenges of microfluidic cell culture in polydimethylsiloxane devices, Biosens. Bioelectron. 63 (2015) 218.
[48] T. Kilic, F. Navaee, F. Stradolini, P. Renaud, S. Carrara, Organs-on-chip monitoring: sensors and other strategies, Microphysiol. Syst. 2 (2018) 1.
[49] O. Hofmann, et al., Monolithically integrated dye-doped PDMS long-pass filters for disposable on-chip fluorescence detection, Lab Chip 6 (2006) 981.
[50] Y.S. Zhang, et al., Multisensor-integrated organs-on-chips platform for automated and continual in situ monitoring of organoid behaviors, Proc. Natl Acad. Sci. U.S.A. 114 (2017) E2293.
[51] S.R. Shin, et al., Aptamer-based microfluidic electrochemical biosensor for monitoring cell-secreted trace cardiac biomarkers, Anal. Chem. 88 (2016) 10019.
[52] S.R. Shin, et al., Label-free and regenerative electrochemical microfluidic biosensors for continual monitoring of cell secretomes, Adv. Sci. 4 (2017) 1600522.
[53] Q. Lyu, et al., Real-time and in-situ monitoring of H_2O_2 release from living cells by a stretchable electrochemical biosensor based on vertically aligned gold nanowires, Anal. Chem. 91 (2019) 13521.
[54] R. Singh, S. Hong, J. Jang, Label-free detection of influenza viruses using a reduced graphene oxide-based electrochemical immunosensor integrated with a microfluidic platform, Sci. Rep. 7 (2017) 42771.
[55] B. Ibarlucea, C. Fernández-Sánchez, S. Demming, S. Büttgenbach, A. Llobera, Selective functionalisation of PDMS-based photonic lab on a chip for biosensing, Analyst 136 (2011) 3496.
[56] A. Gokaltun, M.L. Yarmush, A. Asatekin, O.B. Usta, Recent advances in nonbiofouling PDMS surface modification strategies applicable to microfluidic technology, Technology 5 (2017) 1.
[57] R. Luttge, Nano-and Microfabrication for Industrial and Biomedical Applications, William Andrew, 2016.
[58] S.H. Ahn, L.J. Guo, Large-area roll-to-roll and roll-to-plate nanoimprint lithography: a step toward high-throughput application of continuous nanoimprinting, ACS Nano 3 (2009) 2304.
[59] S. Lee, K.-B. Kim, Y.-I. Kim, Love wave SAW biosensors for detection of antigen-antibody binding and comparison with SPR biosensor, Food Sci. Biotechnol. 20 (2011) 1413.
[60] T. Endo, et al., Reflectometric detection of influenza virus in human saliva using nanoimprint lithography-based flexible two-dimensional photonic crystal biosensor, Sens. Actuators B: Chem. 148 (2010) 269.
[61] S. Ahn, et al., Poly-amide modified copper foam electrodes for enhanced electrochemical reduction of carbon dioxide, ACS Catal. 8 (2018) 4132.

[62] E. Destandau, et al., A novel microfluidic flow-injection analysis device with fluorescence detection for cation sensing. Application to potassium, Anal. Bioanal. Chem. 387 (2007) 2627.
[63] T. Wu, et al., Determination of lead in water by combining precolumn adsorption and fluorimetric detection in a microfluidic device, Anal. Methods 4 (2012) 989.
[64] X. Zhang, et al., Using biochar for remediation of soils contaminated with heavy metals and organic pollutants, Environ. Sci. Pollut. Res. 20 (2013) 8472.
[65] J. Qi, et al., Three-dimensional paper-based microfluidic chip device for multiplexed fluorescence detection of Cu2 + and Hg2 + ions based on ion imprinting technology, Sens. Actuators B: Chem. 251 (2017) 224.
[66] N. Kumar, et al., New sensitive and selective calixarene-based fluorescent sensors for the detection of Cs + in an organoaqueous medium, N. J. Chem. 41 (2017) 7162.
[67] A.K. Yetisen, et al., Paper-based microfluidic system for tear electrolyte analysis, Lab Chip 17 (2017) 1137.
[68] F. Ghaderinezhad, et al., Sensing of electrolytes in urine using a miniaturized paper-based device, Sci. Rep. 10 (2020) 13620.
[69] S. Gómez-de Pedro, et al., Optical microfluidic system based on ionophore modified gold nanoparticles for the continuous monitoring of mercuric ion, Sens. Actuators B: Chem. 194 (2014) 19.
[70] E.A. Luy, S.C. Morgan, J.J. Creelman, B.J. Murphy, V.J. Sieben, Inlaid microfluidic optics: absorbance cells in clear devices applied to nitrite and phosphate detection, J. Micromech. Microeng. 30 (2020) 095001.
[71] L.-M. Fu, Y.-N. Wang, Detection methods and applications of microfluidic paper-based analytical devices, TrAC. Trends Anal. Chem. 107 (2018) 196.
[72] G.G. Morbioli, T. Mazzu-Nascimento, A.M. Stockton, E. Carrilho, Technical aspects and challenges of colorimetric detection with microfluidic paper-based analytical devices (μPADs) – a review, Anal. Chim. Acta 970 (2017) 1.
[73] S. Chaiyo, W. Siangproh, A. Apilux, O. Chailapakul, Highly selective and sensitive paper-based colorimetric sensor using thiosulfate catalytic etching of silver nanoplates for trace determination of copper ions, Anal. Chim. Acta 866 (2015) 75.
[74] T. Komatsu, M. Maeki, A. Ishida, H. Tani, M. Tokeshi, Paper-based device for the facile colorimetric determination of lithium ions in human whole blood, ACS Sens. 5 (2020) 1287.
[75] N. Qi, et al., Surface-enhanced Raman scattering on a zigzag microfluidic chip: towards high-sensitivity detection of As(III) ions, Anal. Methods 6 (2014) 4077.
[76] Y. Wang, M. Shang, Y. Wang, Z. Xu, Droplet-based microfluidic synthesis of (Au nanorod@Ag)–polyaniline Janus nanoparticles and their application as a surface-enhanced Raman scattering nanosensor for mercury detection, Anal. Methods 11 (2019) 3966.
[77] P. Inpota, et al., Chemiluminescence detection with microfluidics for innovative in situ measurement of unbound cobalt ions in dynamic equilibrium with bound ions in binding study with polyethyleneimine and its functionalized nanoparticles, Talanta 188 (2018) 606.
[78] W. Alahmad, K. Uraisin, D. Nacapricha, T. Kaneta, A miniaturized chemiluminescence detection system for a microfluidic paper-based analytical device and its application to the determination of chromium(III), Anal. Methods 8 (2016) 5414.
[79] J. Park, R. Meissner, O. Ducloux, P. Renaud, H. Fujita, A calcium ion-selective electrode array for monitoring the activity of HepG2/C3As in a microchannel, Sens. Actuators B: Chem. 174 (2012) 473.

[80] I. Taurino, et al., Platinum nanopetal-based potassium sensors for acute cell death monitoring, RSC Adv. 6 (2016) 40517.

[81] P. Jothimuthu, et al., Lab-on-a-chip sensor for detection of highly electronegative heavy metals by anodic stripping voltammetry, Biomed. Microdevices 13 (2011) 695.

[82] W. Gao, et al., Fully integrated wearable sensor arrays for multiplexed in situ perspiration analysis, Nature 529 (2016) 509.

[83] G. Xu, et al., Battery-free and wireless epidermal electrochemical system with all-printed stretchable electrode array for multiplexed in situ sweat analysis, Adv. Mater. Technol. 4 (2019) 1800658.

[84] W.-J. Lan, et al., Paper-based potentiometric ion sensing, Anal. Chem. 86 (2014) 9548.

[85] E. Garcia-Cordero, et al., Three-dimensional integrated ultra-low-volume passive microfluidics with ion-sensitive field-effect transistors for multiparameter wearable sweat analyzers, ACS Nano 12 (2018) 12646.

[86] F. Puppo, M. Di Ventra, G. De Micheli, S. Carrara, Memristive sensors for pH measure in dry conditions, Surf. Sci. 624 (2014) 76.

[87] S. Carrara, et al., Memristive-biosensors: a new detection method by using nanofabricated memristors, Sens. Actuators B: Chem. 171–172 (2012) 449.

[88] A. Tarasov, et al., Understanding the electrolyte background for biochemical sensing with ion-sensitive field-effect transistors, ACS Nano 6 (2012) 9291.

[89] N.S. Oliver, C. Toumazou, A.E. Cass, D.G. Johnston, Glucose sensors: a review of current and emerging technology, Diabet. Med. 26 (2009) 197.

[90] X. Jia, et al., Synthesis of palladium/helical carbon nanofiber hybrid nanostructures and their application for hydrogen peroxide and glucose detection, ACS Appl. Mater. Interfaces 5 (2013) 12017.

[91] Z. Yu, N. Jiang, S.G. Kazarian, S. Tasoglu, A.K. Yetisen, Optical sensors for continuous glucose monitoring, Prog. Biomed. Eng. 3 (2021) 022004.

[92] W. Jia, et al., Electrochemical tattoo biosensors for real-time noninvasive lactate monitoring in human perspiration, Anal. Chem. 85 (2013) 6553.

[93] M.S. Mannoor, et al., Graphene-based wireless bacteria detection on tooth enamel, Nat. Commun. 3 (2012) 763.

[94] S. Iguchi, et al., A flexible and wearable biosensor for tear glucose measurement, Biomed. Microdevices 9 (2007) 603.

[95] X. Li, et al., Plasmonic nanohole array biosensor for label-free and real-time analysis of live cell secretion, Lab Chip 17 (2017) 2208.

[96] J.W. Menezes, J. Ferreira, M.J.L. Santos, L. Cescato, A.G. Brolo, Large-area fabrication of periodic arrays of nanoholes in metal films and their application in biosensing and plasmonic-enhanced photovoltaics, Adv. Funct. Mater. 20 (2010) 3918.

[97] Y.-C. Chen, S.-Y. Lu, Y.-T. Liao, A microwatt dual-mode electrochemical sensing current readout with current-reducer ramp waveform generation, IEEE Trans. Biomed. circuits Syst. 13 (2019) 1163.

[98] H. Yin, E. Ashoori, X. Mu, A.J. Mason, A compact low-power current-to-digital readout circuit for amperometric electrochemical sensors, IEEE Trans. Instrum. Meas. 69 (2019) 1972.

[99] A. Sun, A. Venkatesh, D.A. Hall, A multi-technique reconfigurable electrochemical biosensor: enabling personal health monitoring in mobile devices, IEEE Trans. Biomed. Circuits Syst. 10 (2016) 945.

[100] V. Valente, N. Neshatvar, E. Pilavaki, M. Schormans, A. Demosthenous, 1.2-V energy-efficient wireless CMOS potentiostat for amperometric measurements, IEEE Trans. Circuits Syst. II: Express Briefs 67 (2019) 1700.

[101] A. Hassibi, T.H. Lee, A programmable 0.18-$\mu\hbox {m} $ CMOS electrochemical sensor microarray for biomolecular detection, IEEE Sens. J. 6 (2006) 1380.

[102] R.H. Edwards, The neurotransmitter cycle and quantal size, Neuron 55 (2007) 835.

[103] C.I. Dorta-Quiñones, et al., A bidirectional-current CMOS potentiostat for fast-scan cyclic voltammetry detector arrays, IEEE Trans. Biomed. Circuits Syst. 12 (2018) 894.

[104] A.F. Yeknami, et al., A 0.3-V CMOS biofuel-cell-powered wireless glucose/lactate biosensing system, IEEE J. Solid-State Circuits 53 (2018) 3126.

[105] K. Dutta, Potential of impedance spectroscopy towards quantified analysis of gas sensors: a tutorial, IEEE Sens. J. (2021).

[106] N. Neshatvar, P. Langlois, A. Demosthenous, A non-linear feedback current driver with automatic phase compensation for bioimpedance applications, IEEE Trans. Circuits Syst. II: Express Briefs 65 (2018) 1340.

[107] C. Yang, D. Rairigh, A. Mason, in 2007 IEEE Biomedical Circuits and Systems Conference. IEEE, 2007, pp. 21–24.

[108] T.-A. Chen, W.-J. Wu, C.-L. Wei, R.B. Darling, B.-D. Liu, Novel 10-bit impedance-to-digital converter for electrochemical impedance spectroscopy measurements, IEEE Trans. Biomed. Circuits Syst. 11 (2016) 370.

[109] C.-L. Wei, Y.-W. Wang, B.-D. Liu, Wide-range filter-based sinusoidal wave synthesizer for electrochemical impedance spectroscopy measurements, IEEE Trans. Biomed. Circuits Syst. 8 (2013) 442.

[110] C. Yang, S.R. Jadhav, R.M. Worden, A.J. Mason, Compact low-power impedance-to-digital converter for sensor array microsystems, IEEE J. Solid-State Circuits 44 (2009) 2844.

[111] K.-H. Lee, et al., One-chip electronic detection of DNA hybridization using precision impedance-based CMOS array sensor, Biosens. Bioelectron. 26 (2010) 1373.

[112] C. Stagni, et al., CMOS DNA sensor array with integrated A/D conversion based on label-free capacitance measurement, IEEE J. Solid-State Circuits 41 (2006) 2956.

[113] E. Ghafar-Zadeh, M. Sawan, A hybrid microfluidic/CMOS capacitive sensor dedicated to lab-on-chip applications, IEEE Trans. Biomed. Circuits Syst. 1 (2007) 270.

[114] N. Dodel et al., in 2019 IEEE Biomedical Circuits and Systems Conference (BioCAS). IEEE, 2019, pp. 1–4.

[115] M. Schienle, et al., A fully electronic DNA sensor with 128 positions and in-pixel A/D conversion, IEEE J. Solid-State Circuits 39 (2004) 2438.

[116] J. Rodriguez-Manzano, et al., Handheld point-of-care system for rapid detection of SARS-CoV-2 extracted RNA in under 20 min, ACS Cent. Sci. 7 (2021) 307.

[117] Y. Pei-Wen, et al., A device design of an integrated CMOS poly-silicon biosensor-on-chip to enhance performance of biomolecular analytes in serum samples, Biosens. Bioelectron. 61 (2014) 112.

[118] Y. Li et al., in 2021 IEEE International Symposium on Circuits and Systems (ISCAS). IEEE, 2021, pp. 1–5.

[119] D.H. Ta-chien, S. Sorgenfrei, P. Gong, R. Levicky, K.L. Shepard, A 0.18-$\mu $ m CMOS array sensor for integrated time-resolved fluorescence detection, IEEE J. Solid-State Circuits 44 (2009) 1644.

[120] G. Shekhawat, S.-H. Tark, V.P. Dravid, MOSFET-embedded microcantilevers for measuring deflection in biomolecular sensors, Science 311 (2006) 1592.

[121] S. Hwang, et al., CMOS microelectrode array for electrochemical lab-on-a-chip applications, IEEE Sens. J. 9 (2009) 609.

[122] L. Li, H. Yin, A.J. Mason, Epoxy chip-in-carrier integration and screen-printed metalization for multichannel microfluidic lab-on-CMOS microsystems, IEEE Trans. Biomed. circuits Syst. 12 (2018) 416.

[123] H. Jafari, L. Soleymani, R. Genov, 16-channel CMOS impedance spectroscopy DNA analyzer with dual-slope multiplying ADCs, IEEE Trans. Biomed. Circuits Syst. 6 (2012) 468.

[124] Y. Yusof, N. Kazuo, in 2014 IEEE Conference on Biomedical Engineering and Sciences (IECBES). IEEE, 2014, pp. 365–368.

[125] P. Grosso, S. Carrara, C. Stagni, L. Benini, Cancer marker detection in human serum with a point-of-care low-cost system, Sens. Actuators B: Chem. 147 (2010) 475.

[126] G. Brabant, et al., Serum insulin-like growth factor I reference values for an automated chemiluminescence immunoassay system: results from a multicenter study, Hormone Res. Paediatrics 60 (2003) 53.

[127] T. Tanaka, T. Matsunaga, Fully automated chemiluminescence immunoassay of insulin using antibody – protein A – bacterial magnetic particle complexes, Anal. Chem. 72 (2000) 3518.

[128] A. Padoan, C. Cosma, L. Sciacovelli, D. Faggian, M. Plebani, Analytical performances of a chemiluminescence immunoassay for SARS-CoV-2 IgM/IgG and antibody kinetics, Clin. Chem. Laboratory Med. (CCLM) 58 (2020) 1081.

[129] M.R. Romero, F. Ahumada, F. Garay, A.M. Baruzzi, Amperometric biosensor for direct blood lactate detection, Anal. Chem. 82 (2010) 5568.

[130] U. Saxena, M. Chakraborty, P. Goswami, Covalent immobilization of cholesterol oxidase on self-assembled gold nanoparticles for highly sensitive amperometric detection of cholesterol in real samples, Biosens. Bioelectron. 26 (2011) 3037.

[131] S. Chakraborty, C.R. Raj, Amperometric biosensing of glutamate using carbon nanotube based electrode, Electrochem. Commun. 9 (2007) 1323.

[132] C. Baj-Rossi, G.D. Micheli, S. Carrara, Electrochemical detection of anti-breast-cancer agents in human serum by cytochrome P450-coated carbon nanotubes, Sensors 12 (2012) 6520.

[133] S. Carrara, A. Cavallini, V. Erokhin, G. De Micheli, Multi-panel drugs detection in human serum for personalized therapy, Biosens. Bioelectron. 26 (2011) 3914.

[134] F. Stradolini, T. Kilic, I. Taurino, G. De Micheli, S. Carrara, Cleaning strategy for carbon-based electrodes: long-term propofol monitoring in human serum, Sens. Actuators B: Chem. 269 (2018) 304.

[135] F. Stradolini, et al., An IoT solution for online monitoring of anesthetics in human serum based on an integrated fluidic bioelectronic system, IEEE Trans. Biomed. Circuits Syst. 12 (2018) 1056.

[136] S. Aiassa, et al., Smart portable pen for continuous monitoring of anaesthetics in human serum with machine learning, IEEE Trans. Biomed. Circuits Syst. 15 (2021) 294.

[137] R. Thewes et al., in 2002 IEEE International Solid-State Circuits Conference. Digest of Technical Papers (Cat. No. 02CH37315). IEEE, 2002, 1, pp. 350–473.

[138] C. Toumazou, et al., Simultaneous DNA amplification and detection using a pH-sensing semiconductor system, Nat. Methods 10 (2013) 641.

[139] J. Min, et al., Integrated biosensor for rapid and point-of-care sepsis diagnosis, ACS Nano 12 (2018) 3378.

[140] F. Basilotta et al., in 2015 IEEE Biomedical Circuits and Systems Conference (BioCAS). IEEE, 2015, pp. 1–4.

[141] M. Trotter, N. Borst, R. Thewes, F. von Stetten, Electrochemical DNA sensing—principles, commercial systems, and applications, Biosens. Bioelectron. 154 (2020) 112069.

[142] T. Kilic, P.J. Philipp, P. Giavedoni, S. Carrara, Milk allergen detection: sensitive label-free voltammetric immunosensor based on electropolymerization, BioNanoScience 10 (2020) 512.

[143] R. Chauhan, et al., A novel electrochemical piezoelectric label free immunosensor for aflatoxin B1 detection in groundnut, Food Control. 52 (2015) 60.

[144] Y. Pang, Recent development and progress of electrochemical sensors for antibiotic detection, Int. J. Electrochem. Sci. 15 (2020) 5232.

[145] G. Hu, W. Sheng, Y. Zhang, X. Wu, S. Wang, A novel and sensitive fluorescence immunoassay for the detection of fluoroquinolones in animal-derived foods using upconversion nanoparticles as labels, Anal. Bioanal. Chem. 407 (2015) 8487.

[146] F. Bertani, et al., Optical detection of aflatoxins B in grained almonds using fluorescence spectroscopy and machine learning algorithms, Food Control. 112 (2020) 107073.

[147] L. Smeesters, W. Meulebroeck, S. Raeymaekers, H. Thienpont, Optical detection of aflatoxins in maize using one-and two-photon induced fluorescence spectroscopy, Food Control. 51 (2015) 408.

[148] J. Nelis, A. Tsagkaris, M. Dillon, J. Hajslova, C. Elliott, Smartphone-based optical assays in the food safety field, TrAC. Trends Anal. Chem. 129 (2020) 115934.

[149] C. Baj-Rossi, et al., In-vivo validation of fully implantable multi-panel devices for remote monitoring of metabolism, IEEE Trans. Biomed. circuits Syst. 10 (2016) 955.

[150] C. Baj-Rossi, et al., Continuous monitoring of Naproxen by a cytochrome P450-based electrochemical sensor, Biosens. Bioelectron. 53 (2014) 283.

[151] F. Criscuolo, I. Taurino, F. Stradolini, S. Carrara, G. De Micheli, Highly-stable Li + ion-selective electrodes based on noble metal nanostructured layers as solid-contacts, Anal. Chim. Acta 1027 (2018) 22.

[152] E. Mehmeti, T. Kilic, C. Laur, S. Carrara, Electrochemical determination of nicotine in smokers' sweat, Microchem. J. 158 (2020) 105155.

[153] L. Han, et al., Detecting work-related stress with a wearable device, Computers Ind. 90 (2017) 42.

[154] K.M. Szostak, L. Grand, T.G. Constandinou, Neural interfaces for intracortical recording: requirements, fabrication methods, and characteristics, Front. Neurosci. 11 (2017) 665.

[155] A. Amon, F. Alesch, Systems for deep brain stimulation: review of technical features, J. Neural Transm. 124 (2017) 1083.

[156] S. Carrara, Body dust: Well beyond wearable and implantable sensors, IEEE Sens. J. 21 (2020) 12398.

[157] J.M. Kahn, R.H. Katz, K.S. Pister, in Proceedings of the 5th annual ACM/IEEE international conference on Mobile computing and networking, 1999, pp. 271–278.

[158] V.E. Balas, V.K. Solanki, R. Kumar, M.A.R. Ahad, A Handbook of Internet of Things in Biomedical and Cyber Physical System, Springer, 2020.
[159] D. Seo, J.M. Carmena, J.M. Rabaey, E. Alon, M.M. Maharbiz, Neural dust: an ultrasonic, low power solution for chronic brain-machine interfaces. arXiv preprint arXiv:1307.2196, 2013.
[160] S. Carrara, P. Georgiou, Body dust: Miniaturized highly-integrated low power sensing for remotely powered drinkable CMOS bioelectronics. arXiv preprint arXiv:1805.05840, 2018.
[161] D.K. Piech, et al., A wireless millimetre-scale implantable neural stimulator with ultrasonically powered bidirectional communication, Nat. Biomed. Eng. 4 (2020) 207.
[162] J. Snoeijs, P. Georgiou, S. Carrara, in 2017 IEEE Biomedical Circuits and Systems Conference (BioCAS). IEEE, 2017, pp. 1−4.
[163] G.L. Barbruni, C. Bielli, D. Demarchi, S. Carrara, in SMACD/PRIME 2021; International Conference on SMACD and 16th Conference on PRIME. VDE, 2021, pp. 1−4.
[164] N.-C. Kuo, B. Zhao, A.M. Niknejad, Equation-based optimization for inductive power transfer to a miniature CMOS rectenna, IEEE Trans. Microw. Theory Tech. 66 (2018) 2393.

CHAPTER

Microfluidic-based plasmonic biosensors 10

Yanting Liu and Xuming Zhang

Department of Applied Physics, The Hong Kong Polytechnic University, Hong Kong, P.R. China

10.1 Introduction

10.1.1 Basics of microfluidics and biosensor technology

Microfluidics refers to micro total analysis systems for the control and manipulation of fluids, which can process or manipulate small amounts of fluids ($10^{-9}-10^{-18}$ L) using submillimeter dimension channels [1]. The microfluidic devices and systems provide an ideal platform for the ultra-sensitive detection of molecules and the manipulation of cells, or small-scale particles, along with demonstrating great potential for applications in biomedical-related research [2]. The microfluidic systems with enhanced capabilities, such as high-throughput processing and less sample dosage, can greatly reduce the complexity and time of assay with a miniature operating platform, replacing traditional macroscale assays [3,4]. Furthermore, the microfluidic devices with simple construction can combine with various sensing techniques. Microfluidic devices integrated with on-chip sensors have proven useful for biological and biomedical assays [5,6]. Biosensors are powerful analytical tools used for the detection of specific biomolecules, including enzymes, aptamers, immune proteins, nucleic acids, microorganisms, and cancer or tumor cells [7,8]. Biosensors consists of a physicochemical transducer or transducing microsystems integrated with immobilized biological recognition molecules or elements. It is capable of converting biological or chemical molecules' responses into different signals, such as electrical chemical, optical (including visible color), or mass change. Biosensing techniques can help detect target biomolecules with high sensitivity and multiplex analysis ability [9–12]. The advances in microfluidic techniques like high throughput, miniaturization, and low-cost features can further provide interesting opportunities for biosensor platforms. However, there are still several challenges remain that need to be addressed for the integration of lab-on-chip-based biosensor technology because it requires the combination of numerous disciplines and interdisciplinary areas.

Microfluidic Biosensors. DOI: https://doi.org/10.1016/B978-0-12-823846-2.00017-1

10.1.2 Biosensors and classification

Generally, biosensors can be classified as mechanical, electrochemical, and optical biosensors depending on the types of transducers or biorecognition elements [13,14]. Mechanical biosensors can detect and measure the signal through the changes of mass on the surface of the transducer caused by the biochemical interaction [15,16]. As for the electrochemical biosensors, with simple manufacturing methods, they can measure electrochemical signal changes caused by the biochemical interaction between molecules via an electrochemical transducer (such as enzyme electrodes for the commercially available glucose biosensors) [17,18]. Optical biosensors are one of the most important classes of biosensors that measure changes in optical properties as a result of biomolecular reactions or affinity interactions by integrating biorecognition sensing elements in the optical signal detection system [19,20]. Overall, biosensors can provide quantitative analytical information through biorecognition components using different types of integrated transducers [21].

10.1.3 Optical biosensor

Compared with other biosensing methods, optical biosensors exhibit unique advantages for nondestructive analysis with less interference, low detection limit, high sensitivity, low background noise, and fast signal monitoring [20,22,23]. Optical biosensors can directly detect biochemical interaction through various optical techniques and platforms such as optical fibers, optical waveguides, and surface plasmon resonance (SPR) [24–26]. It can measure the variations of an optical transducer, including intensity, wavelength, refractive index (RI), and polarization properties of light [27]. Besides, the optical biosensor can also indirectly detect analytes through optically labeled probes, such as fluorescence and luminescence [28–30]. Furthermore, the optical biosensor can accomplish multiplex monitoring and analysis of bioreceptor molecules by measuring different signals from multiple analytes [31].

Particularly, optical biosensors that are based on plasmonic micro/nano-structures and surface plasmon phenomena stand out due to the advances in optical properties of sensing materials and sensor instrumentation [32]. Plasmonic biosensors are a kind of label-free optical biosensors, and such a sensing method has promoted the rapid development of biosensing technology. It is usually composed of sensing elements that comprise metal or metal–dielectric micro/nano-structures supported by the unique electromagnetic wave surface plasmon-polaritons-to-probe interactions, and bio/chem molecular recognition elements that are immobilized on the surface of the sensor, which can selectively bind a target analyte [33,34].

Given the advantages of the plasmonic biosensors, microfluidic chip integration gives rise to promoting ultrasensitive analysis of plasmonic biosensing systems for simultaneous detection of multiple analytes with high-throughput multiplex processing ability. Therefore, microfluidic-based plasmonic biosensing systems with integrated sample handling features have shown promise in improving the performance of biomedical-related detection applications [35–37]. In this chapter, different types of microfluidic-integrated plasmonic biosensors are summarized and represented, which demonstrates the advantages for biomedical detection applications. This chapter is focused on the intersection of plasmonic biosensing technology with the microfluidics platform illustrated in Fig. 10.1. Particularly, patterned plasmonic nanostructure array-based plasmonic biosensors in microfluidic systems have been reviewed comprehensively. The following sections will provide insights into (1) plasmonic biosensors; (2) patterned plasmonic nanostructure arrays for enhanced plasmonic biosensors; and (3) different types of

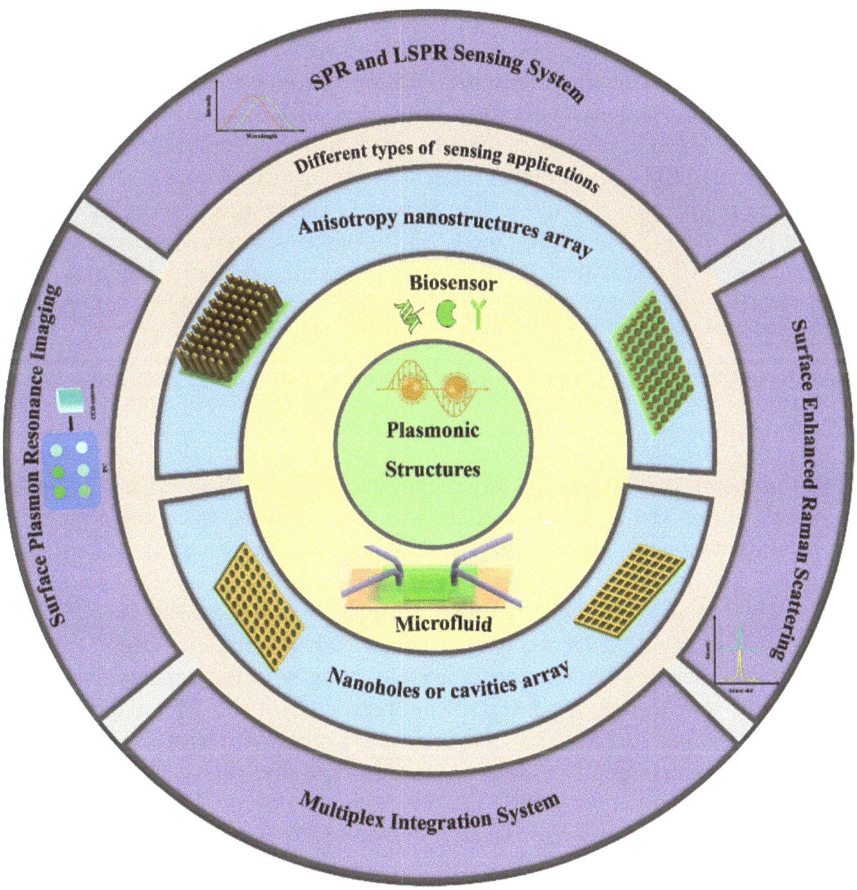

FIGURE 10.1

Scheme of microfluidic-integrated various plasmonic patterned structure-based plasmonic biosensor systems with different types of sensing approaches.

integrated microfluidic plasmonic biosensing approaches. We hope this chapter would further guide application of biosensing techniques in various promising biomedical-related fields by presenting the versatility and perspective of the microfluidic-integrated plasmonic biosensing systems.

10.2 Plasmonic biosensors

10.2.1 Principle of surface plasmons

Plasmons are usually defined as a quantum of collective electron oscillations in metals, which originate either from the bulk and surface of the metal, or confinement of a surface plasmon in a nanoparticle [38]. As for the SPR mode at the surface of a metal, it can be excited at the surface of a metal and propagate along the metal–dielectric interface. Furthermore, its evanescent field could penetrate the surrounding media [39]. In the case of localized SPR (LSPR), it is confined and excited on sub-wavelength-scale metal nanoparticles with a specific frequency [40]. Due to the SPR near the metal surface and their evanescent field being highly sensitive to the RI changes, it can be used as a refractometric sensing platform [41]. Similarly, the LSPR of plasmonic nanostructures is sensitive to changes in the local dielectric environment, and thus it can act as a refractometric sensor by measuring the LSPR wavelength shift of plasmonic nanoparticles [42].

10.2.2 Conventional surface plasmon resonance biosensors

SPR biosensors have attracted increasing attention and has become the most commonly used and commercialized plasmonic sensors since the first SPR-based refractometric sensors were developed by Nylander et al. in 1983 [43,44]. Typically, for SPR biosensors, the incident light could excite a special mode of electromagnetic field-surface plasmon, when the SPR matches with the component of the wave vector of light, which is parallel to the metal sur face. So, the surface plasmon can propagate along with the generation of electromagnetic waves along the interface between a flat metal film and dielectric, as shown in Fig. 10.2A [43,44]. There are several different light coupling approaches to enhancing the incident light for exciting the special SPR mode, including (1) prism coupling [46], (2) waveguide coupling [47], and (3) grating coupling [44] as shown in Fig. 10.2B [48]. For SPR biosensing, any change in RI caused by the biological–chemical molecule interactions that occur in the proximity of the interface between metal and dielectric could induce a change in the SPR propagation condition. The above changes in the propagation can be detected and then transformed into measurable and quantifiable signals in real time [41].

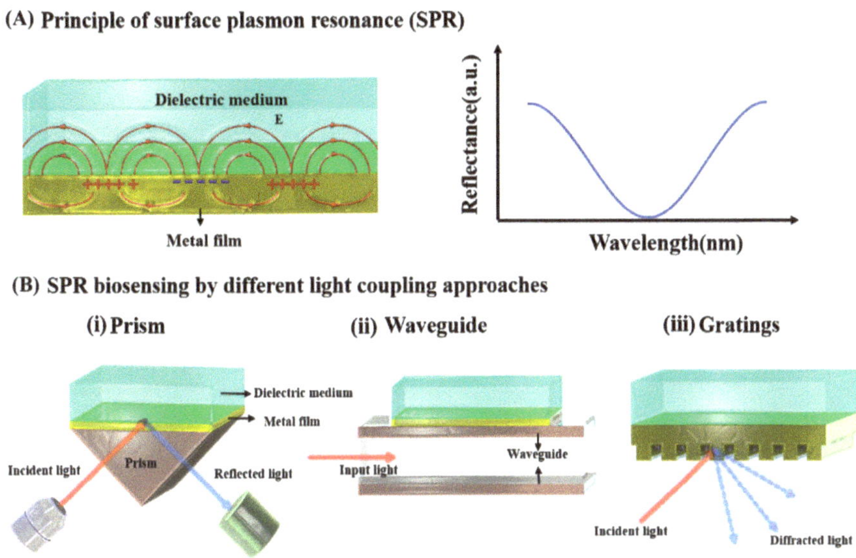

FIGURE 10.2

Scheme of (A) the working principle of SPR at the interface between a flat metal film and dielectric. (B) SPR biosensing methods using different light coupling approaches, including (i) prism coupling, (ii) the waveguide coupling, and (iii) light diffraction grating coupling [45]. *SPR*, Surface plasmon resonance.

10.2.3 LSPR biosensors

Over the last few decades, nanotechnology has been in use for preparing and assembling nanomaterials using either bottom-up or top-down methods [49]. Hence, nanotechnology and nanomaterial have been widely applied and utilized in various fields, including biosensing-related fields. The sensitivity and selectivity of biosensors can be greatly improved and enhanced using the nanomaterials as biorecognition transmitters for biomolecular detection, especially the noble metal nanomaterials, which stand out among all nanomaterials due to their unique optical plasmonic properties. When incident light interacts with plasmonic nanostructures, such as the noble metal nanoparticles, a collective oscillation of conduction electrons known as LSPR can be produced, and this resonant oscillation can also produce large wavelength-selective enhancements in the absorption, scattering, and electromagnetic field adjacent to the nanostructure surface [50]. Furthermore, the LSPR characteristics of the plasmonic nanostructures are explicitly dependent on the surrounding dielectric, their morphology, size, component, and distance between adjacent nanostructures [51]. Among differently shaped nanostructures, the electron oscillation of sphere-shaped plasmonic nanostructure can produce a single dipole of surface plasmon mode and only show one LSPR band in the optical spectrum, as shown in Fig. 10.3A(i) [52]. Compared with spherical structure, the anisotropic

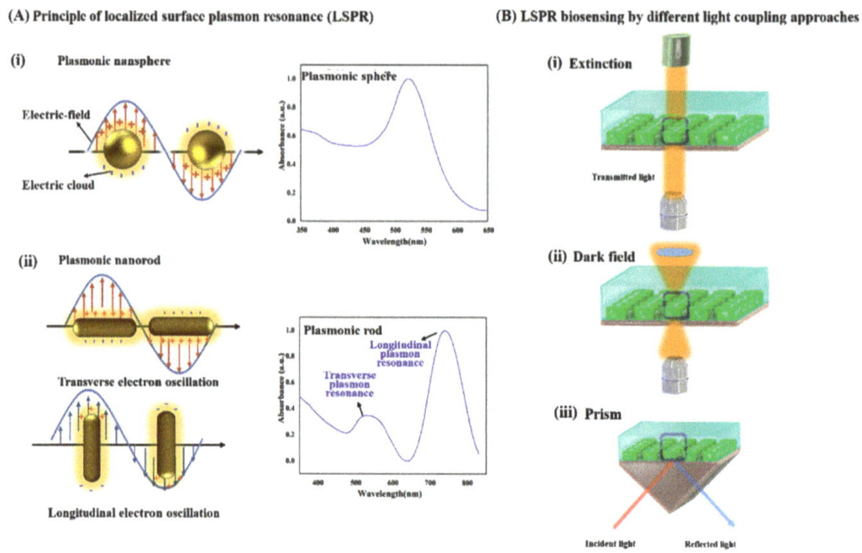

FIGURE 10.3

Scheme of (A) the working principle of LSPR with (i) plasmonic nanosphere and with (ii) plasmonic nanorod. (B) LSPR biosensing using different light coupling approaches on the plasmonic nanostructured surface, including (i) extinction, (ii) dark-field, and (iii) prism coupler [45]. *LSPR*, Localized surface plasmon resonance.

morphology like nanorod will show more LSPR modes due to the high aspect ratio and surface curvature of geometry [53]. The SPR surface plasmons of nanorod structure are split into two distinct modes, including the TPRLPR, as shown in Fig. 10.3A(ii) [48,54]. The plasmonic nanoparticles also process several superior properties, enabling them to be excellent candidates for biosensors, such as high stability, good biocompatibility, large surface energy, and strong signal amplification [55,56]. Besides, it also has specific advantages drawn out of the advances in unique nanoscale plasmonic components, such as myriad nanostructures and sophisticated array assemblies of nanostructure. Therefore, the LSPR biosensing has become a benchmark for label-free detection techniques with rapid, miniaturized, multiplexing, and high-resolution detection and analysis by integrating the conventional coupling mode of the light [57].

The LSPR is at the origin of the strong absorption/scattering of the incident light, and the sum of light absorption and scattering is called extinction [42]. LSPR can be characterized by extinction, which is dependent on the intrinsic parameters of plasmonic nanostructures as mentioned above [58]. For the LSPR refractometric biosensors, the wavelength peak value of LSPR is extremely sensitive to the surrounding media's RI changes, so it can be monitored by extinction spectra [59]. The RI changes when binding an analyte on the surface of plasmonic nanostructures will result in the shift of the extinction peak [60,61]. Similar to SPR biosensing, by modulating the light transmission or reflection configurations, the LSPR biosensing methods can be classified based on different approaches, as shown in Fig. 10.3B: (1) extinction, (2) dark-field, and (3) prism on a surface of plasmonic nanostructure [48,62,63]. The extinction mode is the most straightforward way by measuring the extinction spectrum changes induced by specific binding of the analyte [64]. When involving the analytes located in small regions or even single nanoparticles, the dark-field light-scattering configuration method may prove to be an extremely powerful measurement tool. Furthermore, the light-illuminating and light-collecting lenses allow fine-tuning of the spot size of the plasmonic nanostructured region in the sensing system [64,65], while the prism coupler in plasmonic nanostructure array mode is similar to the SPR prism configuration [66].

10.3 Patterned metallic structure array for enhanced plasmonic biosensors

Generally, plasmonic metallic nanostructures act as a core part of biosensing applications. Recent advancements and innovations in low-cost and large-area nanofabrication techniques have endowed researchers with tools to prepare various plasmonic nanostructure arrays [67–69]. Compared to nanoparticles in colloids, plasmonic nanostructure arrays exhibit unique advantages, such as long-range order and superior plasmonic optical modes for biosensing applications [70]. So, in this section, we draw our attention to plasmonic biosensor-integrated microfluidic chips, which are based on the patterned metallic nanostructure arrays, including anisotropy nanostructure arrays and nanohole/cavity array substrates. Due to the different optical responses of SPR, there are two different plasmonic sensing measurements for anisotropy nanostructure arrays and nanohole/cavity arrays, respectively [71]. The two types of sensing strategies of patterned metallic structure arrays are shown in Fig. 10.4. The anisotropy nanostructure arrays can measure the extinction of LSPR for sensing strategies, where the LSPR arises from the interaction of adjacent nanostructures on the substrate with incident light in a random, noninteracting manner [72,73]. For nanohole/cavity arrays, they measure the extraordinary optical transmission (EOT), where the EOT arises from the interaction of nanohole/cavity with incident light in a long-range periodic or short-range ordering manner [70].

10.3.1 Anisotropy nanostructure arrays

The anisotropy nanostructure arrays can avoid the cluster and agglomeration as sensing substrates compared with colloidal plasmonic nanoparticle solutions for plasmonic biosensors [74]. Furthermore, nanostructured array-based sensing can provide superior reproducibility of the analysis because it can have better control over the LSPR of the nanostructured array. On integration with a microfluidic chip, multiplexed and ultrasensitive analysis in different biomedical detection

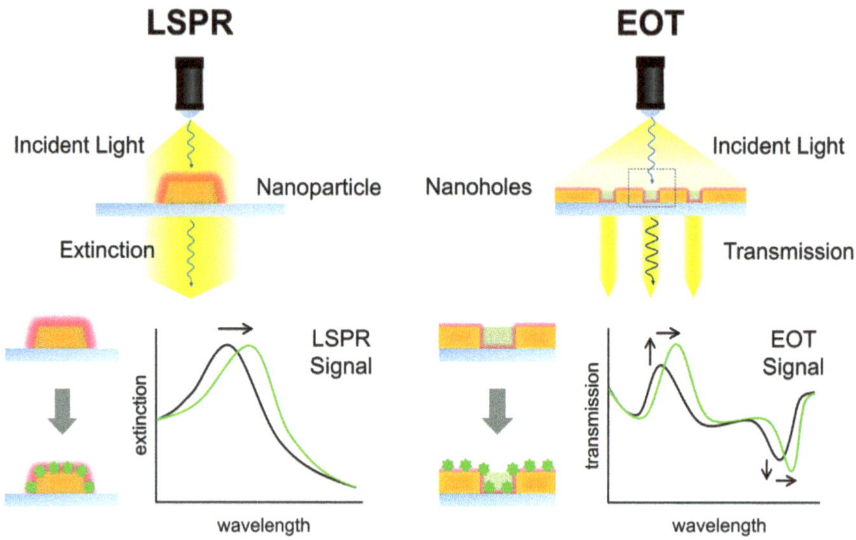

FIGURE 10.4

Scheme of patterned metallic structure array-based plasmonic biosensors with different sensing strategies. The anisotropy nanostructure array sensing is done through measuring the LSPR extinction spectrum, while for nanohole/cavity array sensing is done through measuring an EOT spectrum. *EOT*, Extraordinary optical transmission.

Reproduced with permission from J.A. Jackman, A.R. Ferhan, N.-J. Cho, Nanoplasmonic sensors for biointerfacial science, J. Chem. Soc. Rev. 46 (12) (2017) 3615–3660. ©2017 Royal Society of Chemistry.

applications could be realized using anisotropic patterned nanostructure array-based biosensors [75–79].

As shown in Fig. 10.5A, a patterning microfluidic-based LSPR biosensor device with gold nanorod (Au NRs) microarrays conjugated can realize a high-throughput cytokine antibody detection. The system can offer six cytokine detection parallel immunoassays with lower concentrations down to the pg/mL scale in a serum medium. The scattering light intensity across the ensemble of Au NR microarrays can be measured through the dark-field imaging scanning technique [80]. While a vertically aligned and densely packed gold nanorods arrays based plasmonic biosensor integrated with the microfluidic channel has been developed by Baraban group as shown in Fig. 10.5B [81]. This system can realize label-free detection of the immobilization, backfilling, and hybridization of short DNA sequences by investigating biochemical reactions of specific binding the target DNA molecules on these nanorods based array sensors. In some cases, they may contain extremely large regions of dielectric materials in LSPR sensing devices. These regions should be carefully blocked to avoid compromising the accuracy of measurements because it could induce the nonspecific adsorption of the analyte. To overcome this problem, Aćimović's group has developed vertical gold nanodisk arrays based on plasmonic biosensors.

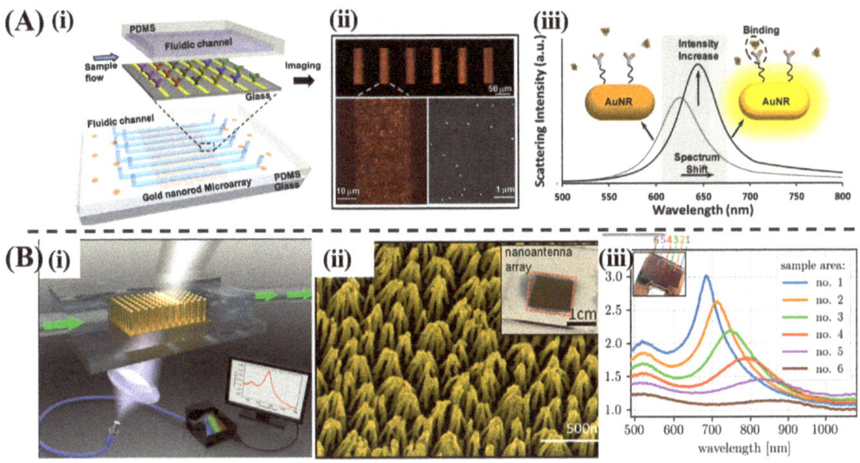

FIGURE 10.5

Anisotropy nanostructure array-based plasmonic sensing platforms. (A) Au NR microarray-based LSPR biosensor with microfluidic patterning for cytokine detection: (i) the LSPR microarray chip's scheme; (ii) dark-field images of Au NR microarrays for multiplex antibodies detection; and (iii) LSPR shift of the Au NR microarray system with bending. (B) Vertical arrays of gold nanorods-based plasmonic biosensor: (i) Scheme of microfluidic incorporated vertical arrays of gold nanorods-based plasmonic biosensing system. (ii) SEM and optical images of vertically aligned nanorods substate. (iii) Extinction spectra of vertically aligned nanorods-based chips. *Au NR*, Gold nanorod; *LSPR*, localized surface plasmon resonance; *PEG*, polyethylene glycol.

(A) Reproduced with permission from P. Chen, M.T. Chung, W. McHugh, R. Nidetz, Y. Li, J. Fu, et al., Multiplex serum cytokine immunoassay using nanoplasmonic biosensor microarrays, ACS Nano 9 (4) (2015) 4173–4181. ©2015 American Chemical Society. (B) Reproduced with permission from Klinghammer, S., Uhlig, T., Patrovsky, F., Böhm, M., Schütt, J., Pütz, N., Baraban, L., Eng, L. M., Cuniberti, G., Plasmonic biosensor based on vertical arrays of gold nanoantennas. ACS Sensors. 3 (7) (2018) 1392-1400. Copyright 2018 American Chemical Society.

This device can markedly reduce the vulnerability to adsorption of the nonspecific substances, while, at the same time, retaining a high effective sensing surface by the vertical nanodisk structure [82]. Furthermore, such an approach offers a unique array design for label-free LSPR biosensing systems, and the combination with a microfluidic chip could provide minimal sample consumption and high-throughput merits.

10.3.2 Nanohole/cavity arrays

The nanoholes/cavities with a sub-wavelength-scale pattern array can be fabricated on thick metal films, which can have tunable plasmonic properties [83]. Usually, the dimension of the periodic structure is less than the wavelength of incident light propagating through it [84]. By grating coupling of the incident light, the EM field near the nanoholes can be greatly enhanced by the excitation of nanoholes' plasmons [70]. The ordered nanohole/cavity arrays can modulate the intensity and the spectral position of SPR mode due to their periodic arrangement. The plasmonic properties like the electromagnetic field distribution can also be properly engineered and fine-tuned by changing the periodicity, size of the hole, and metal composition [85]. The nanohole/cavity array-based plasmonic biosensors can be the proper choice for RI sensing by detecting a shift of wavelength in the EOT spectrum [86]. In addition, when integrated with microfluidic chips, the pattern nanohole or cavity array-based plasmonic biosensors can also achieve high-throughput and high-sensitivity multiplexing analysis for various kinds of the biological and biomedical-related assays [87–91].

As shown in Fig. 10.6A, Altug' group has developed a patterned plasmonic nanohole array for a label-free biosensing platform integrated with an optofluidic chip [92]. This sensing platform can utilize the EOT spectrum to monitor the group-specific antibodies to directly detect intact viruses from biological media. The plasmonic nanohole array-based sensing platform can detect genetically derived VSV-pseudotyped Ebola viruses by using antiviral immunoglobulins immobilized at the sensor surface for specific capturing of the pseudo-Ebola virions. Furthermore, their sensing platform is also capable of quantifying virus concentrations and detecting different types of viruses. The Altug group has also realized the selective detection and quantification for simultaneous detection of the two most common bacterial infections: Chlamydia trachomatis and Neisseria gonorrhoeae with similar nanoholes array-based plasmonic biosensing system as shown in Fig. 10.6B [93]. This platform can realize quantified detect the levels of the two bacteria in a one-step assay, without the need for DNA extraction or amplification techniques. Besides, Feng Liang et al. have developed a photonic–-plasmonic antenna hybrid nanocavity biosensor for label-free dynamic detection of single-molecule DNA–protein interaction [94]. The photonic–plasmonic hybrid antenna-in-a-nanocavity system is capable of correctly reflecting molecular dynamic measurements with millisecond time resolution. And it can also realize the current characterization of the DNA–protein interaction. Such a system is suitable studying dynamic processes compared with fluorescent label methods.

10.4 The applications of microfluidic-based plasmonic biosensors

Enormous progress has been made in plasmonic techniques and micro/nano-optic components. Ease of integration of microfluidic techniques can offer miniaturized and low-cost chips with the multiplex channel for the plasmonic biosensor to be applied in bio-related fields. Microfluidic-based plasmonic biosensors endowed researchers with a tool for developing user-friendly, automatized, and

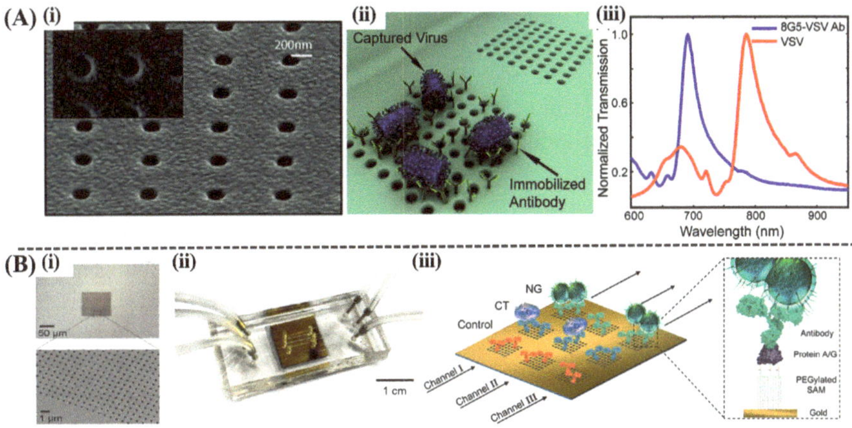

FIGURE 10.6

Nanohole/cavity array-based plasmonic sensing platforms. (A) A label-free patterned nanohole array-based sensing system for directly detecting the live viruses in optofluidic device: (i) SEM images of nanohole array array-based substrate; (ii) scheme of plasmonic nanohole array biosensing; and (iii) EOT transmission spectrum of nanohole array-based biosensor conjected with antibodies. (B) A nanoholes array-based plasmonic biosensing system for simultaneous detection of bacterial infections: (i) Optical and SEM images of the nanoholes-based plasmonic chip. (ii) The picture of the plasmonic nanohole arrays integrated with PDMS microfluidic chip. (iii) Schematic illustration of surface biofunctionalized plasmonic biosensor for different bacterial detection. *EOT*, Extraordinary optical transmission.

(A) Reproduced with permission from A.A. Yanik, M. Huang, O. Kamohara, A. Artar, T.W. Geisbert, J.H. Connor, et al., An optofluidic nanoplasmonic biosensor for direct detection of live viruses from biological media, Nano Lett. 10 (12) (2010) 4962–4969. ©2010 American Chemical Society. (B) Reproduced with permission from Soler, M., Belushkin, A., Cavallini, A., Kebbi-Beghdadi, C., Greub, G., Altug, H., Multiplexed nanoplasmonic biosensor for one-step simultaneous detection of Chlamydia trachomatis and Neisseria gonorrhoeae in urine. Biosens. Bioelectron. 94 (2017) 560-567. Copyright 2017, Elsevier.

portable devices and systems from the laboratory proof-of-concept prototypes and device integration and commercialization to meet clinical practice requirements for point-of-care applications, In addition to the simplest RI change-based SPR/LSPR biosensing techniques, there are several advanced applications of plasmonic biosensing in microfluidics. Special application scenarios have led to the development of microfluidic-based plasmonic biosensing technology, such as the integration of SPR imaging, SERS, or multiplex sensors.

10.4.1 Integrated surface plasmon resonance and LSPR sensing systems

The SPR- and LSPR-based biosensing system-related application cases have been presented in Section 10.3 in detail. Integrating microfluidic chips with plasmonic biosensing systems brings with it the advantages of lab-chip platforms with the benefits of SPR biosensing technology. Especially, the portable and multiplex detection of biomolecules with microfluidics integrated is an integral part of an SPR biosensing platform. Onur Tokel and the coworkers developed an inexpensive and disposable microfluidic-integrated SPR biosensing platform for pathogen detection. The microfluidic chip comprises a gold-coated substrate embedded in the middle of a single microchannel, which has an inlet and an outlet port. The chip is functionalized with antibodies for efficient, selective, and specific capture of the pathogen. The SPR sensing system is based on the Kretschmann configuration. LED can illuminate the surface of the microchip, so reflected light can be captured by a CMOS sensor. The light source, CMOS sensor, and the associated optical and electrical components can be packaged on a portable scale as shown in Fig. 10.7A and B [95]. Furthermore, the sensing platform can be quantified to rapidly detect *Escherichia coli* in a wide range of concentrations ranging from 10^5 to 3.2×10^7 CFUs/mL. This platform also has specificity, selectivity, and multiplexing capability with a mixture of two types of bacteria, *E. coli* and *Staphylococcus aureus*, as shown in Fig. 10.7C. Therefore, such

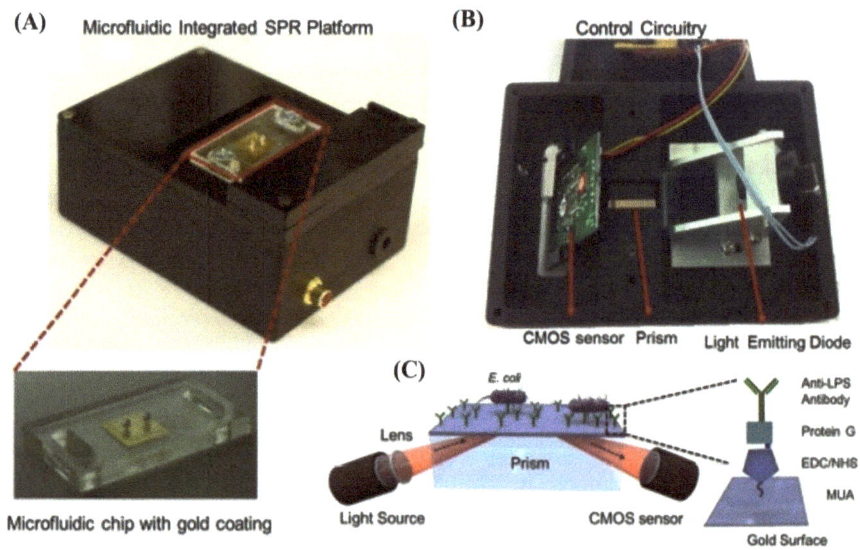

FIGURE 10.7

Portable microfluidic-integrated SPR biosensing platform for pathogen multiplex detection: (A) photograph of a surface-activated disposable microfluidic chip device; (B) control circuitry setup of the device; and (C) scheme of the multiplex microfluidic-integrated SPR biosensing system. *SPR*, Surface plasmon resonance.

Reproduced with permission from O. Tokel, U.H. Yildiz, F. Inci, N.G. Durmus, O.O. Ekiz, B. Turker, et al., Portable microfluidic integrated plasmonic platform for pathogen detection, Sci. Rep. 5 (1) (2015) 9152.

portable microfluidic integrated SPR biosensing systems can enable extending detection to other biomarkers for the development of point-of-care platforms.

In addition to extinction measurements, some alternative microfluidic-based LSPR sensing platforms will be discussed below. By employing a disconnected pattern of the thin metal film, LSPR can also generate a nonpropagating plasmon mode, called transmission configuration (T-LSPR) [96]. The T-LSPR can be readily measured from the transmitted light when it is excited by the passing-through light, thus greatly simplifying the detection system [97]. For example, Carlotta Guiducci et al. have developed a portable, palm-sized T-LSPR device. The T-LSPR microfluidic plasmonic chip comprises white light LED as an excitation source and a CMOS image sensor as a light detector. The light is collected on the CMOS detector by a plano-convex lens, which reduces the focal distance between the plasmonic substrate and the CMOS sensor. The plasmon shift caused by surface binding can be calculated in terms of both plasmon peak location and hue, while the latter can allow faster data elaboration and real-time display of the results with the CMOS image sensor. This T-LSPR sensing device is coupled with DNA-based aptamers for the specific detection of the antibiotic tobramycin [98]. Furthermore, the T-LSPR sensing mode can offer real-time label-free detection at extremely low concentrations.

10.4.2 Surface plasmon resonance imaging

In biomedical detection and assay-related applications, the multiplexed analysis of the plasmonic biosensors is on demand to meet the requirements. SPR imaging (SPRi) is the simplest way to offer high-throughput and multiplex detection. This optical technique was first developed by Benno Rothenhäusler and Wolfgang Knoll in 1988 [99]. The SPRi approach can spatially monitor regional differences in the reflectivity of incident light. The differences are caused by the plasmonic structure interacting with the analyte via prism coupling [100]. The prism coupling method can excite the SPR in an extended area, which is realized by a collimated monochromatic light beam. Naturally, by using a 2D charge-coupled device (CCD) camera, the intensity variations of the reflected light caused by RI changes can be measured. The multiplex analysis of various analytes that are immobilized at different regions of the array substrate surface can be carried out by evaluations of the 2D reflected intensity pattern. So, the signal, including SPR images of the array substrate surface and sensograms (resonance signal vs time), can be detected by the CCD camera. The SPR imaging is capable of visualizing the whole biochip via a video CCD camera in a label-free way. When coupled with microfluidic chip design, it forms a multichannel array of biosensing chips. Thus, it enables high-throughput analysis of the various analyte interactions simultaneously through SPR information measurements of each active site in the sensing system [101]. Tremendous efforts have been made to develop and exploit the SPRi technique-based biosensing detection system in recent years [102–104]. Furthermore, wavelength, phase, and angular polarization in addition to reflected intensity can also be used for interrogation with incident light. Besides, the microfluidic-assisted SPRi biosensing systems can provide further advantages like time monitoring with small sample consumption via high-throughput channels [105–110].

A 50 parallel channels of real-time full-spectral-based SPRi biosensing microfluidic platform has been developed by the Sang-Hyun Oh group for protein–lipid binding interaction detection using large-area metallic nanohole arrays as shown in Fig. 10.8A [111]. This system consists of large-area ($\sim cm^2$) metallic nanohole arrays as SPR sensing substrates integrated on a 50 parallel channel microfluidic chip with a 70 μm period over a $3.5 \times 2\ mm^2$ area, a broadband light source, and then coupled with an imaging spectrometer with a cooled CCD camera. So, this SPRi-based biosensing device can monitor receptor–ligand binding kinetics affinities [cholera toxin b (CTX-b) and ganglioside GM1), simultaneously. To acquire high-throughput and real-time acquisition of full spectra SPRi assay, they designed the individual functionalization of each channel with a

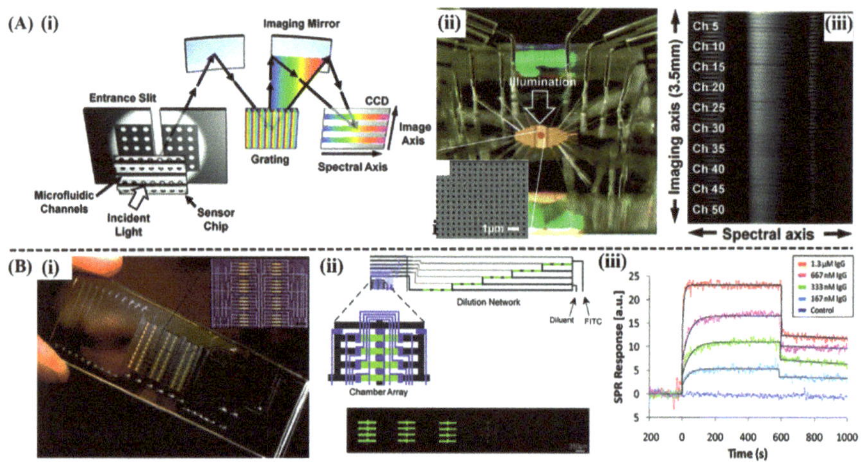

FIGURE 10.8

SPRi-based plasmonic biosensing platforms. (A) 50 microfluidic channel-based real-time full-spectral SPRi sensing platform for affinity measurements: (i) schematic illustration of the spectral imaging system setup; (ii) photograph of inlet and outlet ports of the microfluidic chip (inset is the SEM image of an area of the nanohole chip); and (iii) 50 parallel channels of full spectral content imaging. (B) Parallel microfluidic SPRi arrays: (i) photograph of the microfluidic chip bonded to a gold-patterned glass slide; and (ii and iii) schematic diagram and fluorescence images of the serial dilution network of the chaotic advection micromixers on the chip. *SPRi*, Surface plasmon resonance-imaging.

(A) Reproduced with permission from S.H. Lee, N.C. Lindquist, N.J. Wittenberg, L.R. Jordan, S.-H. Oh, Real-time full-spectral imaging and affinity measurements from 50 microfluidic channels using nanohole surface plasmon resonance, Lab Chip 12 (20) (2012) 3882–3890. ©2012 Royal Society of Chemistry. (B) Reproduced with permission from E. Ouellet, C. Lausted, T. Lin, C.W.T. Yang, L. Hood, E.T. Lagally, Parallel microfluidic surface plasmon resonance imaging arrays, Lab Chip 10 (5) (2010) 581–588. ©2010 Royal Society of Chemistry.

wide range of receptors and high-throughput proteomics. The platform can extract spectral information from every channel in real-time with an RI resolution of 7.7×10^{-6} RI units. Then, the affinity signal can be transformed into high-throughput full-spectral imaging. Such a system can realize the precise and quantified measurement of a wide range of biomolecular interactions. Eric Ouellet et al. presented a high-throughput parallel microfluidic array-based SPRi sensing device, in which a microfluidic chip was fabricated by the polydimethylsiloxane (PDMS)-based soft lithography techniques shown in Fig. 10.8B [112]. The device consists of a PDMS microvalve-based microfluidic chip, p-polarized LED light source, and a customized SPRi instrument to detect the biomolecular reactions within the chambers. It can be used for human α-thrombin immunoassay analysis by detecting the binding affinities of antibodies against the protein. The parallel array microchip system consists of 264 element-addressable chambers isolated by microvalves, so it can realize 264 different immobilized ligands interrogating binding events against multiple analytes in a single experiment. Such a microfluidic-based SPRi platform displays the versatility accommodation for multiple ligands to multiple analytes simultaneously.

10.4.3 Surface-enhanced Raman scattering

Surface-enhanced Raman scattering (SERS) technology is one of the applications of the plasmonic coupling effect, which combines the unique optical properties of plasmonic nanostructures with the laser-based Raman spectroscopy technique [113]. SERS can amplify the Raman spectroscopic signals of analyte's molecule vibration interaction with plasmonic structure and offer high-level structural information of the target analyte [114]. That is, by using the SERS technique, the limits of ultrasensitive detection can be improved significantly by plasmonic-based Raman enhancement. So, such an enhancement can facilitate trace detection of biomolecular analysis down to the single-molecule level. A series of advancements and progress in SERS technology have made it a viable option for biosensing applications. Furthermore, the SERS sensing method enables fast and reliable biomolecular marker assays by the identification of fingerprinting spectra from molecules similar in structure and function, which can eliminate extra time-consuming steps like sample preparations in other immunoassay techniques [115]. According to the sources of SERS signals, it can be divided into two different types of sensing strategies: (1) direct detection of target molecules by the characteristic fingerprint peak recognition in Raman spectra; and (2) indirect detection of target analytes with the specific molecular conjugation by additional surface functionalization steps [116]. Similarly, the SERS technique can also integrate with microfluidic devices to realize high-throughput and ultrasensitive detection and analysis [79,117,118]. The integration with a microfluidic chip could facilitate manipulating the analytes and measuring signal because it can achieve full attachment and interaction of the analytes with SERS-based plasmonic sensor substrate in microfluidic channels [119]. Furthermore, special topographical morphologies of the SERS-active plasmonic structure-based sensing substrate are capable of affecting the detection performances in the microfluidic device [79,117,119].

As shown in Fig. 10.9A, Jaebum Choo et al. reported a SERS-based biosensing platform for anthrax biomarker detection in the solenoid-embedded microfluidic chip [120]. The integrated system is composed of a solenoid-embedded PDMS dual-channel microfluidic chip, a CCD camera for bright field image capture, and a 633-nm Raman microscope system. The SERS-biosensing system can offer highly sensitive, rapid, quantitative detection of anthrax biomarkers called poly-γ-D-glutamic acid (PGA). The dual-channel microfluidic device has two parallel channel compartments, including the PGA sensing and control measurements, which allow performing PGA immunoassay in an automated manner.

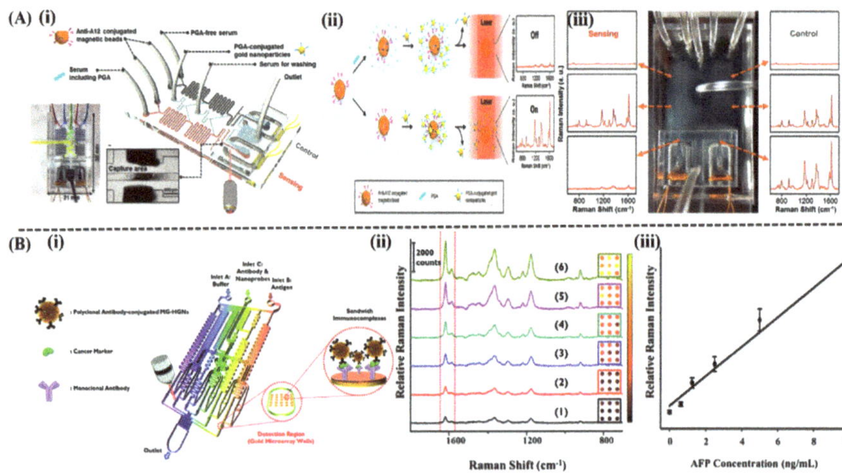

FIGURE 10.9

SERS-based plasmonic biosensing microfluidic platforms. (A) SERS-based plasmonic sensor for anthrax biomarker detection in the solenoid microfluidic device: (i) scheme of SERS-based plasmonic biosensor system for competitive immunoassay in the special designed microfluidic; (ii) scheme of the SERS spectra-based anthrax marker competitive immunoassay in serum; and (iii) SERS spectra at six different channel positions detected in the solenoid-embedded microfluidic chip. (B) Gold-patterned microarray wells embedded SERS-based immunoassay in the gradient microfluidic chip: (i) scheme of gold-patterned microarray well embedded gradient microfluidic chip, and (ii and iii) quantitative SERS spectra of an immunoassay for alpha-fetoprotein (AFP). *AFP*, Alpha-fetoprotein; *SERS*, surface-enhanced Raman scattering.

(A) Reproduced with permission from R. Gao, J. Ko, K. Cha, J. Ho Jeon, G.-E. Rhie, J. Choi, et al., Fast and sensitive detection of an anthrax biomarker using SERS-based solenoid microfluidic sensor, Biosens. Bioelectron. 72 (2015) 230–236. ©2015 Elsevier. (B) Reproduced with permission from M. Lee, K. Lee, K. H. Kim, K.W. Oh, J. Choo, SERS-based immunoassay using a gold array-embedded gradient microfluidic chip, Lab Chip 12 (19) (2012) 3720–3727. ©2012 Royal Society of Chemistry.

Furthermore, solenoid-embedded microfluidic structure design can enable direct acquisition and analysis of SERS spectra via the competitive reaction (PGA and PGA-conjugated gold nanoparticles versus anti-PGA-immobilized magnetic beads). The limit of detection of PGA in serum by this microfluidic SERS-based biosensor system can be estimated at 100 pg/mL. Another type of SERS-based gold-patterned microarray well embedded gradient microfluidic sensing chip has been developed by the Jaebum Choo group for the immunoassay of alpha-fetoprotein (AFP) model protein marker (cancer biomarkers) detection shown in Fig. 10.9B [121]. The device consists of three panels: the top and middle PDMS panels of the microfluidic circuits, and the bottom glass panel embedded with 30-μm gold patterned microarray wells (5 × 5). Such a network of gradient channel chip design with embedded gold microarray wells can offer programmable and fully automatic generation of various concentrations of antigen solutions for AFP immunoassay. Furthermore, the gradient microfluidic chip can eliminate the tedious manual dilution process so that it takes less than 60 minutes for the total immunoassay assay from serial dilution to SERS detection. This platform also provides a convenient and reproducible SERS-based quantitative immunoassay for AFP within the concentration range of 0–10 ng/mL and the limit of detection down to 1 ng/mL^{-1}.

10.4.4 Plasmonic sensor integrated with other sensors for multiplex biosensing

In addition to the aforementioned plasmonic biosensing techniques in microchips, the deployment of plasmonic biosensors as analytical platforms has not been fully accomplished for bio-medical-related applications. Researchers have also reported and focused on the combination of multiple biosensors. Although these biosensors process different operation principles, the integration of multiple sensors can bring a more portable and powerful system by combining the advantages of both sensors for potential applications in biomedical fields [29,122].

As shown in Fig. 10.10A, Liang Dong and their coworkers have integrated SPR and the electrochemical sensor for dual-modality target biomarker molecule detection in the microfluidic chip [122]. The substrate chip is made up of a patterned periodic pillar array coated with a gold-graphene oxide (Au-GO) layer, which can either serve as the working electrode of an electrochemical sensor or serve as the nanopatterned substrate of a plasmonic sensor. The dual-modality integrated biosensor can couple incident light and operate voltage sources simultaneously to detect the specific immunoassay of the target breast cancer biomarker (ErbB2) and anti-ErbB2 molecules. Sensitivities of the electrochemical and the SPR measurements to ErbB2 are 20.47 $\mu A/(\mu M\ cm^2)$ in a range from 1 fM to 0.1 μM and 1.35 nm/μM in a range from 10 pM to 1 nM, respectively. Furthermore, the integrated dual-modality design can improve the detection reliability with higher sensitivity through the acquired SPR and electrochemical

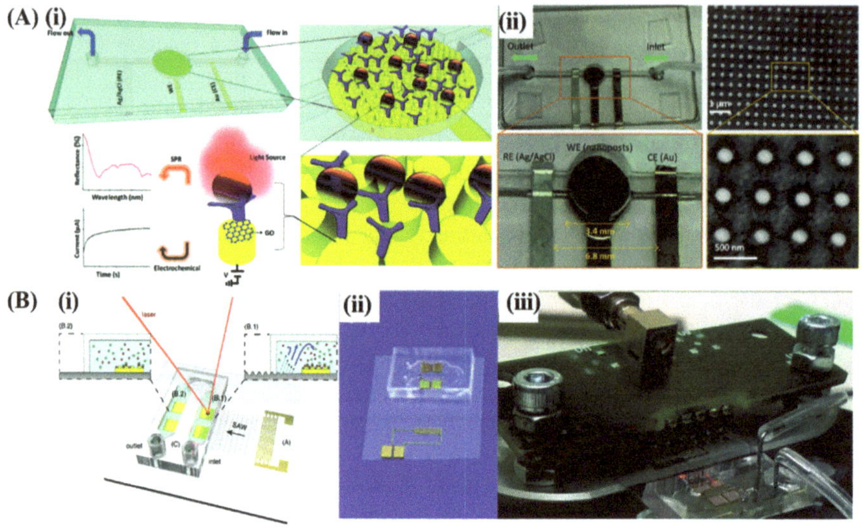

FIGURE 10.10

Multiplex sensor integration biosensing microfluidic platform. (A) Dual-modality SPR and electrochemical integrated biosensor: (i) scheme of the dual-modality sensors integrated microfluidic biosensing system for cancer biomarker detection; and (ii) the photograph of the dual-modality integrated biosensing microchip device and the SEM images of the patterned periodic arrays substrate. (B) Phase-interrogation SPR coupled with SAW enhanced grating biosensing microfluidic system: (i) scheme of phase-interrogation SPR coupled with SAW enhanced grating biosensing microfluidic system; (ii) the photograph of for acoustic–plasmonic sensor chip's configuration; and (iii) the set-up of SAW-driven SPR detection biosensing device. *SAW*, Surface acoustic wave; *SPR*, surface plasmon resonance **[123]**.

(A) Reproduced with permission from M.A. Ali, S. Tabassum, Q. Wang, Y. Wang, R. Kumar, L. Dong, Integrated dual-modality microfluidic sensor for biomarker detection using lithographic plasmonic crystal, Lab Chip 18 (5) (2018) 803–817. ©2018 Royal Society of Chemistry. (B) Reproduced with permission from A. Sonato, M. Agostini, G. Ruffato, E. Gazzola, D. Liuni, G. Greco, et al., A surface acoustic wave (SAW)-enhanced grating-coupling phase-interrogation surface plasmon resonance (SPR) microfluidic biosensor, Lab Chip 16 (7) (2016) 1224–1233. ©2016 Royal Society of Chemistry.

signals, and enable dynamic tracking of the interaction by the specific antigen–antibody affinity reaction.

Sonato et al. have demonstrated an integrated biosensing microfluidic system by coupling the surface acoustic wave (SAW) enhanced grating with phase-interrogation SPR, as shown in Fig. 10.10B [124]. The biosensing microchip consists of three main parts: (1) an interdigital transducer for Rayleigh SAW generation; (2) four SPR nanostructured areas on the single lithium niobate (LN) substrate, which can generate phase-interrogation grating-coupling SPR, while only two SPR could be affected by SAW-induced mixing; and (3) a PDMS microchannel with two microchambers. With the azimuthal control, it can realize the high sensitivity of grating-coupling SPR. The sensor response was detected using phase-interrogation SPR; thus, the avidin/biotin binding kinetics assay and thiol-polyethylene glycol (bPEG-SH) adsorption can be monitored and recorded under different experimental conditions. Furthermore, the fluid mixing can be enhanced with the assistance of SAW-generated chaotic advection. Therefore, the SAW-induced integrated SPR biosensing microchip-based system can significantly reduce instances of bioreaction experiment with the improvement of the molecule binding kinetics. The time saturation binding kinetics of polyethylene and avidin adsorption can be reduce by 82% and 24%, respectively.

10.5 Conclusion and outlook

In this chapter, we have systematically summarized the recent developments and the literature work in microfluidic-based plasmonic biosensors. We have also illustrated that it could act as an advanced detection tool for various biomolecule detections with different types of plasmonic structure-based biosensors. Furthermore, we also reveal that the microfluidic-integrated plasmonic biosensors would offer high-throughput and real-time ultrasensitive detection and analysis of biomolecular molecules. This chapter aims to guide application of different strategies toward the development of more optimized plasmonic biosensor devices for potential applications in biomedical fields.

To date, the microfluidic-based plasmonic biosensing techniques have shown great potential for biomolecular detection and can be used to detect a wide variety of analytes, such as proteins, viral pathogens, bacterial, and nucleic acids, especially the development of microfluidic-based plasmonic biosensors for the ultrasensitive detection of cancer biomarkers, which have high significance for medical diagnostics at early stages. Despite that plasmonic microstructure arrays can improve the sensitivity and detection limit of analytes, there still exist several challenges and limitations to overcome. The biomarkers' detection usually constitutes proof-of-concept experiments, which are performed in buffer media rather than in medically relevant samples, for example, the biomarker in blood serum, bodily fluids, or tissue extract environments. These media substances usually have complex compositions, which may be contaminated and corrode the core part of the chip in the device. So, it is important to increase the stability of the sensing system for clinical diagnostics in the future.

Another challenge for plasmonic biosensing is its suitability for point-of-care diagnostics. The portability of the plasmonic biosensor also faces the challenge of establishing its practicality in clinic tests. Great efforts have been made to build more affordable, more accessible, and more applicable plasmonic biosensing devices. Although the microchip integration has facilitated setting up portable plasmonic sensing systems, there are still some engineering roadblocks facing the miniaturization of all components into a single portable platform. For example, the optical instruments like illumination systems and spectrometers may be difficult to be scaled down to the microscale.

Consequently, one future perspective can be aiming at the development of innovative plasmonic biosensor chips with whole system miniaturization to hand palm and portable size. Thus, it can be predicted that the microfluidic-integrated plasmonic biosensor can gradually become an ideal sensing platform for practical point-of-care, smart, and potable clinic detection and diagnostic applications in the future. Here are some perspective scenarios, such as rapid detection of pathogenic species for security and defense against bioterrorism or identifying toxins in the aquatic environment in areas with no access to laboratory facilities. On the other hand, novel human coronavirus 2019 (COVID-19) is usually detected by quantitative real-time polymerase chain reaction from various types of biological specimens currently, which is time-consuming, labor-intensive, and may not be able to deploy in resource-limited regions. Our future perspectives can aim at the development of suitable and robust plasmonic biosensors in a rapid, scalable, and sensitive way to detect COVID-19. Besides, SPRi technology has been utilized for the high-throughput drug screening assay, so integrating with a multichannel microfluidic chip, the SPRi may prove to be a useful tool to accelerate the drug development process with high-throughput processing and less sample dosage. For example, the microfluidic-integrated plasmonic SPRi biosensing platform can extend its application in mRNA drug manufacturing.

References

[1] G.M. Whitesides, The origins and the future of microfluidics, Nature 442 (7101) (2006) 368–373.

[2] E.K. Sackmann, A.L. Fulton, D.J. Beebe, The present and future role of microfluidics in biomedical research, Nature 507 (7491) (2014) 181–189.

[3] A. Webster, J. Greenman, S.J. Haswell, Development of microfluidic devices for biomedical and clinical application, J. Chem. Technol. Biotechnol. 86 (1) (2011) 10–17.

[4] R. Gorkin, J. Park, J. Siegrist, M. Amasia, B.S. Lee, J.-M. Park, et al., Centrifugal microfluidics for biomedical applications, Lab Chip 10 (14) (2010) 1758–1773.

[5] P. Mehrotra, Biosensors and their applications – a review, J. Oral. Biol. Craniofacial Res. 6 (2) (2016) 153–159.

[6] G. Luka, A. Ahmadi, H. Najjaran, E. Alocilja, M. DeRosa, K. Wolthers, et al., Microfluidics integrated biosensors: a leading technology towards lab-on-a-chip and sensing applications, Sensors 15 (2015) 12.

[7] D.R. Thévenot, K. Toth, R.A. Durst, G.S. Wilson, Electrochemical biosensors: recommended definitions and classification International Union of Pure and Applied Chemistry: Physical Chemistry Division, Commission I.7 (Biophysical Chemistry); Analytical Chemistry Division, Commission V.5 (Electroanalytical Chemistry)0.1, Biosens. Bioelectron. 16 (1) (2001) 121–131.

[8] D. R., Thevenot, K. Toth, R. A. Durst, G. S. Wilson, Electrochemical biosensors: recommended definitions and classification, Pure Appl. Chem. 71 (12) (1999) 2333–2348.

[9] K. Abe, K. Suzuki, D. Citterio, Inkjet-printed microfluidic multianalyte chemical sensing paper, Anal. Chem. 80 (18) (2008) 6928–6934.
[10] M.A. Unger, H.-P. Chou, T. Thorsen, A. Scherer, S.R. Quake, Monolithic microfabricated valves and pumps by multilayer soft lithography, Science 288 (5463) (2000) 113.
[11] N. Futai, W. Gu, J.W. Song, S. Takayama, Handheld recirculation system and customized media for microfluidic cell culture, Lab Chip 6 (1) (2006) 149–154.
[12] K.W. Oh, Microfluidic devices for biomedical applications: biomedical microfluidic devices 2019, Micromachines 11 (4) (2020).
[13] B. Caballero, L.C. Trugo, P.M. Finglas, Encyclopedia of Food Sciences and Nutrition, Academic, 2003.
[14] M. Schaechter, Encyclopedia of Microbiology, Academic Press:, 2009.
[15] J. Tamayo, P.M. Kosaka, J.J. Ruz, Á. San Paulo, M. Calleja, Biosensors based on nanomechanical systems, Chem. Soc. Rev. 42 (3) (2013) 1287–1311.
[16] J.L. Arlett, E.B. Myers, M.L. Roukes, Comparative advantages of mechanical biosensors, Nat. Nanotechnol. 6 (4) (2011) 203–215.
[17] N.J. Ronkainen, H.B. Halsall, W.R. Heineman, Electrochemical biosensors, Chem. Soc. Rev. 39 (5) (2010) 1747–1763.
[18] J. Wang, Electrochemical biosensors: Towards point-of-care cancer diagnostics, Biosens. Bioelectron. 21 (10) (2006) 1887–1892.
[19] M. Alvarez, G.P. Anderson, J. Bachmann, I. Biran, D.A. Blake, Y.C. Cao, et al., List of contributors, in: F.S.; Ligler, C.R. Taitt (Eds.), Optical Biosensors, *Second ed.*, Elsevier:, Amsterdam, 2008. pp xi-xix.
[20] C. Chen, J. Wang, Optical biosensors: an exhaustive and comprehensive review, Analyst 145 (5) (2020) 1605–1628.
[21] L.D. Mello, L.T. Kubota, Review of the use of biosensors as analytical tools in the food and drink industries, Food Chem. 77 (2) (2002) 237–256.
[22] X. Fan, I.M. White, S.I. Shopova, H. Zhu, J.D. Suter, Y. Sun, Sensitive optical biosensors for unlabeled targets: a review, Anal. Chim. Acta 620 (1) (2008) 8–26.
[23] M.A. Cooper, Optical biosensors in drug discovery, Nat. Rev. Drug Discovery 1 (7) (2002) 515–528.
[24] A. Leung, P.M. Shankar, R. Mutharasan, A review of fiber-optic biosensors, Sens. Actuators B: Chem. 125 (2) (2007) 688–703.
[25] P. Kozma, F. Kehl, E. Ehrentreich-Förster, C. Stamm, F.F. Bier, Integrated planar optical waveguide interferometer biosensors: a comparative review, Biosens. Bioelectron. 58 (2014) 287–307.
[26] X. Guo, Surface plasmon resonance based biosensor technique: a review, J. Biophotonics 5 (7) (2012) 483–501.
[27] C.L. Baird, D.G. Myszka, Current and emerging commercial optical biosensors, J. Mol. Recognit. 14 (5) (2001) 261–268.
[28] H.J. Watts, D. Yeung, H. Parkes, Real-time detection and quantification of DNA hybridization by an optical biosensor, Anal. Chem. 67 (23) (1995) 4283–4289.
[29] F.S. Ligler, Perspective on optical biosensors and integrated sensor systems, Anal. Chem. 81 (2) (2009) 519–526.
[30] A. Roda, M. Mirasoli, E. Michelini, M. Di Fusco, M. Zangheri, L. Cevenini, et al., Progress in chemical luminescence-based biosensors: a critical review, Biosens. Bioelectron. 76 (2016) 164–179.

[31] S. Spindel, K.E. Sapsford, Evaluation of optical detection platforms for multiplexed detection of proteins and the need for point-of-care biosensors for clinical use, Sensors 14 (12) (2014).

[32] B. Špačková, P. Wrobel, M. Bocková, J. Homola, Optical biosensors based on plasmonic nanostructures: a review, Proc. IEEE 104 (12) (2016) 2380–2408.

[33] R.L. Rich, D.G. Myszka, Advances in surface plasmon resonance biosensor analysis, Curr. Opin. Biotechnol. 11 (1) (2000) 54–61.

[34] R.T. Hill, Plasmonic biosensors, Wiley Interdiscip Rev Nanomed Nanobiotechnol. 7 (2) (2015) 152–168.

[35] Z. Liao, Y. Zhang, Y. Li, Y. Miao, S. Gao, F. Lin, et al., Microfluidic chip coupled with optical biosensors for simultaneous detection of multiple analytes: a review, Biosens. Bioelectron. 126 (2019) 697–706.

[36] F.B. Myers, L.P. Lee, Innovations in optical microfluidic technologies for point-of-care diagnostics, Lab Chip 8 (12) (2008) 2015–2031.

[37] A. Bange, H.B. Halsall, W.R. Heineman, Microfluidic immunosensor systems, Biosens. Bioelectron. 20 (12) (2005) 2488–2503.

[38] M. Pelton, J. Aizpurua, G. Bryant, Metal-nanoparticle plasmonics, Laser Photonics Rev. 2 (3) (2008) 136–159.

[39] S.A. Maier, Plasmonics: *Fundamentals* and *Applications*, Springer Science & Business Media, 2007.

[40] S.A. Maier, H.A. Atwater, Plasmonics: localization and guiding of electromagnetic energy in metal/dielectric structures, J. Appl. Phys. 98 (1) (2005) 011101.

[41] J. Homola, M. Piliarik, Surface plasmon resonance (SPR) sensors, *Surface Plasmon Resonance Based Sensors*, Springer, 2006, pp. 45–67.

[42] K.M. Mayer, J.H. Hafner, Localized surface plasmon resonance sensors, Chem. Rev. 111 (6) (2011) 3828–3857.

[43] J. Homola, Present and future of surface plasmon resonance biosensors, Anal. Bioanal. Chem. 377 (3) (2003) 528–539.

[44] Reather, H., Surface plasmons on smooth and rough surfaces and on gratings, in: Springer Tracts in Modern Physics, 1988, 111, 1–3.

[45] Y. Liu, X. Zhang, Microfluidics-based plasmonic biosensing system based on patterned plasmonic nanostructure arrays, Micromachines (Basel) 12 (7) (2021) 826.

[46] E. Kretschmann, Die bestimmung optischer konstanten von metallen durch anregung von oberflächenplasmaschwingungen, Zeitschrift für Physik A Hadrons and Nuclei 241 (4) (1971) 313–324.

[47] R. Kashyap, G.J.J.o S. Nemova, Surface plasmon resonance-based fiber and planar waveguide sensors, J. Sensors 2009 (2009).

[48] Y. Liu, X. Zhang, Microfluidics-based plasmonic biosensing system based on patterned plasmonic nanostructure arrays, Micromachines 12 (7) (2021).

[49] L.A. Kolahalam, I.V. Kasi Viswanath, B.S. Diwakar, B. Govindh, V. Reddy, Y.L.N. Murthy, Review on nanomaterials: synthesis and applications, Mater. Today: Proc. 18 (2019) 2182–2190.

[50] C.F. Bohren, D.R. Huffman, Absorption and *Scattering* of *Light* by *Small Particles*, John Wiley & Sons:, 2008.

[51] L.M. Liz-Marzán, Tailoring surface plasmons through the morphology and assembly of metal nanoparticles, Langmuir 22 (1) (2006) 32–41.

[52] T.K. Sau, A.L. Rogach, F. Jäckel, T.A. Klar, J. Feldmann, Properties and applications of colloidal nonspherical noble metal nanoparticles, Adv. Mater. 22 (16) (2010) 1805–1825.
[53] G.V. Hartland, Optical studies of dynamics in noble metal nanostructures, Chem. Rev. 111 (6) (2011) 3858–3887.
[54] J. Cao, T. Sun, K.T.V. Grattan, Gold nanorod-based localized surface plasmon resonance biosensors: a review, Sens. Actuators B Chem. 195 (2014) 332–351.
[55] L.M. Liz-Marzán, C.J. Murphy, J. Wang, Nanoplasmonics, Chem. Soc. Rev. 43 (11) (2014) 3820–3822.
[56] M.I. Stockman, Nanoplasmonics: past, present, and glimpse into future, Opt. Express 19 (22) (2011) 22029–22106.
[57] S. Unser, I. Bruzas, J. He, L. Sagle, Localized surface plasmon resonance biosensing: current challenges and approaches, Sensors 15 (7) (2015).
[58] L.S. Jung, C.T. Campbell, T.M. Chinowsky, M.N. Mar, S.S. Yee, Quantitative interpretation of the response of surface plasmon resonance sensors to adsorbed films, Langmuir 14 (19) (1998) 5636–5648.
[59] J.N. Anker, W.P. Hall, O. Lyandres, N.C. Shah, J. Zhao, R.P. Van Duyne, Biosensing with plasmonic nanosensors, Nat. Mater. 7 (6) (2008) 442–453.
[60] D. Howes, P. Rana, S, M. Stevens, Plasmonic nanomaterials for biodiagnostics, Chem. Soc. Rev. 43 (11) (2014) 3835–3853.
[61] M.C. Estevez, M.A. Otte, B. Sepulveda, L.M. Lechuga, Trends and challenges of refractometric nanoplasmonic biosensors: a review, Analyt. Chim. Acta 806 (2014) 55–73.
[62] O. Tokel, F. Inci, U. Demirci, Advances in plasmonic technologies for point of care applications, Chem. Rev. 114 (11) (2014) 5728–5752.
[63] M. Soler, C.S. Huertas, L.M. Lechuga, Label-free plasmonic biosensors for point-of-care diagnostics: a review, Expert. Rev. Mol. Diagn. 19 (1) (2019) 71–81.
[64] K.A. Willets, R.P. Van Duyne, Localized surface plasmon resonance spectroscopy and sensing, Annu. Rev. Phys. Chem. 58 (1) (2007) 267–297.
[65] M. Hu, C. Novo, A. Funston, H. Wang, H. Staleva, S. Zou, et al., Dark-field microscopy studies of single metal nanoparticles: understanding the factors that influence the linewidth of the localized surface plasmon resonance, J. Mater. Chem. 18 (17) (2008) 1949–1960.
[66] S. Sharma, R. Kumari, S.K. Varshney, B. Lahiri, Optical biosensing with electromagnetic nanostructures, Rev. Phys. 5 (2020) 100044.
[67] G.A. Lopez, M.C. Estevez, M. Soler, L.M. Lechuga, Recent advances in nanoplasmonic biosensors: applications and lab-on-a-chip integration, Nanophotonics 6 (1) (2017) 123–136.
[68] C. Wang, Y. Cai, A. MacLachlan, P. Chen, Novel nanoplasmonic-structure-based integrated microfluidic biosensors for label-free in situ immune functional analysis: a review of recent progress, IEEE Nanotechnol. Mag. 14 (2) (2020) 46–C3.
[69] M. Soler, O. Calvo-Lozano, M.-C. Estevez, L.M. Lechuga, Nanophotonic biosensors: driving personalized medicine, Opt. Photonics N. 31 (4) (2020) 24–31.
[70] J.A. Jackman, A.R. Ferhan, N.-J. Cho, Nanoplasmonic sensors for biointerfacial science, Chem. Soc. Rev. 46 (12) (2017) 3615–3660.
[71] J.A. Jackman, A.R. Ferhan. N.-J. Cho, Nanoplasmonic sensors for biointerfacial science, J. Chem. Soc. Rev. 46 (12) (2017) 3615–3660.
[72] A. Curry, G. Nusz, A. Chilkoti, A. Wax, Analysis of total uncertainty in spectral peak measurements for plasmonic nanoparticle-based biosensors, Appl. Opt. 46 (10) (2007) 1931–1939.

[73] R. Gordon, D. Sinton, K.L. Kavanagh, A.G. Brolo, A new generation of sensors based on extraordinary optical transmission, Acc. Chem. Res. 41 (8) (2008) 1049–1057.
[74] J.N. Anker, W.P. Hall, O. Lyandres, N.C. Shah, J. Zhao, R.P. Van Duyne, Biosensing with plasmonic nanosensors, Nanoscience and Technology, Co-Published with Macmillan Publishers Ltd, UK, 2009, pp. 308–319.
[75] X. Lv, Z. Geng, Y. Su, Z. Fan, S. Wang, W. Fang, et al., Label-free exosome detection based on a low-cost plasmonic biosensor array integrated with microfluidics, Langmuir 35 (30) (2019) 9816–9824.
[76] V. Singh, Ultrasensitive quantum dot-coupled-surface plasmon microfluidic aptasensor array for serum insulin detection, Talanta 219 (2020) 121314.
[77] A. Horrer, J. Haas, K. Freudenberger, G. Gauglitz, D.P. Kern, M. Fleischer, Compact plasmonic optical biosensors based on nanostructured gradient index lenses integrated into microfluidic cells, Nanoscale 9 (44) (2017) 17378–17386.
[78] X. Hu, G. Xu, L. Wen, H. Wang, Y. Zhao, Y. Zhang, et al., Metamaterial absorber integrated microfluidic terahertz sensors, Laser & Photonics Rev. 10 (6) (2016) 962–969.
[79] L. Ming, Z. Fusheng, Z. Jianbo, Q. Ji, L. Jing, S. Wei-Chuan, Microfluidic surface-enhanced Raman scattering sensor with monolithically integrated nanoporous gold disk arrays for rapid and label-free biomolecular detection, J. Biomed. Opt. 19 (11) (2014) 1–8.
[80] P. Chen, M.T. Chung, W. McHugh, R. Nidetz, Y. Li, J. Fu, et al., Multiplex serum cytokine immunoassay using nanoplasmonic biosensor microarrays, ACS Nano 9 (4) (2015) 4173–4181.
[81] S. Klinghammer, T. Uhlig, F. Patrovsky, M. Böhm, J. Schütt, N. Pütz, L. Baraban, L. M. Eng, G. Cuniberti, Plasmonic biosensor based on vertical arrays of gold nanoantennas, ACS Sensors 3 (7) (2018) 1392–1400.
[82] S.S. Aćimović, H. Šípová, G. Emilsson, A.B. Dahlin, T.J. Antosiewicz, M. Käll, Superior LSPR substrates based on electromagnetic decoupling for on-a-chip high-throughput label-free biosensing, Light: Sci. Appl. 6 (8) (2017) e17042.
[83] J. Henzie, M.H. Lee, T.W. Odom, Multiscale patterning of plasmonic metamaterials, Nat. Nanotechnol. 2 (9) (2007) 549–554.
[84] D. Sinton, R. Gordon, A.G. Brolo, Nanohole arrays in metal films as optofluidic elements: progress and potential, Microfluidics Nanofluidics 4 (1) (2008) 107–116.
[85] D. Kotlarek, S. Fossati, P. Venugopalan, N. Gisbert Quilis, J. Slabý, J. Homola, et al., Actuated plasmonic nanohole arrays for sensing and optical spectroscopy applications, Nanoscale 12 (17) (2020) 9756–9768.
[86] R. Gordon, A.G. Brolo, D. Sinton, K.L. Kavanagh, Resonant optical transmission through hole-arrays in metal films: physics and applications, Laser Photonics Rev. 4 (2) (2010) 311–335.
[87] X. Li, M. Soler, C.I. Özdemir, A. Belushkin, F. Yesilköy, H. Altug, Plasmonic nanohole array biosensor for label-free and real-time analysis of live cell secretion, Lab Chip 17 (13) (2017) 2208–2217.
[88] A. Vázquez-Guardado, S. Barkam, M. Peppler, A. Biswas, W. Dennis, S. Das, et al., Enzyme-free plasmonic biosensor for direct detection of neurotransmitter dopamine from whole blood, Nano Lett. 19 (1) (2019) 449–454.
[89] A.F. Coskun, A.E. Cetin, B.C. Galarreta, D.A. Alvarez, H. Altug, A. Ozcan, Lensfree optofluidic plasmonic sensor for real-time and label-free monitoring of molecular binding events over a wide field-of-view, Sci. Rep. 4 (1) (2014) 6789.

[90] J. Zhou, F. Tao, J. Zhu, S. Lin, Z. Wang, X. Wang, et al., Portable tumor biosensing of serum by plasmonic biochips in combination with nanoimprint and microfluidics, Nanophotonics 8 (2) (2019) 307–316.

[91] A. Hsiao, M.R. Gartia, T.-W. Chang, X. Wang, P. Khumwan, G.L. Liu, Colorimetric plasmon resonance microfluidics on nanohole array sensors, Sens. Bio-Sensing Res. 5 (2015) 24–32.

[92] A.A. Yanik, M. Huang, O. Kamohara, A. Artar, T.W. Geisbert, J.H. Connor, et al., An optofluidic nanoplasmonic biosensor for direct detection of live viruses from biological media, Nano Lett. 10 (12) (2010) 4962–4969.

[93] M. Soler, A. Belushkin, A. Cavallini, C. Kebbi-Beghdadi, G. Greub, H. Altug, Multiplexed nanoplasmonic biosensor for one-step simultaneous detection of Chlamydia trachomatis and Neisseria gonorrhoeae in urine, Biosens. Bioelectron. 94 (2017) 560–567.

[94] F. Liang, Y. Guo, S. Hou, Q. Quan, Photonic-plasmonic hybrid single-molecule nanosensor measures the effect of fluorescent labels on DNA-protein dynamics, Sci. Adv. 3 (5) (2017) e1602991.

[95] O. Tokel, U.H. Yildiz, F. Inci, N.G. Durmus, O.O. Ekiz, B. Turker, et al., Portable microfluidic integrated plasmonic platform for pathogen detection, Sci. Rep. 5 (1) (2015) 9152.

[96] O. Kedem, A.B. Tesler, A. Vaskevich, I. Rubinstein, Sensitivity and optimization of localized surface plasmon resonance transducers, ACS Nano 5 (2) (2011) 748–760.

[97] L. Malic, K. Morton, L. Clime, T. Veres, All-thermoplastic nanoplasmonic microfluidic device for transmission SPR biosensing, Lab Chip 13 (5) (2013) 798–810.

[98] G. Cappi, F.M. Spiga, Y. Moncada, A. Ferretti, M. Beyeler, M. Bianchessi, et al., Label-free detection of tobramycin in serum by transmission-localized surface plasmon resonance, Anal. Chem. 87 (10) (2015) 5278–5285.

[99] B. Rothenhäusler, W. Knoll, Surface–plasmon microscopy, Nature 332 (6165) (1988) 615–617.

[100] G. Steiner, Surface plasmon resonance imaging, Anal. Bioanal. Chem. 379 (3) (2004) 328–331.

[101] C.L. Wong, M. Olivo, Surface plasmon resonance imaging sensors: a review, Plasmonics 9 (4) (2014) 809–824.

[102] G. Spoto, M. Minunni, Surface plasmon resonance imaging: what next? J. Phys. Chem. Lett. 3 (18) (2012) 2682–2691.

[103] M. Bocková, J. Slabý, T. Špringer, J. Homola, Advances in surface plasmon resonance imaging and microscopy and their biological applications, Annu. Rev. Anal. Chem. 12 (1) (2019) 151–176.

[104] D. Wang, J.F. Loo, J. Chen, Y. Yam, S.-C. Chen, H. He, et al., Recent advances in surface plasmon resonance imaging, Sens. Sensors 19 (6) (2019).

[105] S. Wang, X. Shan, U. Patel, X. Huang, J. Lu, J. Li, et al., Label-free imaging, detection, and mass measurement of single viruses by surface plasmon resonance, Proc. Natl. Acad. Sci. U.S.A. 107 (37) (2010) 16028.

[106] H. Guner, E. Ozgur, G. Kokturk, M. Celik, E. Esen, A.E. Topal, et al., A smartphone based surface plasmon resonance imaging (SPRi) platform for on-site biodetection, Sens. Actuators B: Chem. 239 (2017) 571–577.

[107] A. Roda, E. Michelini, M. Zangheri, M. Di Fusco, D. Calabria, P. Simoni, Smartphone-based biosensors: a critical review and perspectives, TrAC. Trends Anal. Chem. 79 (2016) 317–325.

[108] P. Arora, A. Krishnan, On-chip label-free plasmonic based imaging microscopy for microfluidics, J. Phys. Commun. 2 (8) (2018) 085012.
[109] L. Malic, T. Veres, M. Tabrizian, Nanostructured digital microfluidics for enhanced surface plasmon resonance imaging, Biosens. Bioelectron. 26 (5) (2011) 2053–2059.
[110] L. Malic, T. Veres, M. Tabrizian, Biochip functionalization using electrowetting-on-dielectric digital microfluidics for surface plasmon resonance imaging detection of DNA hybridization, Biosens. Bioelectron. 24 (7) (2009) 2218–2224.
[111] S.H. Lee, N.C. Lindquist, N.J. Wittenberg, L.R. Jordan, S.-H. Oh, Real-time full-spectral imaging and affinity measurements from 50 microfluidic channels using nanohole surface plasmon resonance, Lab Chip 12 (20) (2012) 3882–3890.
[112] E. Ouellet, C. Lausted, T. Lin, C.W.T. Yang, L. Hood, E.T. Lagally, Parallel microfluidic surface plasmon resonance imaging arrays, Lab Chip 10 (5) (2010) 581–588.
[113] A. Campion, P. Kambhampati, Surface-enhanced Raman scattering, Chem. Soc. Rev. 27 (4) (1998) 241–250.
[114] A. Otto, I. Mrozek, H. Grabhorn, W. Akemann, Surface-enhanced Raman scattering, J. Physics: Condens. Matter 4 (5) (1992) 1143–1212.
[115] K. Kneipp, H. Kneipp, I. Itzkan, R.R. Dasari, M.S. Feld, Surface-enhanced Raman scattering and biophysics, J. Physics: Condens. Matter 14 (18) (2002) R597–R624.
[116] L. Rodriguez-Lorenzo, L. Fabris, R.A. Alvarez-Puebla, Multiplex optical sensing with surface-enhanced Raman scattering: a critical review, Anal. Chim. Acta 745 (2012) 10–23.
[117] Q. Zhou, T. Kim, Review of microfluidic approaches for surface-enhanced Raman scattering, Sens. Actuators B: Chem. 227 (2016) 504–514.
[118] J. Kneipp, H. Kneipp, K. Kneipp, SERS—a single-molecule and nanoscale tool for bioanalytics, Chem. Soc. Rev. 37 (5) (2008) 1052–1060.
[119] R. Gao, Z. Lv, Y. Mao, L. Yu, X. Bi, S. Xu, et al., SERS-based pump-free microfluidic chip for highly sensitive immunoassay of prostate-specific antigen biomarkers, ACS Sens. 4 (4) (2019) 938–943.
[120] R. Gao, J. Ko, K. Cha, J. Ho Jeon, G.-E. Rhie, J. Choi, et al., Fast and sensitive detection of an anthrax biomarker using SERS-based solenoid microfluidic sensor, Biosens. Bioelectron. 72 (2015) 230–236.
[121] M. Lee, K. Lee, K.H. Kim, K.W. Oh, J. Choo, SERS-based immunoassay using a gold array-embedded gradient microfluidic chip, Lab Chip 12 (19) (2012) 3720–3727.
[122] C.M. Miyazaki, D.J. Kinahan, R. Mishra, F. Mangwanya, N. Kilcawley, M. Ferreira, et al., Label-free, spatially multiplexed SPR detection of immunoassays on a highly integrated centrifugal Lab-on-a-Disc platform, Biosens. Bioelectron. 119 (2018) 86–93.
[123] M.A. Ali, S. Tabassum, Q. Wang, Y. Wang, R. Kumar, L. Dong, Integrated dual-modality microfluidic sensor for biomarker detection using lithographic plasmonic crystal, Lab Chip 18 (5) (2018) 803–817.
[124] A. Sonato, M. Agostini, G. Ruffato, E. Gazzola, D. Liuni, G. Greco, et al., A surface acoustic wave (SAW)-enhanced grating-coupling phase-interrogation surface plasmon resonance (SPR) microfluidic biosensor, Lab Chip 16 (7) (2016) 1224–1233.

CHAPTER

Fiber-optic devices for sensing, manipulating, and imaging cells in vitro and in vivo

11

Shaodi Zhu[1,2,3], Shuwen Zeng[2,3], Yuye Wang[1,4], Aaron Ho Pui Ho[1] and Wu Yuan[1]

[1]*Department of Biomedical Engineering, The Chinese University of Hong Kong, Shatin, Hong Kong*

[2]*CNRS, XLIM Research Institute, UMR 7252, University of Limoges, Limoges, France*

[3]*Light, Nanomaterials and Nanotechnologies (L2n), CNRS-ERL 7004, Université de Technologie de Troyes, Troyes, France*

[4]*Laboratory of Biomedical Microsystems and Nano Devices, Bionic Sensing and Intelligence Center, Institute of Biomedical and Health Engineering, Shenzhen Institute of Advanced Technology, Chinese Academy of Sciences, Shenzhen, P.R. China*

11.1 Introduction

The cell is the basic structural and functional unit of living organisms. Understanding the intracellular processes and the interaction between cells is essential for revealing the fundamental factors of physiological activities and pathogenic processes. Nowadays, available methods of cell analysis, i.e., characterizing the genotypes and phenotypes of living cells, are surging, accompanied by the recent advancements in biosensing and bioimaging technologies [1,2]. The mainstream analytical techniques include flow-cytometry-based cell sorting, fluorescent-labeling-assisted cell imaging, sequencing-based gene expression profiling, and mass-spectroscopy-based organic compound analysis [1].

Fiber optics has attracted increasing interest in the field of cell analysis since it possesses the advantages of compactness, flexibility, immunity to electromagnetic (EM) interference (EMI), and versatile application possibilities. For instance, the sub-millimeter scale diameter and the flexibility allow the fiber optics to propagate light into narrow spaces, which enables the remote cell analysis at almost arbitrary positions. In addition, the EMI immunity enables the fiber-optic analytical devices to be more stable during the operation in comparison with devices utilizing electric signals, such as electrochemical sensors. Moreover, the optical fiber can be further processed into various functional structures that make the light sensitive to surrounding refractive index (RI) or shape the output light, e.g., fiber gratings, interferometers, and focusing lenses [3]. The RI-sensitive structures can serve as highly

Microfluidic Biosensors. DOI: https://doi.org/10.1016/B978-0-12-823846-2.00019-5

sensitive label-free biosensors. The output shaping structures can serve as optical tweezers and endoscopes. These fiber-optic devices have been successfully utilized in various applications of cell analysis, such as biomarker detection [4–9], cell counting and/or sorting [10–13], and cell imaging [14–17].

This chapter will provide an overview of essential components for fiber-optic-based compact devices and their applications in cell sensing, manipulation, and imaging. To begin with, the basic principles of fiber-optic devices are introduced. Then, in vitro and in vivo applications of cell analysis in three scenarios are reviewed. For bulk-cell samples, both direct and indirect approaches to detecting specific cells and monitoring cell behaviors using fiber-optic biosensors are summarized. For single cells, cell/organelle manipulation and intracellular analysis with fiber-optic probes are introduced. For intravital cells, the in vivo cell detection in the circulation system and the minimally invasive cell imaging in deep tissues are presented.

11.2 Basic principles

This section will give a brief overview of utilizing structured fibers for cell analysis. These structured fibers are employed to perform two main functions: (1) transducing the environmental RI to spectral contents and phase, and (2) focusing the output beam to generate optical gradient force. Based on the functions, three analytical techniques will be introduced accordingly, i.e., plasmonic sensing, interferometric sensing, and optical trapping.

11.2.1 Plasmonic sensing

Plasmonic sensing is a commonly used mechanism of fiber-optic sensors [18–20]. The conventional plasmonic sensing configurations were established by Kretschmann and Otto on prisms [21,22]. The evanescent field from the total reflected incident beam can excite the collective oscillation of surface electrons in the adjacent metal film, which is also known as the surface plasmon polaritons (SPPs). Such surface plasmon resonance (SPR) would happen only if the wavenumber of the evanescent field (k_{ev}) equals the wavenumber of excited SPPs (k_{spp}), which is given as follows:

$$k_{ev} = \frac{2\pi}{\lambda}\sin\theta\sqrt{\varepsilon_p} \tag{11.1}$$

$$k_{spp} = \frac{2\pi}{\lambda}\sqrt{\frac{\varepsilon_m\varepsilon_d}{\varepsilon_m + \varepsilon_d}}. \tag{11.2}$$

In Eqs. (11.1) and (11.2), λ denotes the wavelength of the incident light, θ is the incident angle, and $\varepsilon_p, \varepsilon_m$, and ε_d are the relative electric permittivities of the glass prism, the metal film, and the dielectric analyte, respectively. Note that the

RI, n, is commonly used in bioanalysis. As for nonmagnetic biomaterials, n equals $\sqrt{\varepsilon}$. Therefore, the RI change of the analyte would result in a different k_{spp}, which matches a different k_{ev}. This manifests as the shift of SPR absorption angle θ when the incident light wavelength λ is fixed or the variation of absorption wavelength when a light beam of broad spectral bandwidth is incident at a certain angle θ.

Similar to SPR, localized SPR (LSPR) denotes that the excited collective oscillation of electrons is constrained in metallic nanoparticles. Unlike SPR, the excitation of oscillating electrons in nanoparticles does not require the exact matching of wavenumbers through an evanescent field [23]. For example, the resonance angular frequency ω_{LSPR} of spherical particles can be estimated using the following equation:

$$\omega_{\mathrm{LSPR}} = \frac{\omega_{\mathrm{p}}}{\sqrt{1 + 2\varepsilon_{\mathrm{d}}}}. \tag{11.3}$$

The aforementioned equation is deduced from the Drude model [24], where ω_{p} and ε_{d} represent the plasma frequency of metallic nanoparticles and the relative permittivity of surrounding dielectrics, respectively. Note that ω_{LSPR} equals $2\pi c/\lambda$, with c being the speed of light in free space. Thus, the change of relative permittivity ε_{d} or refract index n would result in the shift of ω_{LSPR}, inducing the change of absorption wavelength λ.

It is worthwhile to mention that the local EM field of SPPs would be dramatically enhanced at resonance [8,23]. This enhanced EM field has been demonstrated to provide sensitivity amplification in conventional SPR sensing, fluorescence signals, and Raman scattering signals [25,26].

As for the fiber-optic-based plasmonic sensors (Fig. 11.1A), the unidimensional structure of optical fibers does not allow the convenient measurement of the incident angle of light, i.e., θ, as in the prism configurations. Hence, measuring the shift of absorption wavelength has become the mainstream interrogation approach [3,27,28]. As shown in Fig. 11.1B, a variety of fiber optic structures have been proposed for plasmonic sensing to achieve high sensitivity. These plasmonic sensors are implemented using different fiber techniques, including heterocore splicing [6,8,29], core exposing [9,30–32], side polishing [33], U-shaped bending [5,34], and fiber Bragg grating [4,35–37], to name a few.

11.2.2 Interferometric sensing

Fiber-optic-based interferometry is another widely used optical sensing technique. As illustrated in Fig. 11.2, a number of different optical fiber sensors have been proposed to enable a variety of interferometric technologies, such as the Fabry--Pérot interferometer(FPI), the Mach–Zehnder interferometer (MZI), the Michelson interferometer (MI), and the Sagnac interferometer (SI).

The FPI fiber sensors are usually fabricated by introducing an in-line optical cavity consisting of two closely spaced reflective surfaces into optical fiber

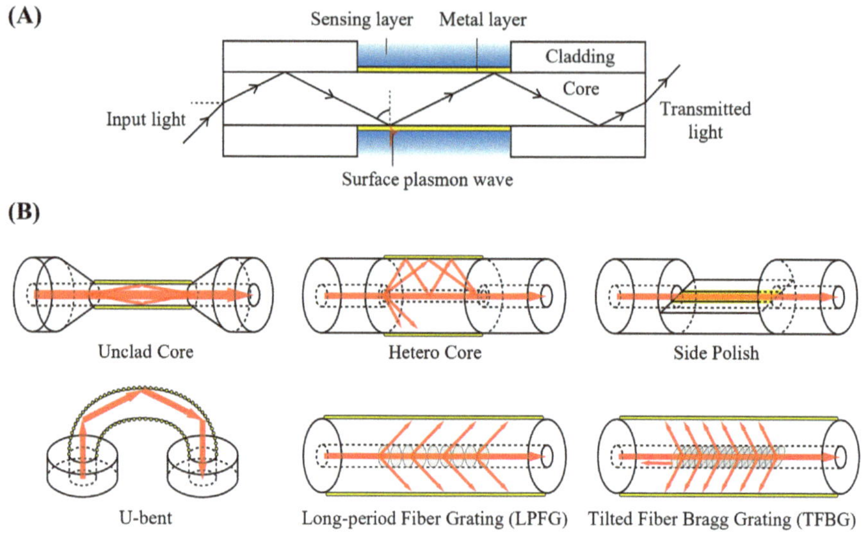

FIGURE 11.1

(A) Schematic showing the principle of the fiber-optic-based plasmonic sensor. (B) Typical fiber structures for plasmonic sensing.

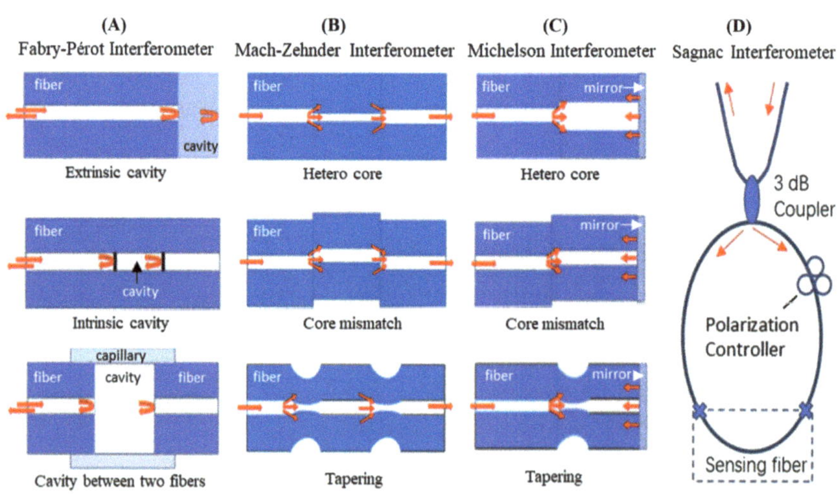

FIGURE 11.2

The typical fiber-optic-based interferometric sensing structures. (A) Fabry–Pérot interferometer (B) Mach-Zehnder interferometer (C) Michelson interferometer (D) Sagnac interferometer. Red arrows indicate the propagation direction of light [38].

Adapted from [38]. © 2021 the author.

through a fusion splicing procedure or other methods [39,40]. The RI change of the analyte within the cavity or attached to cavity walls will lead to a variation in the difference in optical path lengths (OPLs) between two light beams reflected from cavity surfaces, resulting in the shift of the peaks and valleys of the interference pattern.

The MZI fiber sensors split the propagation light in fiber into a reference beam and a sensing beam. The reference beam is usually confined in the fiber core, whereas the sensing beam will propagate in the fiber cladding region and interact with the ambient analyte. This type of fiber sensor can be fabricated through different methods, including fiber tapers, heterocore fiber splicing, and side polishing of optical fibers [7,41]. Since the OPL of the sensing beam is determined by the exposed cladding modes, which are sensitive to the RI of the surrounding analytes, the resulting peaks and valleys of the interference pattern will shift accordingly if the analytes' RI changes. Similar to the MZI counterparts, the MI fiber sensors also consist of reference and sensing beams. MI fiber sensors usually have a reflective mirror on the fiber end facet and measure the beam interference in the reflection mode [38,42]. In contrast, the MZI fiber sensors generally work in the transmission mode (Fig. 11.2).

The SI fiber sensors utilize a Sagnac loop, in which two orthogonal principal polarization states in the optical fiber are oppositely propagated as reference and sensing beams. The polarization states are tuned by a polarization controller, and the sensing part of the SI fiber consists of a polarization-maintaining fiber. The polarized sensing beam is exposed to surrounding analytes through etching or side polishing of the optical fibers. The RI change of the analyte will only change the OPL of the polarized sensing beam and eventually lead to the shift of the peaks and valleys of the interference pattern between the polarized sensing and reference beams [43,44].

11.2.3 Optical cell trapping

The method of trapping particles using the optical gradient force of a tightly focused laser beam, also known as an optical tweezer, was established by Ashkin et al. [45]. As shown in Fig. 11.3A, the optical force on Rayleigh particles can be decomposed into an optical scattering force pushing the particle away and an optical gradient force induced by particle polarization toward focus [46]. Given the light intensity of I_0, the optical forces on a Rayleigh particle can be calculated using the following equations:

$$F_{\mathrm{scat}} = \frac{\sigma n_{\mathrm{m}}}{c} I_0 \tag{11.4}$$

$$F_{\mathrm{grad}} = \frac{2\pi\alpha}{c n_{\mathrm{m}}^2} \nabla I_0. \tag{11.5}$$

In Eqs. (11.4) and (11.5), σ is the scattering cross-section of a particle, α represents the polarizability of a particle, n_m denotes the RI of a particle, and c is

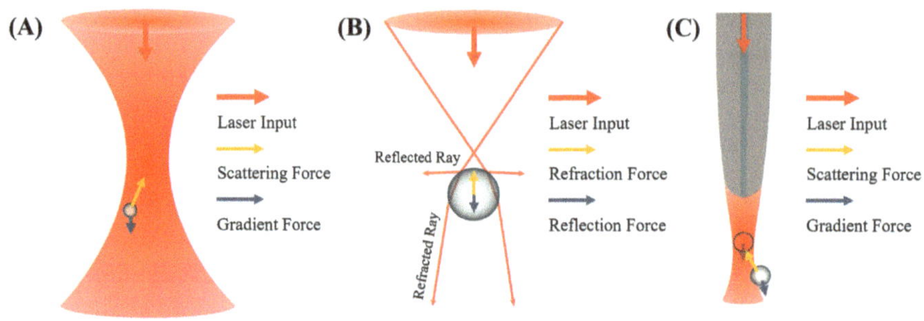

FIGURE 11.3

Schematics showing the optical trapping principle for (A) subwavelength Rayleigh particles and (B) cellular scale particles. (C) Schematic of the optical fiber tweezer.

the speed of light in free space. As for particles larger than the light wavelength or cellular-scale particles, the optical forces exerted by the tightly focused laser beam on cellular-scale particles can be simply interpreted using the reflective and refractive rays and their induced momentums on the particle. The refractive optical force pulls the particle/cell toward the focus, whereas the reflective optical force on the particle/cell surface leads to repulsion [47,48].

As for the fiber optical tweezer shown in Fig. 11.3C, the exerted optical forces on cells cannot be straightforwardly estimated using the above-mentioned ray method due to the loosely focused laser beam exiting from the fiber tip. In this case, electrodynamic simulation has been widely used to calculate the inhomogeneous distribution of the EM field and the resulting time-averaged Maxwell stress tension $\overline{T_M}$ on cells induced by the laser beam. Then, the total optical force on cells can be conveniently estimated using an integral of $\overline{T_M}$ over the cell surface S [49,50]

$$F_{\text{optical}} = \int \overline{T_M} \bullet \vec{n} \quad dS \tag{11.6}$$

where $\vec{n}$ denotes the normal vector on the cell surface [50,51].

As for fiber-optic-based cell imaging, a variety of working principles have been successfully implemented, and the selected cases will be introduced in Section 11.5.2.

11.3 Bulk-cell analysis

The bulk-cell analysis in this chapter refers to the analytical measurement of the common properties in a sample containing a large number of cells. The bulk-cell sample can be obtained by culturing cells or processing biopsy samples. Three properties can be measured by fiber-optic biosensors in this scenario. The first is the number of cells that express specific membrane markers [4–6]. Their membrane marker expression level can also be measured accordingly [5,6]. The second is the concentration of cell metabolites, such as nucleotides [7], proteins [8], and extracellular vesicles (EVs) [9]. The third is the cell responses against environmental stimuli, such as the antigen-induced mast cell activation [30,35], as well as the fetal bovine serum (FBS)-induced nutrient uptake, sodium-azide-induced metabolism inhibition, and trypsin-induced cell detachment [35].

Recently, various fiber-optic biosensors have been proposed to provide a highly sensitive, label-free, and real-time approach for detecting specific cells and profiling their metabolites and responses. Thanks to their unique advantages of compact size and one-step-detection capability, they have become competitive choices in clinical applications, especially for point-of-care testing.

11.3.1 Detecting cell

Cell detection using fiber optics is typically an affinity assay targeting the biomarkers expressed on the cell membrane. Accordingly, it can also serve as an approach to measure the membrane biomarker expression level. The detection of mutated cells in bulk samples, such as blood and single-cell suspension, plays a crucial role in cancer diagnostics. For example, the presence of circulating tumor cells (CTCs) in blood shedding from solid tumors is believed to be associated with hematogenous metastasis. Therefore, it can be used as a biomarker to determine the progression of cancer. Moreover, the membrane biomarker expression level, such as *N*-glycan expression for glucose transporters (GLUTs), can be used as an indicator for monitoring the growth of different cancer cells after incubation with the external stimuli of inhibitors.

Loyez et al. proposed an aptamer-functionalized tilted fiber Bragg grating SPR (TFBG-SPR) fiber sensor for detecting breast-originated CTCs [4]. As shown in Fig. 11.4, the sensor was fabricated by coating a 50-nm-thick gold film on TFBG and subsequently functionalized by MAMA2 aptamers. These aptamers are short oligonucleotides with a high affinity toward mammaglobin (MAM) on the MDA-MB-415 cell line [51]. Besides, the aptamer-functionalized sensor also showed significant specificity in the control experiments of MAM-negative cells. Furthermore, this work demonstrated the nonlabel CTC detection limit in phosphate-buffered saline (PBS) at 49 cells/mL, by measuring the extinction spectrum of the most sensitive SPR mode. In addition, the experimental limit of detection (LOD) can be further reduced to 10 cells/mL by using aptamer-functionalized gold nanoparticles (AuNPs) of 10 nm in diameter.

The combination of cytosensing and evaluation of membrane marker expression has also attracted academic interest. Luo et al. proposed a cytosensing system based on a U-shaped LSPR fiber sensor with concanavalin A functionalization [5]. As displayed in Fig. 11.5, the sensor was fabricated through the cladding etching and U-bending of a fiber with a 600-μm core. A layer of AuNPs was subsequently coated. The RI sensitivity was optimized by adjusting the outer bending diameter to 4 mm and the AuNPs' size to 32 nm.

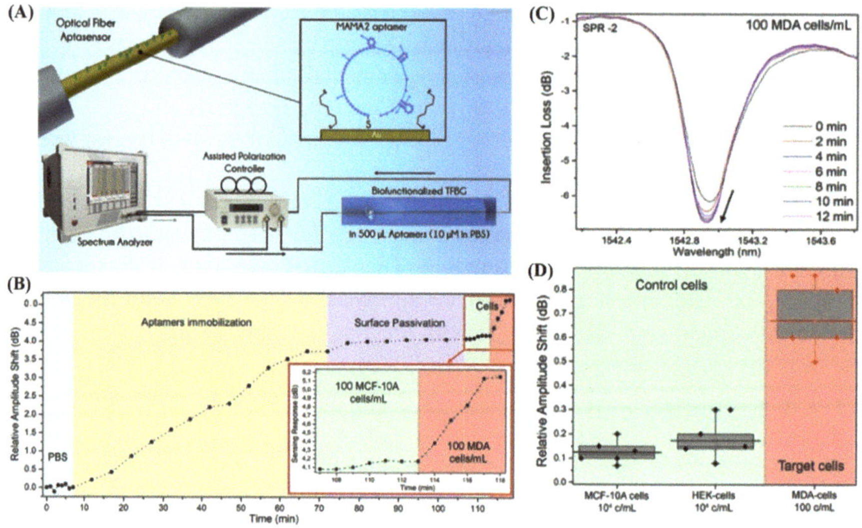

FIGURE 11.4

(A) Optical interrogation setup for the TFBG-SPR cytosensor and the schematic of sensor functionalization. (B) Sensorgram during the functionalization, blocking, and cell detection. The analysis was performed in phosphate-buffered saline (PBS). (C) Extinction spectra of an SPR-sensitive mode acquired during a 12-minute binding process of the target cell. (D) Boxplots of the biosensing shifts between the control and the target. *TFBG-SPR*, Tilted fiber Bragg grating surface plasmon resonance.

Adapted with permission from [4]. © 2020 American Chemical Society.

The detection of MDA-MB-231 cancer cells has been demonstrated through the affinity binding between *N*-glycan and concanavalin A. The calculated LOD of the optimized sensor was 30 cells/mL. Furthermore, the authors also examined the *N*-glycan expression level on six different cell lines through tunicamycin (TM) inhibitor treatment. The resulting LSPR absorption is proportional to the *N*-glycan expression level and would progressively decrease with the increased concentration of TM, suggesting the sensor as a promising tool to study *N*-glycan-related biological processes.

Apart from the transmissive optical interrogation configurations mentioned above, several fiber-optic biosensors adopted a reflective design [6,9,30]. Singh et al. proposed a heterocore reflective LSPR cytosensor by splicing a 1-mm etched multicore fiber (MCF) into a single-mode fiber (SMF) [6]. The sensing layer was composed of sequentially coated graphene oxide (GO), AuNPs, and copper oxide nanoflowers (CuO-NFs) (Fig. 11.6A). In this design, the GO and CuO-NFs were reported to enhance surface biocompatibility [52,53], and AuNPs were regularly employed as LSPR sensing agents. Moreover, the biosensor was further functionalized by 2-deoxy-D-glucose targeting GLUTs overexpressed on

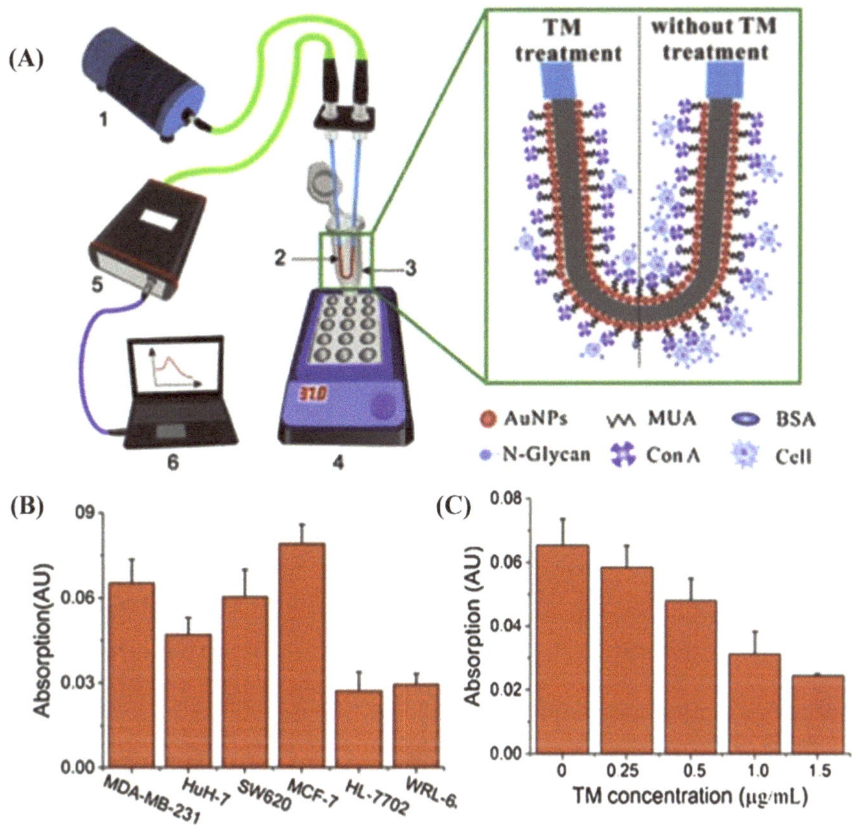

FIGURE 11.5

(A) Experimental setup for a U-shaped fiber-optic cytosensor and schematic of sensor functionalization. (B) LSPR absorption across different cells, cell suspension concentration 1×10^5 cells/mL. (C) LSPR absorption versus TM concentration. Cell line MDA-MB-231, concentration 1×10^5 cells/mL. *LSPR*, Localized surface plasmon resonance; *TM*, tunicamycin. *Adapted with permission from* [5]. © *2019 Elsevier*

cancer cells. Based on such affinity, the authors demonstrated the cytosensing of four different cancer cell lines, i.e., Hepa1-6, A549, MCF-7, and HepG. The calculated LODs were 2, 2, 2, and 3 cells/mL, respectively. Similar to the work of Luo et al. [5], the correlation between the GLUT expression level and LSPR absorption has also been confirmed in this work (Fig. 11.6B–G).

11.3.2 Detecting cell metabolites

Cell metabolites, such as nucleic acids, proteins, and EVs, are also important biomarkers for profiling specific cells. The metabolites are released into the blood,

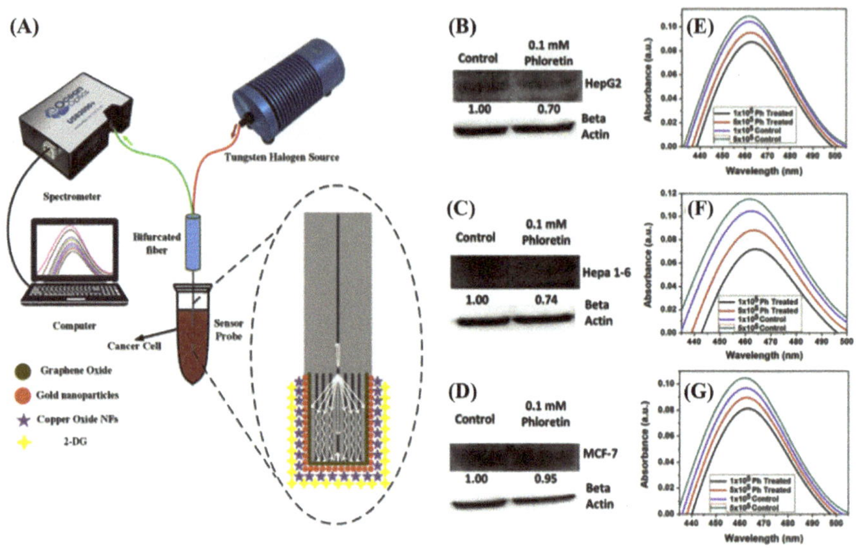

FIGURE 11.6

(A) Experimental setup and the schematic of probe functionalization of the heterocore reflective LSPR cytosensor. (B–D) Western blot showing GLUT expression between control and 0.1 mM phloretin inhibitor treated cancer cell lines. (E–G) LSPR absorption spectra at different cell concentrations (cells/mL). Cell lines: (B and E) HePG2, (C, F) Hepa1-6, and (D, G) MCF-7. *LSPR*, Localized surface plasmon resonance; *GLUT*, glucose transporter.

Adapted with permission from [6]. © 2020 Elsevier.

saliva, urine, or other body fluids through exocytosis and apoptosis, carrying the metabolic and genomic information of cells.

Lu et al. proposed a AuNP-enhanced MZI sensor for nucleic acid detection [7]. The microfiber was fabricated by tapering the SMF, as shown in Fig. 11.7. The taper region allows for the excitation of higher-order modes and the recombination with core modes to form interference. Since the surrounding RI strongly influences the phase of higher-order modes propagating along the fiber cladding, the RI change can result in an interference shift [54,55]. The authors functionalized the microfiber with a predesigned oligo sequence for targeted capture and employed a complementary oligo decorated with AuNPs for signal amplification, also known as the "sandwich" configuration [56]. In this work, the authors improved the detection limit of oligos to the order of 10^{-15} M and showed excellent selectivity against the mismatched oligo sequences.

Xia et al. reported another approach of combining plasmonic nanoparticles and the RI sensor for label-free and ultrasensitive immunoassay [8]. They proposed that the stacking of nanoparticles with nanorods can enhance the EM field

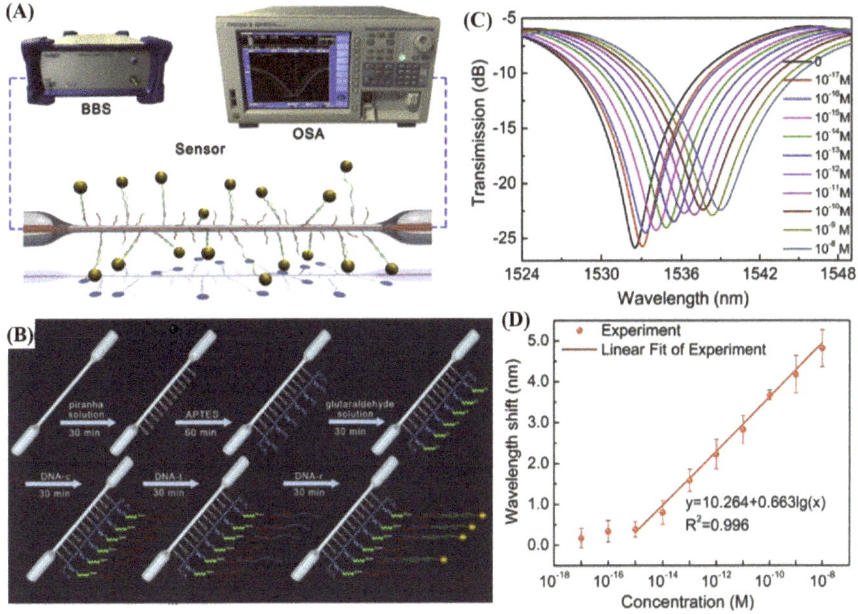

FIGURE 11.7

(A) Schematic of the experimental setup and surface functionalization of the MZI microfiber sensor. (B) Schematic of functionalization and DNA hybridization process. (C) Transmission spectra of DNA-t hybridization with AuNP amplification. (D) The shift of the dip wavelength versus DNA-t concentration. *MZI*, Mach–Zehnder interferometer.

Adapted from [7]. © 2021 Optical Society of America under the terms of the OSA Open Access.

of LSPR and improve the sensitivity of immunoassay. As shown in Fig. 11.8, a heterocore sensor was developed by coating a layer of 50-nm gold and two additional layers of gold particles and nanorods on a photonic crystal fiber. A sensitivity of 25,642 nm/RIU was identified by measuring solutions of different refractive indices. Furthermore, the immunosensing performance was evaluated by further functionalizing the sensor with GO and Immunoglobuline G (IgG) antibodies. The LOD of human IgG is 4.6 ng/mL, which is in the order of 10^{-11} M [57].

In addition to biomolecules, larger metabolites, like EVs, are easier to be detected with an RI sensor. Yildizhan et al. reported a novel EV detection approach with a commercialized fiber-optic SPR probing system [9] (Fox Biosystems Inc.). This work explored the target enrichment and detection with different antibodies to selectively maximize the signal of EVs from specific cancer cells, as shown in Fig. 11.9. Through recombinant sample calibration, the LOD of EVs in complex matrices, such as plasma and cell culture medium, is in the order of 10^7 counts/mL, which is 10^4 times lower than the concentration in a typical cancer patient's plasma [58].

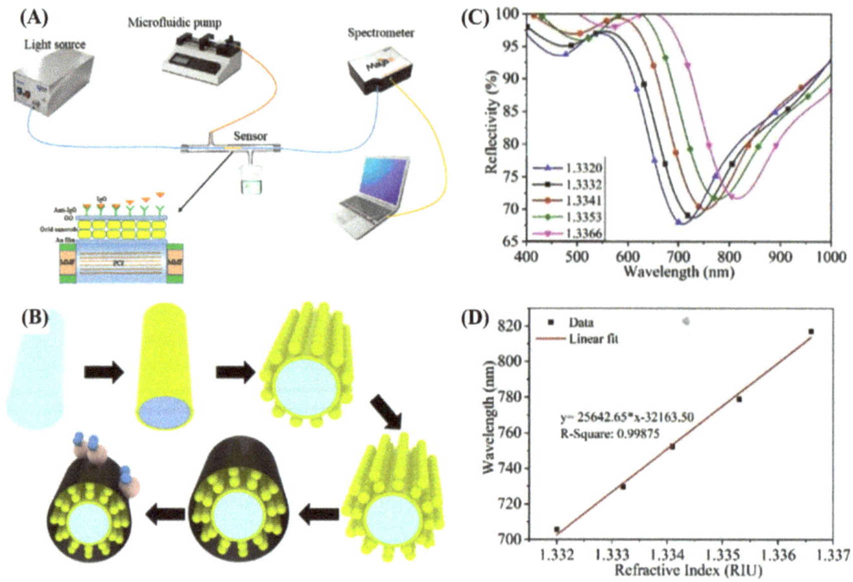

FIGURE 11.8

(A) Schematic of the immunoassay setup. (B) Flowchart showing the fabrication of nanoparticles–nanorods sensor. Note that the nanoparticles are not in scale for clarity. (C) LSPR spectra of the nanorod immobilized sensor in different refractive solutions. (D) The sensitivity of the nanorod immobilized sensor. *LSPR*, Localized surface plasmon resonance.

Adapted with permission from [8]. © 2020 Elsevier.

11.3.3 Monitoring cell behavior

Other than detecting membrane markers and metabolites of cells, fiber-optic sensors have also been used to study cell behaviors against stimuli through detecting the subtle RI change. For example, Yanase et al. proposed an SPR fiber optic sensor to detect the antigen-induced activation of mast cells [30]. The sensor was fabricated by coating a 50-nm gold sensing layer on an exposed fiber of 200 μm in diameter. Moreover, an aluminum layer coated on the fiber tip allows for the interrogation of the reflected signal. The RBL-2H3 mast cells can be immobilized on the sensor surface with biotin functionalization. As shown in Fig. 11.10, through dinitrophenol (DNP) antibody sensitization, the authors have demonstrated in situ detection of the cellular immune reaction induced by DNP-conjugated human serum albumin [59].

As shown in Fig. 11.11, Shevchenko et al. reported another cell behavior monitoring platform based on a TFBG-SPR fiber sensor [35]. In this work, the intensity of the SPR signal was proportional to the ambient RI, which consists of the contributions from the background medium and the density and morphology of

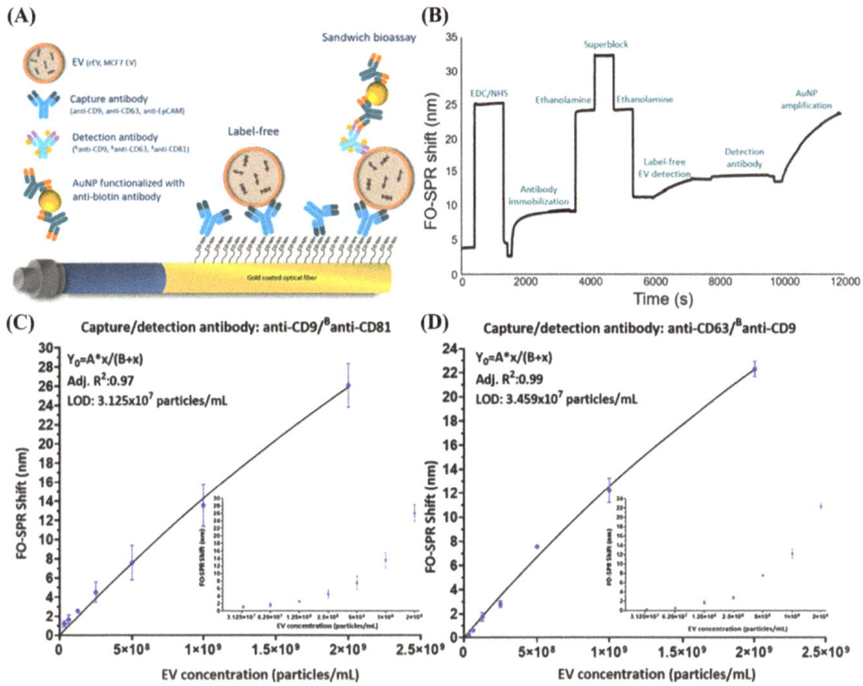

FIGURE 11.9

(A) Schematic of surface functionalization and EV-targeted immunoassay. (B) SPR dip wavelength shift during the immunoassay. (C, D) SPR dip wavelength shift versus EV concentration when using different detection antibodies. *SPR*, Surface plasmon resonance.

Adapted from [9]. © 2021 the authors.

cells attached to the sensing fiber. Therefore, with rigorously controlled experiments, cell behaviors can be recognized from the change of the SPR signal and RI. The authors examined the fibronectin immobilized NIH-3T3 fibroblast under various stimuli and confirmed the RI change resulting from the cell activities, including FBS uptake, trypsin-induced detachment, and sodium-azide-induced metabolism inhibition.

11.4 Single-cell analysis

The single-cell analysis includes the studies of genomics, transcriptomics, proteomics, metabolomics, and cell−cell interactions on individual cells [2], which can provide valuable information on intracellular biological dynamics and the cell-to-cell variation within a cell population [60]. Fiber-tip-based manipulation by

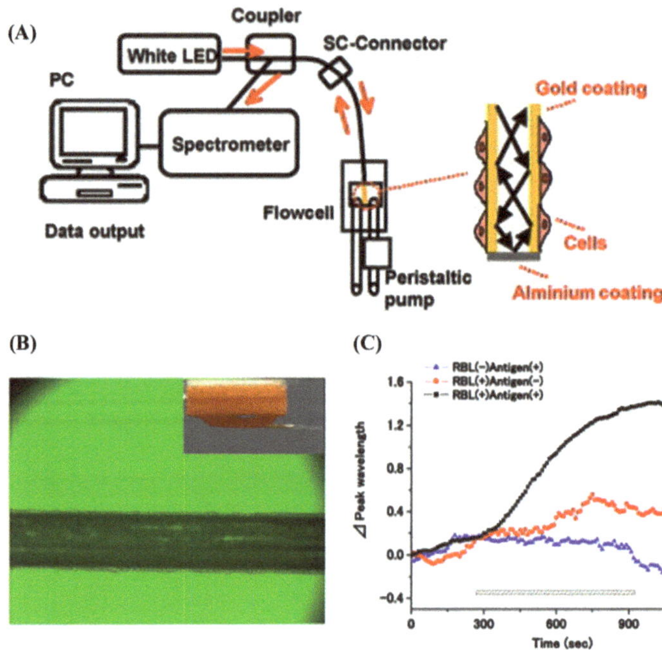

FIGURE 11.10

(A) Optical interrogation setup and the schematic of the probe structure for monitoring the cell's responses to stimuli. (B) An image of the immobilized mast cells on the fiber surface acquired with a phase-contrast microscope. (C) Refractive indices change of DNP-IgE activated (+) and inactivated (–) RBL-2H3 cells with DNP-conjugated human serum albumin stimulation.

Adapted with permission from [30]. © 2010 Elsevier.

optical gradient force [10,61,62] and fiber-optic intracellular sensing [31,32,36] are important analytical techniques in this scenario and provide the needed flexibility and multifunctionality for evaluating the mechanical and biological properties of the cell.

11.4.1 Manipulating cells and organelles

There are three types of single-cell analysis based on cellular and subcellular manipulation. The first type is to position and sort cells and organelles based on a templated assembly process. The second type is to measure the mechanical properties at cellular and subcellular levels, such as the velocity of organelle movement and the power of molecular motors. The third type is to perform a microsurgery and realize the gene delivery into the cell.

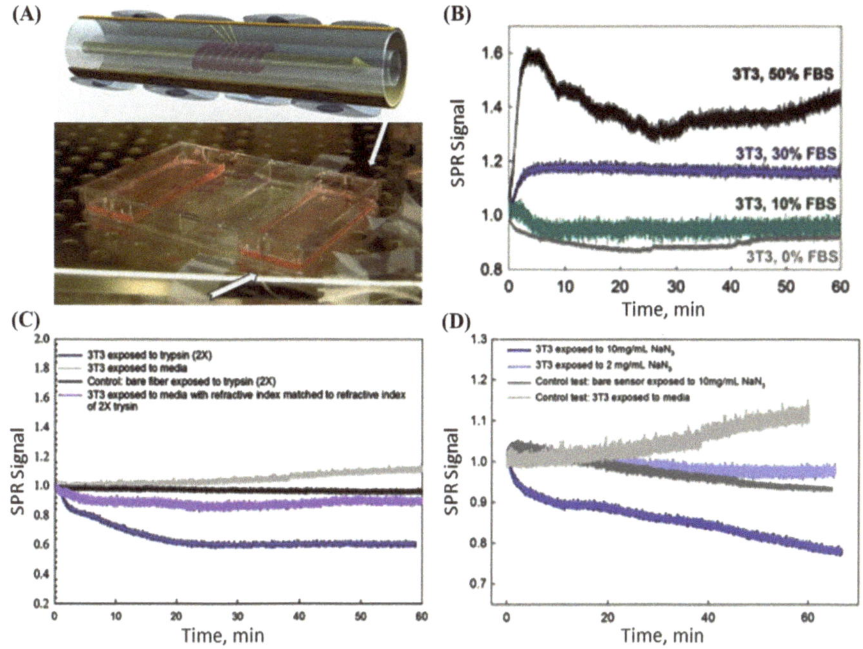

FIGURE 11.11

(A) Schematic of the proposed TFBG-SPR fiber sensor and the picture showing the sensing setup. (B) SPR signal monitoring the cellular uptake of FBS. (C) SPR signal of cell detachment through trypsin. (D) SPR signal monitoring the cellular metabolism inhibition through sodium azide. *FBS*, Fetal bovine serum; *TFBG-SPR*, tilted fiber Bragg grating surface plasmon resonance.

Adapted with permission from [35].

Liu et al. proposed an optical fiber tweezer (OFT)-based cell sorting chip for selectively separating *Escherichia coli* cells from red blood cells (RBCs) by leveraging their difference in morphology [10]. A tapered fiber tip (Fig. 11.12) was designed to generate the proper optical gradient force distribution for trapping *E. coli* while pushing the RBCs away. By using a 1064-nm laser of 10 mW, the authors have demonstrated the capability of trapping *E. coli* cells in a 20 μm/s dynamic flow with a trapping distance of less than 8.8 μm. More interestingly, *E. coli* cells can be further arranged in a cell chain on a fiber tip and the length can be extended with a higher laser power [63]. In addition to *E. coli*, the authors have also discussed the feasibility of cancer cell separation through simulation, showing the versatility of the OFT sorting technique.

Li et al. reported the operation of OFT at the subcellular scale [61]. Different from the work of Liu et al. [10], the fiber tip was designed to deliver the trapping center further away from the fiber tip to offer a noncontact in vivo manipulation

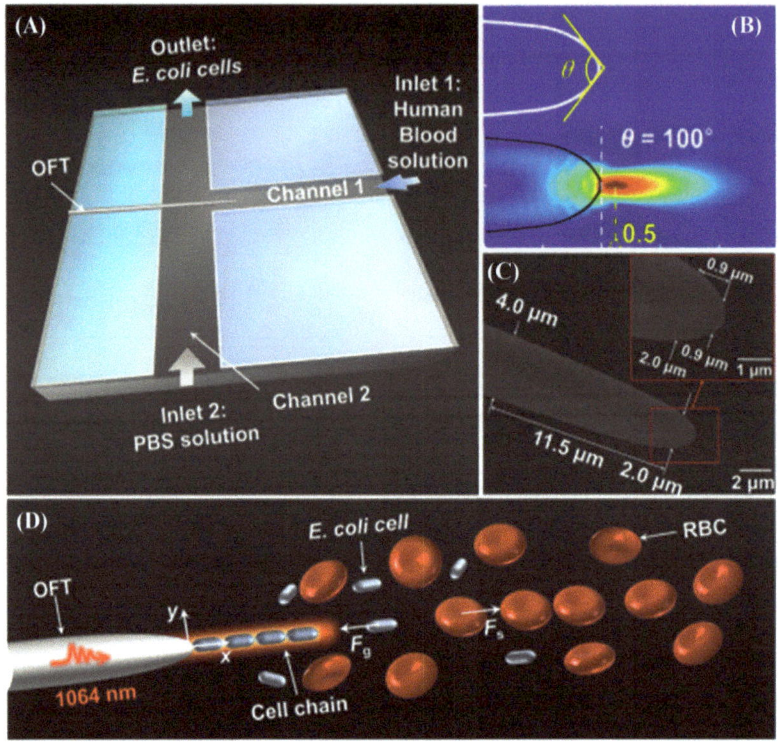

FIGURE 11.12

(A) Schematic of the OFT-based cell sorting setup. (B) Numerical simulation of electromagnetic field on the fiber tip. (C) Scanning electron microscopy (SEM) picture of the fiber tip. (D) Artistic illustration of using OFT to separate *E. coli* cells from RBCs. *OFT*, Optical fiber tweezer; *RBCs*, red blood cells.

Adapted with permission from [10]. © 2019 Optical Society of America.

force (Fig. 11.13). The authors have demonstrated the moving and arranging of chloroplasts on *Hydrilla verticillata* leaves by launching a 980-nm laser with an optical power of 20–50 mW. The maximum velocity of intracellular chloroplasts transported by OFT is 7.5 ± 0.2 μm/s at 35 mW power output. This work has further explored the cytoplasmic viscosity measurement to indicate the integrity of the cell matrix.

Zhao et al. further widened the application of OFT in subcellular operation [62]. They proposed a technique that combined thermoplastics and optical trapping for in situ intracellular organelle manipulation and single-cell microsurgery (Fig. 11.14). In this work, the photothermal effect from internalized gold nanorods was demonstrated to induce localized breakdown and repair of the cell membrane. With an 808-nm laser of 15 mW and a 5-minute irradiation, the perforated hole size was found to be 8.7 μm. Higher optical power and longer irradiation

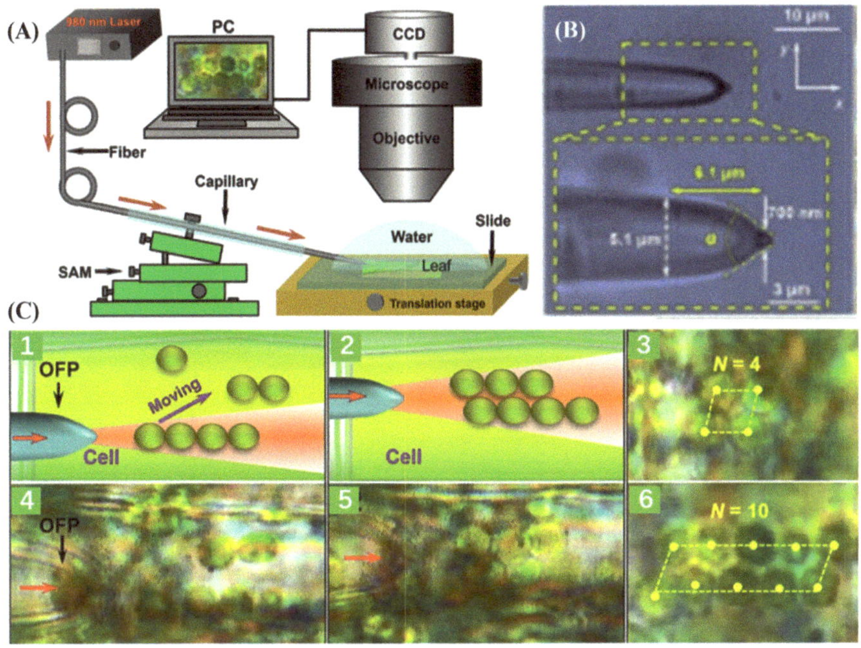

FIGURE 11.13

(A) Schematic of the experimental setup of OFT. (B) Microscopy picture showing the details of the fiber tip. (C) Schematic and pictures of binding multiple chloroplasts. *OFT*, Optical fiber tweezer.

Adapted from [62]. © 2015 the authors.

time will result in larger wound size and potential intracellular damage, whereas a lower power of less than 14 mW will lead to a wound that can be repaired by a heat-facilitated Brownian diffusion reversely. The authors have demonstrated the mitochondrial extraction from cells with minor damage.

11.4.2 Probing intracellular process

In addition to the optical tweezers implemented on fiber tips, nanoscale fiber tip bioprobes are also emerging in recent years. These probes opened a precise window for understanding intracellular dynamics, such as the protein uptake [31], oxidative stress [32], and apoptosis [36]. For example, Hong et al. proposed a label-free LSPR sensor on fiber tip for protein uptake analysis [31]. The fiber tip was fabricated with a precisely monitored wet etching process, which can help achieve a tip of a diameter less than 100 nm, minimizing the invasiveness in cell. A gold nanorod as an LSPR agent was sequentially picked from a dispersion droplet and attached to the fiber tip. The authors have demonstrated the detection

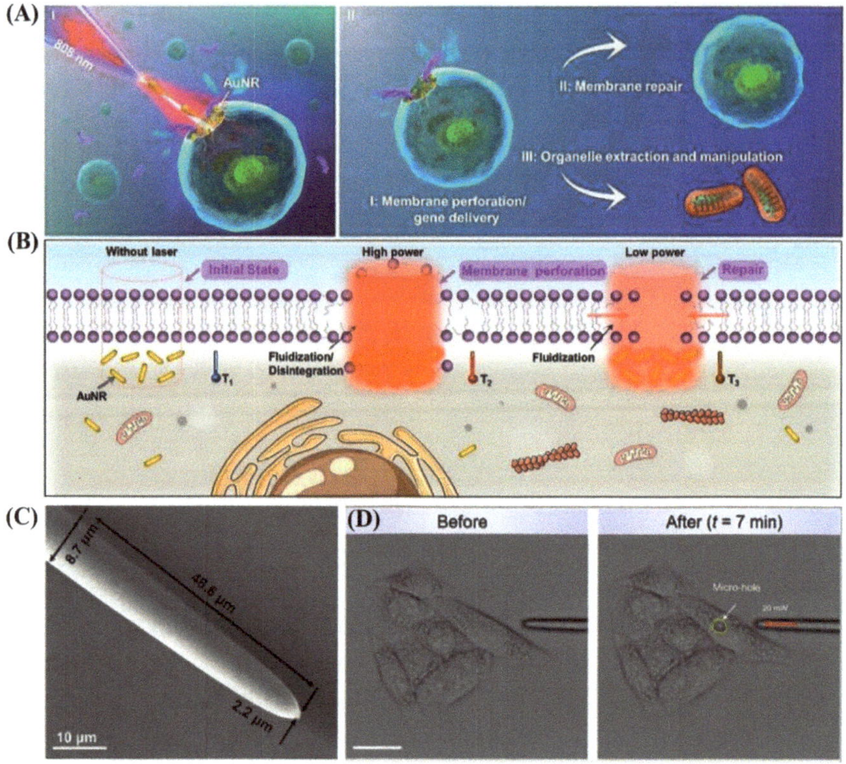

FIGURE 11.14

(A) Artistic illustration of the microsurgery process on the cell. (B) Schematic of membrane perforating and repairing of the cell. (C) Scanning electron microscope (SEM) picture showing the fiber tip details. (D) Microscopy picture of cell membrane perforation.

Adapted with permission from [63]. © 2022 American Chemical Society.

of tumor suppressor protein p53 with sc-126 antibody functionalization. The resulting LOD was less than 1 nM. Using this probe, the p53 uptake process in HeLa cells has been investigated under various therapeutic stimuli, such as UV radiation and neocarzinostatin exposure (Fig. 11.15).

The cellular oxidative stress induced by reactive oxygen species (ROS) has also been explored for many years since the intracellular redox process can help reveal the key factors of aging and metabolic diseases [64,65]. Chen et al. reported a fiber-tip-based surface-enhanced Raman spectroscopy (SERS) sensor for intracellular monitoring of ROS and the reduced form of nicotinamide adenine dinucleotide (NADH) [32]. The sensor was fabricated through a wet etching method similar to that used by Hong et al. [31], except that silver nanoparticles were sequentially deposited as an SERS substrate in this work. The authors have examined the cellular response with oxidative stress inducer phorbol 12-myristate

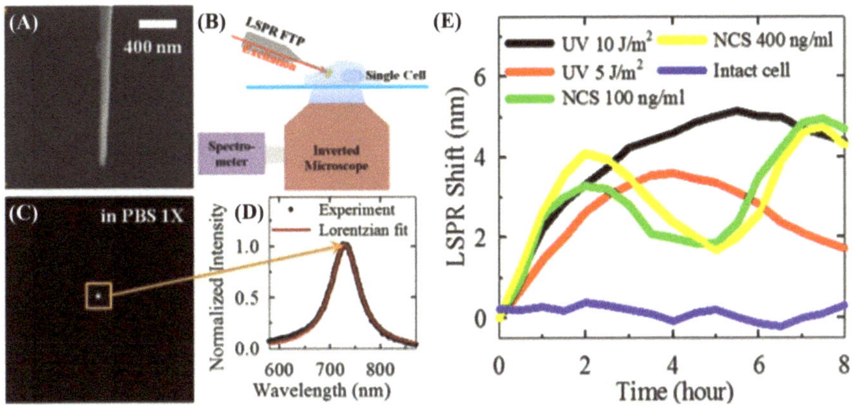

FIGURE 11.15

(A) Scanning electron microscope (SEM) picture of the nanoscale fiber tip with a gold nanorod attached. (B) The experimental setup for intracellular sensing. (C) Microscopic picture showing the scattering from the fiber tip. (D) The spectrum of the scattered light. (E) LSPR spectra shift illustrating the p53 dynamic in HeLa cells. *LSPR*, Localized surface plasmon resonance.

Adapted from [31]. © 2014 the authors.

13-acetate and demonstrated the oxidation events from ROS generation as well as the reduction from NADH. This process has been monitored through the transform between Raman active 4-mercaptophenol and Raman inactive 4-mercaptocyclohexa-2,5-dienone functionalized on the fiber tip (Fig. 11.16).

For single-cell analysis, the necessity of a high-quality microscopic device for signal collection has limited the in vivo detection in deep tissues. Recently, Li et al have reported an all-fiber subcellular sensor for intracellular RI sensing, as shown in Fig. 11.17 [36]. The sensor was fabricated by integrating a zinc oxide (ZnO) nanograting onto a fiber taper tip. Light coupled in ZnO nanograting will be partially reflected ($\sim$30% at 655 nm), and the central wavelength of the reflected spectrum was determined by the surrounding RI. The diameter of nanograting is 800 nm, which would significantly reduce the damage to cells. The authors have investigated the intracellular RI of HeLa cells and demonstrated the RI change of the nucleus during apoptosis.

11.5 Intravital-cell analysis

The flexibility and low profile of optical fibers enable the successful translation of a bulky microscopy system to a miniaturized fiber optic imaging device [66–70]. In vivo flow cytometry and endomicroscopy are the most widely adopted techniques for fiber-based intravital-cell analysis. In vivo flow cytometry

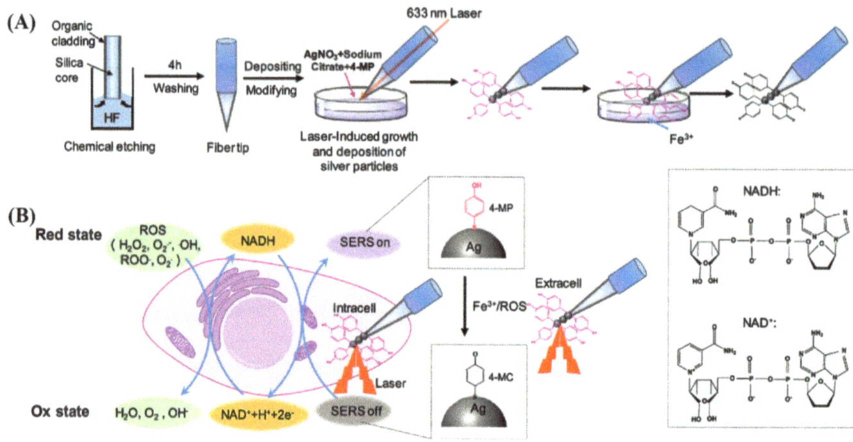

FIGURE 11.16

(A) Schematic of fabricating the fiber tip SERS sensor. (B) Schematics showing the principle of intracellular redox reaction detection. *SERS*, Surface-enhanced Raman spectroscopy.

Reproduced with permission from [32]. © 2021 American Chemical Society.

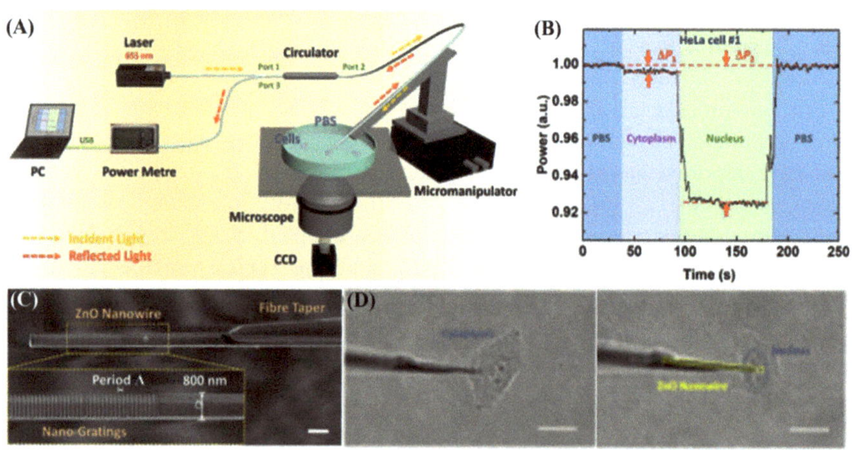

FIGURE 11.17

(A) Schematic of the experimental setup. (B) Reflection intensity showing the intracellular RI difference. (C) Details of ZnO nanograting attached on the tapered optical fiber. (D) Pictures of intracellular RI sensing.

Adapted from [36]. © 2022 the authors.

provides rapid cell screening in the circulation system, whereas endomicroscopy offers morphological and physiological details of cells in tissues.

11.5.1 In vivo flow cytometry

In vivo flow cytometry was first proposed by Georgakoudi et al. in 2004 to introduce a new cell counting approach through an intravital fluorescent microscope [71]. After that, a variety of techniques have been developed in this field as an important preclinical research tool [72]. For instance, Pera et al. reported a multimode-fiber-bundle-based in vivo cytometry for detecting scattered fluorescence from labeled CTCs, named diffuse fluorescence flow cytometer (DFFC) [11]. The fiber probe is composed of a central fiber delivering an excitation laser beam through a 635-nm bandpass filter and eight surrounding fibers for collecting fluorescent signals through 650-nm long-pass filters. The backscattered fluorescent signals from the labeled CTCs in superficial blood vessels were then detected by photomultiplier tube (PMT) and indicated as intensity spikes during the continuous recording of DFFC. The detection of multiple circulating myeloma cells has been demonstrated in the tails of nude mice; the count rate was found to be roughly proportional to the concentration of the injected CTCs.

Recently, the same group has introduced a dual-probe DFFC configuration to further enhance the sensitivity, reduce false-positive signals, and improve the penetration depth ($\sim$2 mm) in tissues, as shown in Fig. 11.18 [12]. The dual probe was placed along the paired ventral caudal artery and vein, where the labeled CTCs traveled in the opposite direction. There, the cell passing event can be detected as a matched peak pair to remove the random fluorescent signal. The consistent flow speed of cells allowed the convenient conversion of the count rate to the CTC concentration. The false alarm rate from the control sample was found to be 0.014 count/min, indicating a detection limit of 0.35 cells/mL.

Despite the wide applications of fluorescent dyes in animal models, their safety issues in clinical applications require further study. Hence, label-free modality, such as photoacoustic (PA) technology, has attracted great interest. Juratli et al. proposed an acoustic-resolution PA probe for detecting the formation of intravascular blood cell emboli (Fig. 11.19) [13], which is also known as circulating blood clots (CBCs). Using a 1064-nm laser, the selective absorption of hemoglobin would provide a stable ultrasound signal, which was detected by a transducer focusing inside the blood vessel (with a resolution 45–80 μm). If a high-absorptive erythrocyte-rich CBC or a low-absorptive leukocytes/ platelet-rich CBC passes by, different absorption properties will result in a positive or negative peak, respectively, while the CBC with mixed cells generate a combined contrast. The authors have examined the vessel injury induced coagulation and demonstrated the detection of emboli generation in both rat tail vein (depth 0.3 mm) and carotid artery (depth 3–4 mm). Their previous work has also demonstrated the detection of embolus with an optical-resolution PA, which provided a higher resolution and a shallower penetration depth of 0.1–0.15 mm [73].

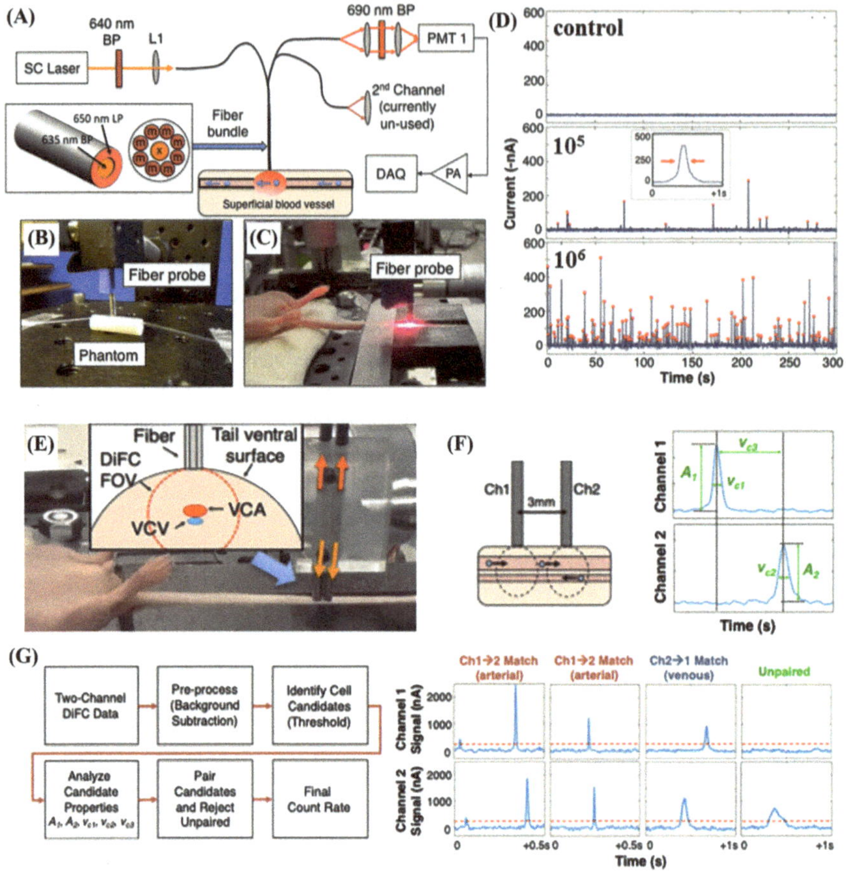

FIGURE 11.18

(A) Schematic of the DFFC system. The detection in flow phantom (B) and on mouse tail (C). (D) Data trace from mice injected with 10^5 and 10^6 cancer cells. The intensity peaks indicate the fluorescently labeled passing tumor cells. (E) Dual-probe DFFC system. (F) Cell passing event detected by two channels. (G) Dual-probe cell detection algorithm. *DFFC*, Diffuse fluorescence flow cytometer.

(A–D) Adapted from [11]. © 2017 the authors. (E–G) Adapted from [12]. © 2019 the authors.

11.5.2 Endomicroscopy

In contrast to the single-point detection of in vivo flow cytometry, endomicroscopy performs in vivo tissue imaging with a cellular resolution and a large field of view. The temporal resolution of endomicroscopy is usually lower than that of in vivo flow cytometry. Such a modality fills the gap between 2D histological analysis and 3D pathological imaging of magnetic resonance imaging (MRI), computerized tomography (CT), and ultrasound imaging.

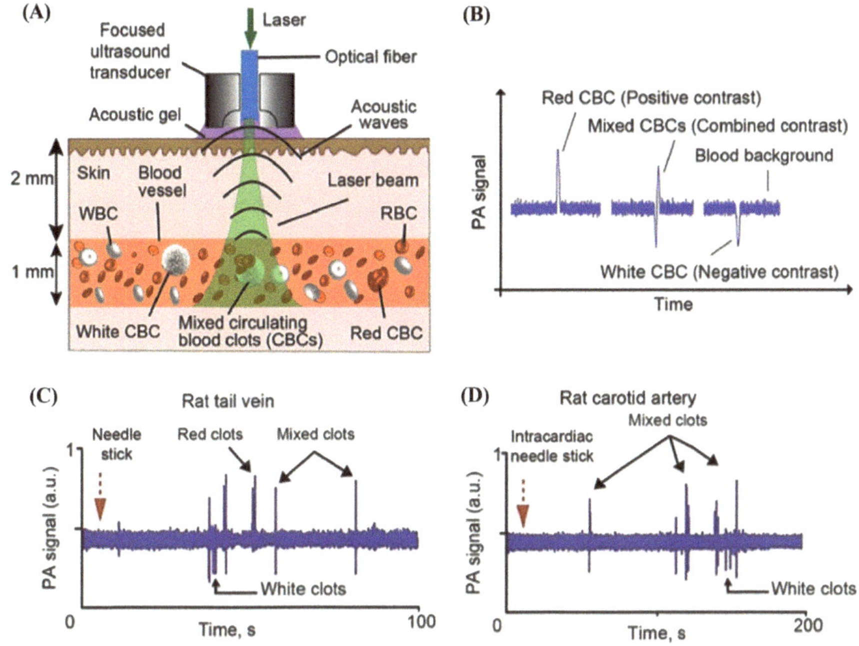

FIGURE 11.19

(A) Schematic of PA probe for CBC detection. (B) Data trace of different types of CBCs. The blood vessel injury induced CBC forming in (C) rat tail vein and (D) carotid artery. *CBC*, Circulating blood clots; *PA*, photoacoustic.

Typical endomicroscopy has been modified from the conventional microscopy through minimizing and fiberizing the bulky optics [14–16]. Liang et al. reported a compact nonlinear endomicroscopy probe of ~2 mm in diameter, offering a subcellular resolution of ~0.7 μm in the lateral direction and ~6.5 μm in the axial direction [16]. This probe was composed of a miniature objective lens and a customized double-clad fiber driven by a piezoelectric scanner, providing a working distance of 200 μm. The laser transmitted in the fiber core was used to excite nonlinear signals of tissue, including two-photon fluorescence and second harmonic scattering, which were subsequently collected back into the inner cladding of the fiber. As illustrated in Fig. 11.20, the nonlinear endomicroscopic images clearly delineated the subcellular structure and extracellular matrix of mouse tissues both in vitro and in vivo. As a label-free imaging modality, this probe has great clinical potential for in situ optical biopsy.

As for intravital inspection, accessibility to deep tissue is desirable. A variety of novel endomicroscopy, such as endoscopic PA imaging [74] and endoscopic optical

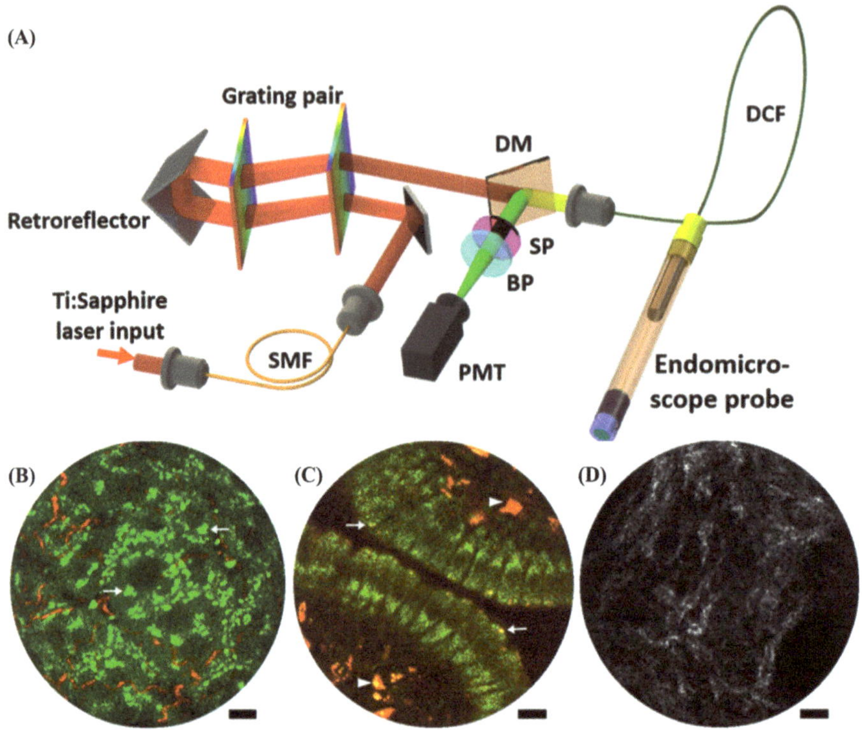

FIGURE 11.20

(A) Schematic of the experimental setup of two-photon endomicroscopy. (B) In vitro imaging of the mouse liver overlapping 2PF (green) with SHG (red). (C) In vivo 2PF image of autofluorescence of NADH (green) and Flavin adenine dinucleotide (FAD, red) from the mucosa of the mouse small intestine. (D) In vitro SHG image of the mouse cervical collagen fiber network. *2PF*, Two-photon fluorescent; *NADH*, nicotinamide adenine dinucleotide; *SHG*, second-harmonic generation signal.

Adapted from [16]. © 2017 the authors.

coherence tomography [75], are emerging. However, these endoscopic modalities are mainly used for in vivo anatomical analysis due to their limited cell contrast [76,77]. Instead, using minimally invasive distal optics provides a promising endomicroscopic approach to help access the deep tissue. For example, Vasquez-Lopez et al. reported a minimally invasive endomicroscopy by embedding a multimode fiber (MMF) in the deep brain of a mouse [17]. In this endomicroscopy, the wavefront modulation enabled by a spatial light modulator was used to perform the 3D imaging of brain tissues through the MMF [78,79]. In this work, the resolution was $\sim 1.27\ \mu m$ in the lateral direction and $\sim 20.2\ \mu m$ in the axial direction. The capability of intravital neuron imaging in the deep brain has been demonstrated on a mouse model, as shown in Fig. 11.21. Through tracking the intensity change of the

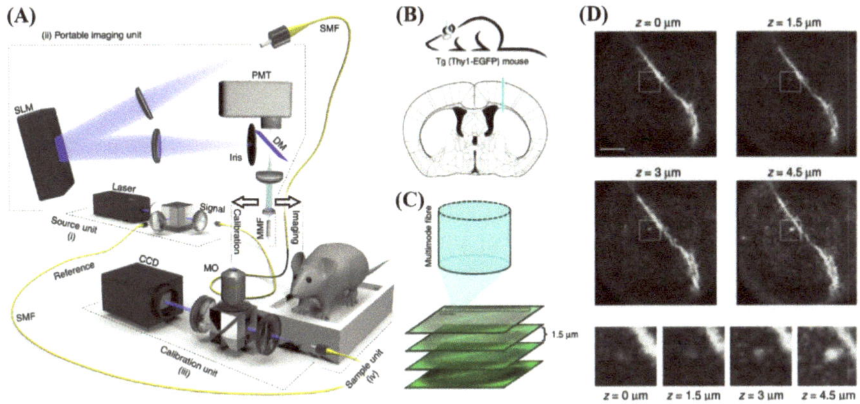

FIGURE 11.21

(A) Schematic of a minimally invasive endomicroscopy system. (B) The placement of an MMF in the mouse brain. (C) Optical 3D deep brain imaging via the MMF using a spatial light modulator. (D) Fluorescent images acquired at different depths in the mouse brain. *MMF*, Multimode fiber.

Adapted from [17]. © 2018 the authors.

fluorescent Ca^{2+} reporter in the auditory thalamus with the sound stimuli, this system was able to study the neuron dynamics in vivo in the deep brain.

11.6 Conclusion

Fiber-optic-based sensing, manipulation, and imaging techniques for cellular analysis have advanced the fundamental understanding of basic cell biology and led to new diagnostic methods. This chapter summarizes the recent development of fiber-optic devices for bulk-cell analysis, single-cell analysis, and intravital-cell analysis.

As for bulk-cell analysis, RI sensing with fiber-optics devices provides a highly sensitive, label-free, and real-time approach for detecting cancer cells, sensing the cell metabolites, and tracking the cellular behaviors. Although great advancement has been made in the sensitivity of fiber-optic-based RI sensors, there are still two challenges in cell detection and cell response monitoring. First, optical fiber is not an ideal platform to enrich the rare target cells due to its limited size. Combining the sensitive cell detection of fiber-optic RI sensors with the flexible analyte processing capability of microfluidic devices maybe a feasible solution. Second, it is challenging for fiber-optic sensors to monitor complex cellular responses involving multiple reactions simultaneously, such as the triggering of several signaling pathways in cells. A rigorous design of the control experiment with an intelligent method is needed to distinguish multiple cellular responses from the collective sensing signals of intracellular RI changes.

As for single-cell analysis, two types of fiber-optic-based devices have been introduced. The first type is based on OFTs for cellular and subcellular manipulations, including cell positioning and sorting, measurements of mechanical properties of cells, and cell microsurgery. OFTs can apply force in the order of pico-Newton and measure the velocity in the range of a few micrometers per second. The second type is for profiling of intracellular biochemical dynamics, such as molecular uptake, oxidative stress events, and apoptosis. These fiber biosensors usually have a sub-micrometer fiber tip, offering label-free and minimally invasive probing of cells without affecting their viability. It is promising to develop fiber-optic devices integrating cellular and subcellular manipulations with biosensing capabilities in the future.

As for intravital-cell analysis, fiber-optic-based in vivo flow cytometry and endomicroscopy have been developed. In vivo flow cytometry provides accurate, rapid, and real-time detection of target cells, such as CTCs, or cell clusters, such as CBCs, in blood vessels. However, it is imperative to further advance in vivo flow cytometry for high throughput. On the other hand, fiber-optics-based endomicroscopy offers cellular resolution and a large field of view to clearly delineate the tissue structures and monitor the activities of living cells in deep tissues. For future clinical use of endomicroscopy, it is desirable to further improve the spatial and temporal resolutions, increase the field of view, and reduce the size to minimize the invasiveness.

Reference

[1] J.R. Heath, et al., Single-cell analysis tools for drug discovery and development, Nat. Rev. Drug Discov. 15 (2016) 204–216.

[2] D. Wang, S. Bodovitz, Single cell analysis: the new frontier in 'omics', Trends Biotechnol. 28 (2010) 281–290.

[3] C. Caucheteur, et al., Review of plasmonic fiber optic biochemical sensors: improving the limit of detection, Anal. Bioanal. Chem. 407 (2015) 3883–3897.

[4] M. Loyez, et al., Rapid detection of circulating breast cancer cells using a multiresonant optical fiber aptasensor with plasmonic amplification, ACS Sens. 5 (2020) 454–463.

[5] Z. Luo, et al., Ultrasensitive U-shaped fiber optic LSPR cytosensing for label-free and in situ evaluation of cell surface *N*-glycan expression, Sens. Actuators B: Chem. 284 (2019) 582–588.

[6] R. Singh, et al., Etched multicore fiber sensor using copper oxide and gold nanoparticles decorated graphene oxide structure for cancer cells detection, Biosens. Bioelectron. 168 (2020) 112557.

[7] H. Lu, et al., Au-NPs signal amplification ultra-sensitivity optical microfiber interferometric biosensor, Opt. Express 29 (2021) 13937–13948.

[8] F. Xia, et al., Ultra-high sensitivity SPR fiber sensor based on multilayer nanoparticle and Au film coupling enhancement, Measurement 164 (2020) 108083.

[9] Y. Yildizhan, et al., FO-SPR biosensor calibrated with recombinant extracellular vesicles enables specific and sensitive detection directly in complex matrices, J. Extracell. Vesicles 10 (2021) e12059.

[10] S. Liu, et al., Miniaturized optical fiber tweezers for cell separation by optical force, Opt. Lett. 44 (2019) 1868–1871.

[11] V.E. Pera, et al., Diffuse fluorescence fiber probe for in vivo detection of circulating cells, J. Biomed. Opt. 22 (2017) 037004.

[12] X. Tan, et al., In vivo flow cytometry of extremely rare circulating cells, Sci. Rep. 9 (2019) 1–11.

[13] M.A. Juratli, et al., Noninvasive label-free detection of circulating white and red blood clots in deep vessels with a focused photoacoustic probe, Biomed. Opt. Express 9 (2018) 5667–5677.

[14] J. Ahn, et al., In vivo longitudinal cellular imaging of small intestine by side-view endomicroscopy, Biomed. Opt. Express 6 (2015) 3963–3972.

[15] E.J. Jun, et al., In vivo fluorescence microendoscopic monitoring of stent-induced fibroblast cell proliferation in an esophageal mouse model, J. Vasc. Intervent. Radiol. 29 (2018) 1756–1763.

[16] W. Liang, et al., Nonlinear optical endomicroscopy for label-free functional histology in vivo, Light: Sci. Appl. 6 (2017) e17082.

[17] S.A. Vasquez-Lopez, et al., Subcellular spatial resolution achieved for deep-brain imaging in vivo using a minimally invasive multimode fiber, Light: Sci. Appl. 7 (2018) 1–6.

[18] X. Yu, et al., Plasmonic enhanced fluorescence spectroscopy using side-polished microstructured optical fiber, Sens. Actuators B: Chem. 160 (2011) 196–201.

[19] X. Yu, et al., An efficient approach for investigating surface plasmon resonance in asymmetric optical fibers based on birefringence analysis, Opt. Express 18 (2010) 17950–17957.

[20] D.J.J. Hu, H.P. Ho, Recent advances in plasmonic photonic crystal fibers: design, fabrication and applications, Adv. Opt. Photonics 9 (2017) 257–314.

[21] E. Kretschmann, H. Raether, Radiative decay of non radiative surface plasmons excited by light, Z. für Naturforschung A 23 (1968) 2135–2136.

[22] A. Otto, Excitation of nonradiative surface plasma waves in silver by the method of frustrated total reflection, Z. für Phys. A Hadron. Nucl. 216 (1968) 398–410.

[23] K.M. Mayer, J.H. Hafner, Localized surface plasmon resonance sensors, Chem. Rev. 111 (2011) 3828–3857.

[24] M. Shabaninezhad, G. Ramakrishna, Theoretical investigation of size, shape, and aspect ratio effect on the LSPR sensitivity of hollow-gold nanoshells, J. Chem. Phys. 150 (2019) 144116.

[25] G.M. Das, et al., Biosensing using SERS active gold nanostructures, Nanomaterials 11 (2021) 2679.

[26] M.A. Badshah, et al., Recent developments in plasmonic nanostructures for metal enhanced fluorescence-based biosensing, Nanomaterials 10 (2020) 1749.

[27] D. Yi, et al., Interrogation technique analyses of a hybrid fiber optic sensor based on SPR and MMI, Opt. Express 28 (2020) 20764–20772.

[28] G. Liang, et al., Fiber optic surface plasmon resonance–based biosensor technique: fabrication, advancement, and application, Crit. Rev. Anal. Chem. 46 (2016) 213–223.

[29] Q. Wu, et al., Singlemode-multimode-singlemode fiber structures for sensing applications—a review, IEEE Sens. J. 21 (2020) 12734–12751.

[30] Y. Yanase, et al., Development of an optical fiber SPR sensor for living cell activation, Biosens. Bioelectron. 25 (2010) 1244–1247.

[31] W. Hong, et al., Nanoscale label-free bioprobes to detect intracellular proteins in single living cells, Sci. Rep. 4 (2014) 1–5.

[32] J. Chen, et al., Single-cell oxidative stress events revealed by a renewable SERS nanotip, ACS Sens. 6 (2021) 1663–1670.

[33] Y. Ying, et al., Recent research progress of optical fiber sensors based on D-shaped structure, Opt. Laser Technol. 90 (2017) 149–157.

[34] A.J.Y. Tan, et al., Trends and applications of U-shaped fiber optic sensors: a review, IEEE Sens. J. 21 (2020) 120–131.

[35] Y. Shevchenko, et al., Surface plasmon resonance fiber sensor for real-time and label-free monitoring of cellular behavior, Biosens. Bioelectron. 56 (2014) 359–367.

[36] D. Li, et al., Label-free fiber nanograting sensor for real-time in situ early monitoring of cellular apoptosis, Adv. Photonics 4 (2022) 016001.

[37] T. Guo, et al., Plasmonic optical fiber-grating immunosensing: a review, Sensors 17 (2017) 2732.

[38] A. Miliou, In-fiber interferometric-based sensors: overview and recent advances, Photonics (2021) 265.

[39] D. Rho, et al., Demonstration of a low-cost and portable optical cavity-based sensor through refractive index measurements, Sensors 19 (2019) 2193.

[40] H. Wu, et al., An ultra-low detection-limit optofluidic biosensor based on all glass Fabry-Perot cavity, Opt. Express 22 (2014) 31977–31983.

[41] W. Yu, et al., Label-free fiber optic biosensor based on thin-core modal interferometer, Sens. Actuators B: Chem. 228 (2016) 322–329.

[42] Y. Li, et al., Immobilized optical fiber microprobe for selective and high sensitive glucose detection, Sens. Actuators B: Chem. 255 (2018) 3004–3010.

[43] X. Li, et al., High-sensitivity Sagnac-interferometer biosensor based on exposed core microstructured optical fiber, Sens. Actuators B: Chem. 269 (2018) 103–109.

[44] G. An, et al., Glucose sensor realized with photonic crystal fiber-based Sagnac interferometer, Opt. Commun. 405 (2017) 143–146.

[45] A. Ashkin, et al., Observation of a single-beam gradient force optical trap for dielectric particles, Opt. Lett. 11 (1986) 288–290.

[46] C.J. Bustamante, et al., Optical tweezers in single-molecule biophysics, Nat. Rev. Methods Prim. 1 (2021) 1–29.

[47] J. Chen, et al., Photochemically synthesized silver nanostructures on tapered fiber as plasmonic tweezers for surface enhanced Raman scattering applications, Vacuum 118 (2015) 171–176.

[48] J. Chen, et al., Plasmonic random nanostructures on fiber tip for trapping live cells and colloidal particles, Opt. Lett. 40 (2015) 3926–3929.

[49] G. Pesce, et al., Optical tweezers: theory and practice, Eur. Phys. J. Plus 135 (2020) 1–38.

[50] V.V. Datsyuk, O.R. Pavlyniuk, Maxwell stress on a small dielectric sphere in a dielectric, Phys. Rev. A 91 (2015) 023826.

[51] E.M. Hassan, et al., In vitro selections of mammaglobin A and mammaglobin B aptamers for the recognition of circulating breast tumor cells, Sci. Rep. 7 (2017) 1–18.

[52] M. Willander, et al., Recent progress on growth and device development of ZnO and CuO nanostructures and graphene nanosheets, J. Mater. Chem. 22 (2012).
[53] V. Georgakilas, et al., Noncovalent functionalization of graphene and graphene oxide for energy materials, biosensing, catalytic, and biomedical applications, Chem. Rev. 116 (2016) 5464–5519.
[54] M. Ding, et al., Mesoporous nanospheres functionalized optical microfiber biosensor for low concentration neurotransmitter detection, Opt. Express 24 (2016) 27152–27159.
[55] W. Luo, et al., Recent progress in microfiber-optic sensors, Photonic Sens. 11 (2021) 45–68.
[56] K. L. Cox, et al., Immunoassay methods, in: Assay Guidance Manual [Internet], 2019.
[57] C. A. Janeway Jr, et al., The structure of a typical antibody molecule, in: Immunobiology: The Immune System in Health and Disease, fifth ed., Garland Science, 2001.
[58] S. Ferguson, R. Weissleder, Modeling EV kinetics for use in early cancer detection, Adv. Biosyst. 4 (2020) 1900305.
[59] J. Kitaura, Tuning IgE: IgE-associating molecules and their effects on IgE-dependent mast cell reactions, Cells 10 (2021).
[60] E. Hodzic, Single-cell analysis: Advances and future perspectives, Bosn. J. Basic Med. Sci. 16 (2016) 313.
[61] Y. Li, et al., Non-contact intracellular binding of chloroplasts in vivo, Sci. Rep. 5 (2015) 1–9.
[62] X. Zhao, et al., In situ single-cell surgery and intracellular organelle manipulation via thermoplasmonics combined optical trapping, Nano Lett. (2021).
[63] H. Xin, et al., Escherichia coli-based biophotonic waveguides, Nano Lett. 13 (2013) 3408–3413.
[64] S. Yang, G. Lian, ROS and diseases: role in metabolism and energy supply, Mol. Cell Biochem. 467 (2020) 1–12.
[65] C.T. Madreiter-Sokolowski, et al., Interrelation between ROS and Ca2+ in aging and age-related diseases, Redox Biol. 36 (2020) 101678.
[66] W. Yuan, et al., Optimal operational conditions for supercontinuum-based ultrahigh-resolution endoscopic OCT imaging, Opt. Lett. 41 (2016) 250–253.
[67] W. Yuan, et al., Super-achromatic monolithic microprobe for ultrahigh-resolution endoscopic optical coherence tomography at 800 nm, Nat. Commun. 8 (2017) 1–9.
[68] W. Yuan, et al., Theranostic OCT microneedle for fast ultrahigh-resolution deep-brain imaging and efficient laser ablation in vivo, Sci. Adv. 6 (2020) eaaz9664.
[69] J. Thiboutot, et al., Visualization and validation of the microstructures in the airway wall in vivo using diffractive optical coherence tomography, Acad. Radiol. (2022).
[70] W. Yuan, et al., In vivo assessment of inflammatory bowel disease in rats with ultrahigh-resolution colonoscopic OCT, Biomed. Opt. Express 13 (2022) 2091–2102.
[71] I. Georgakoudi, et al., In vivo flow cytometry: a new method for enumerating circulating cancer cells, Cancer Res. 64 (2004) 5044–5047.
[72] Y. Suo, et al., Advances of in vivo flow cytometry on cancer studies, Cytometry Part A 97 (2020) 15–23.
[73] M.A. Juratli, et al., Real-time label-free embolus detection using in vivo photoacoustic flow cytometry, PLoS One 11 (2016) e0156269.

[74] Y. Li, et al., Advances in endoscopic photoacoustic imaging, in: *Photonics*, 2021, p. 281.

[75] P. Si, et al., Optical microscopy and coherence tomography of cancer in living subjects, Trends Cancer 6 (2020) 205–222.

[76] J. Hu, et al., Gold nanoshells: contrast agents for cell imaging by cardiovascular optical coherence tomography, Nano Res. 11 (2018) 676–685.

[77] J.E. Lemaster, et al., Gadolinium doping enhances the photoacoustic signal of synthetic melanin nanoparticles: a dual modality contrast agent for stem cell imaging, Chem. Mater. 31 (2018) 251–259.

[78] D. Stellinga, et al., Time-of-flight 3D imaging through multimode optical fibers, Science 374 (2021) 1395–1399.

[79] M. Kim, et al., Transmission matrix of a scattering medium and its applications in biophotonics, Opt. Express 23 (2015) 12648–12668.

Index

Note: Page numbers followed by "*f*" and "*t*" refer to figures and tables, respectively.

G

H

I

T

W

Z

FSC
www.fsc.org
MIX
Papier | Fördert gute Waldnutzung
FSC® C083411

Kolibri 360 GmbH
Ferdinand-Jühlke-Straße 7
99095 Erfurt, Deutschland
produktsicherheit@kolibri360.de